PILBARA SEED ATLAS AND FIELD GUIDE

PILBARA SEED ATLAS AND FIELD GUIDE

Plant Restoration in Australia's Arid Northwest

Todd E. Erickson, Russell L. Barrett, David J. Merritt and Kingsley W. Dixon

CSIRO PUBLISHING

National Library of Australia Cataloguing-in-Publication entry

Pilbara seed atlas and field guide : plant restoration in
Australia's arid northwest / editors: Todd E Erickson,
Russell L Barrett, David J Merritt and Kingsley W Dixon.

9781486305520 (paperback)
9781486305537 (epdf)
9781486305544 (epub)

Includes index.

Revegetation – Western Australia – Pilbara.
Restoration ecology – Western Australia – Pilbara.
Plants – Western Australia – Pilbara.
Plant conservation – Western Australia – Pilbara.

Erickson, Todd E., editor.
Barrett, R. L. (Russell L.), editor.
Merritt, David J., 1956– editor.
Dixon, Kingsley W. (Kingsley Wayne), 1954– editor.

631.64099413

Published by
CSIRO Publishing
36 Gardiner Road, Clayton VIC 3168
Private Bag 10, Clayton South VIC 3169
Australia

Telephone: [+613] 9545 8555
Email: csiropublishing@csiro.au
Website: www.publishing.csiro.au

Front cover: Seeds of *Grevillea refracta* subsp. *refracta*, *Acacia bivenosa*, *Sida echinocarpa* (within the fruit), *Mirbelia viminalis*, *Goodenia stobbsiana*, *Hibbertia glaberrima*, *Gomphrena cunninghamii* and *Triodia wiseana* (photos by David R. Symons and arranged in order from top left to bottom right).

Back cover: An iconic view of the Pilbara landscape where diverse species assemblages, dominated by Triodia grasslands (foreground), overlap with the large-scale active mining operations of the region (background) (photo by Todd E. Erickson).

Set in Adobe Garamond 10/12
Cover by James Kelly
Layout and design by Russell Barrett and Todd Erickson
Printed by Ingram Lightning Source

Kings Park & Botanic Garden
Botanic Gardens & Parks Authority

bhpbilliton

Mar26_RP_ILS

Contents

Foreword

BHP Billiton Iron Ore is proud to join the Botanic Gardens and Parks Authority and the University of Western Australia to develop the *Pilbara Seed Atlas and Field Guide*.

The atlas, which culminates more than five years of collaborative research into seed management for rehabilitation in the Pilbara, will inform best practice for mine-site rehabilitation and for broader rehabilitation activities and the seed collection industry and is a key component of BHP Billiton Iron Ore's drive to maximise seed collection and use.

The development of this reference will deliver sustainable and tangible environmental and economic benefits. It is not only a tremendous field guide for plant and seed identification and collection, but provides a step change in seed management. The atlas provides a one-stop resource from species identification all the way through the life of seed management. The atlas will be instrumental in informing changes to BHP Billiton Iron Ore's processes and make a valuable contribution to positive ecological and rehabilitation outcomes.

I congratulate the efforts of all those involved in the development of the *Pilbara Seed Atlas and Field Guide*.

Gavin Price, Head of Environment, BHP Billiton Iron Ore

Foreword

When I first ventured into Australia's deserts in 1980, the biology of arid zone plants was poorly researched by western scientists. Of course, an intimate and detailed knowledge existed among Aboriginal people. However, few botanists had sought to work collaboratively and learn from first Australian teachers, and a great reservoir of profound understanding accumulated over tens of thousands of years lay largely unknown outside Aboriginal communities. As a consequence, there had been a long need for western desert science to catch up with Aboriginal knowledge, and with biological science that had been prosecuted in the better-studied southern and coastal regions of the continent.

By the time I synthesised what I had learnt about desert biology in 2005[1], a revolution in Australian scientific work had commenced, spurred on by a mining boom and modern requirements for restoration of exhausted mine sites with native plants and animals. The scientists at Kings Park and Botanic Garden played a significant part in this uplift in exploration of desert biology and restoration ecology. I was lucky enough to join that organisation as Director in 1992, and help foster the growth of the small team led by Kingsley Dixon into what today is recognised internationally as one of the world leaders among botanic garden research groups. Much of that growth occurred through novel and productive collaboration with the mining industry and the University of Western Australia, based on a model of industry funds going towards PhD research projects that were prosecuted by higher degree students working in the laboratories at Kings Park and Botanic Garden.

This book beautifully exemplifies the resultant growth in scientific knowledge. Focusing on the Pilbara, where the desert meets the sea in tropical northwestern Australia, the book is brimful of the outcomes of elegant scientific work, from the latest taxonomic exploration, providing a vital underpinning for all biological studies, to cutting edge discoveries in seed biology. A great strength of the work presented is its clear translation into applications in restoration ecology and mine-site rehabilitation. For the first time, readers can engage with state of the art knowledge and images of a sample of important plant genera from the Australian arid zone, packaged in a way that will undoubtedly help conserve a unique biological heritage, and stimulate interest in further scientific research on one of the world's most intriguing desert regions.

The surge in ongoing discovery for the Pilbara is evident in botanical inventory. Just five years ago, the Pilbara had 1,521 recorded native plant taxa, and 103 introduced weeds[2,3]. Herein, we learn that there are now 1,674 native plants and 120 weeds. This represents a rate of new records equal to that for any of the world's most poorly explored botanical regions, including tropical rainforests.

As to seed biology, readers of the book will enjoy some of the finest discoveries being made globally. The investment made by mining companies, the Western Australian government and the State's universities could not have been more handsomely rewarded by the scientific outputs and practical outcomes achieved thus far. Of course, there remains much more to be done, with urgency given the threats faced by poorly known biodiversity, but a significant start is summarised herein.

I congratulate all involved in the production of this book, which sets a high standard indeed for its successors.

Professor Stephen D. Hopper AC, Centre of Excellence in Natural Resource Management and School of Plant Biology, The University of Western Australia, Albany

1. Hopper S.D. (2005) Deserts through time. In P. Nikulinsky and S.D. Hopper, *Soul of the desert*, pp. 10–23. Fremantle Arts Centre Press, Fremantle.
2. Keighery G.J. (2010) The naturalised vascular plants of the Pilbara region, Western Australia. *Records of the Western Australian Museum, Supplement* **78**, 299–311.
3. McKenzie N., van Leeuwen S., Pinder A. (2009) Introduction to the Pilbara biodiversity survey, 2002–2007. *Records of the Western Australian Museum, Supplement* **78**, 3–89.

Contributing authors

Russell L. Barrett

Botanic Gardens and Parks Authority, Kings Park and Botanic Garden, West Perth, Western Australia 6005; Western Australian Herbarium, Department of Parks and Wildlife, Locked Bag 104, Bentley Delivery Centre, Western Australia 6983; School of Plant Biology, Faculty of Science, The University of Western Australia, Crawley, 6009, Western Australia; Current address: Australian National Herbarium, Centre for Australian National Biodiversity Research, National Research Collections Australia, GPO Box 1600, Canberra, Australian Capital Territory 2601

Steven J. Dillon

Western Australian Herbarium, Department of Parks and Wildlife, Locked Bag 104, Bentley Delivery Centre, Western Australia 6983

Kingsley W. Dixon

Department of Environment and Agriculture, Curtin University, Building 311, Bentley Campus, Kent St, Bentley, Western Australia 6102

Todd E. Erickson

School of Plant Biology, Faculty of Science, The University of Western Australia, Crawley, 6009, Western Australia; Botanic Gardens and Parks Authority, Kings Park and Botanic Garden, West Perth, Western Australia 6005

Peter J. Golos

School of Plant Biology, Faculty of Science, The University of Western Australia, Crawley, 6009, Western Australia; Botanic Gardens and Parks Authority, Kings Park and Botanic Garden, West Perth, Western Australia 6005

Stephen D. Hopper

Centre of Excellence in Natural Resource Management and School of Plant Biology, The University of Western Australia, Albany, Western Australia 6330

David J. Merritt

Botanic Gardens and Parks Authority, Kings Park and Botanic Garden, West Perth, Western Australia 6005; School of Plant Biology, Faculty of Science, The University of Western Australia, Crawley, Western Australia 6009

Gavin Price

Head of Environment, BHP Billiton Iron Ore, 225 St Georges Terrace, Perth, Western Australia 6000

David R. Symons

Botanic Gardens and Parks Authority, Kings Park and Botanic Garden, West Perth, Western Australia 6005; School of Plant Biology, Faculty of Science, The University of Western Australia, Crawley, Western Australia 6009

Shane R. Turner

Botanic Gardens and Parks Authority, Kings Park and Botanic Garden, West Perth, Western Australia 6005; School of Plant Biology, Faculty of Science, The University of Western Australia, Crawley, Western Australia 6009

Acknowledgements

When the *Pilbara Seed Atlas and Field Guide* was first conceived, we knew we were entering on a long and arduous journey. Facing an arid region of unique diversity and environmental extremes, many field trips were carried out to locate, photograph, collect, and study seeds of many species throughout the landscape. This could have not occurred without fieldwork and financial support from BHP Billiton Iron Ore's Rehabilitation and Environmental teams. The thank you list is long one, but a direct mention and thanks must go to Joanne Heyes, Paul Simper, Ellissa Tomich, Brad Stokes, Tara Read, Stephen White, David Kaljuste (a.k.a DK), Rosie Turley, Jacqui Roberts, Eliza Cummins, and Ellie Ridley. Without the field time and freedom to collect seeds from obscure corners of the Pilbara, this book would not have been possible. We would also like to thank Steve Dillon from the Department of Parks and Wildlife for assistance with plant identification and processing our herbarium vouchers.

Kevin Thiele from the Department of Parks and Wildlife gave permission to adapt existing descriptions in floras developed by the Western Australian herbarium for this book and provided encouragement and support for the project. Robyn Barker and Ben Anderson are thanked for providing previously unpublished descriptions.

Photographic credits: We are very grateful to the following photographers for allowing us to include their images in this book: Belinda Barnett, Darren Brearley, Geoff and Ruth Byrne, Rob Davis, Murray Fagg, Giuseppe Messina, Alexandra Rouillard and Eddy Wajon.

To the staff and students from the Botanic Gardens and Parks Authority (Kings Park) and The University of Western Australia for direct assistance in the laboratory components of the project, we thank you. Special mention must go to co-authors Shane Turner, David Symons, and Peter Golos, for their contributions to individual chapters, and to Sam Mueller, Wolfgang Lewandrowski, Emma Dalziell, Conor Lawlor, and Adam Cross for direct assistance in the laboratory and field.

We would also like to thank the very helpful and supportive staff at CSIRO Publishing for the final production of this book.

We hope this atlas to the plants and seeds of the Pilbara will be a valuable resource to have in your bookshelf.

Chapter 1
Introduction to plant diversity of the Pilbara

Todd E. Erickson and David J. Merritt

Regional characteristics of the 'The Pilbara'

Landscape and climate

In the northwestern corner of Australia's arid zone lies the Pilbara biogeographical region ('the Pilbara'). This region represents a unique, ancient landscape that consists of distinct geology, climate, topography, and vegetation (Figure 1.1)[1]. Situated largely in the summer rainfall zone of the Australian tropics, the majority of the rainfall occurs when temperatures are at their peak between December and March (Figure 1.2)[2]. Maximum temperatures consistently exceed 40–45°C in parts of the Pilbara during summer. In addition to these extreme temperatures, boom and bust cycles of moisture availability can be highly variable and dependent on summer thunderstorms, tropical cyclones and sporadic winter rains. Therefore, the flora of this arid region is highly attuned to survive harsh seasonal fluctuations in temperature and moisture availability and this is strongly reflected in the biology of the seeds found throughout the Pilbara.

Within the Pilbara, four floristic subregions are recognised: the Hamersley, Fortescue, Chichester, and Roebourne subregions (Figure 1.1)[1]. The elevated ranges, ridges, and mesa outcrops of the Hamersley Ranges located in the south (Figure 1.3) are the landscape features most characteristically associated with the Pilbara. Some of the world's oldest rocks, at least 3.5 billion years old, can be found in these landscapes. Feeding out from, and winding through these hills, are deep gullies, drainage lines, and creeks that flow into the many rivers, drainage flats, and floodplain networks of the Fortescue Marsh to the north and the Gascoyne region to the south.

The Fortescue subregion bisects the Pilbara in an east–west direction and forms a distinct topographical barrier between the southern mountains of the Hamersley Range, and the northern regions that contain more isolated rocky ranges (e.g. Chichester Range), granitic outcrops, extensive plains, coarse river systems (e.g. the De Grey, Fortescue, and Yule Rivers), and the fringing coastal zone. The extensive area of Fortescue Marsh provides a stong biogeographic boundary or barrier for many Pilbara species, both plants and animals. Ecological studies of geckos have highlighted that this extensive topographical divide has created a long-standing genetic divergence between the north and south[3]. Yet, outside of these studies, very little is known about the unique evolutionary history of the Pilbara's flora and fauna and further studies should be undertaken across a wide range of organisms to shed light on this history.

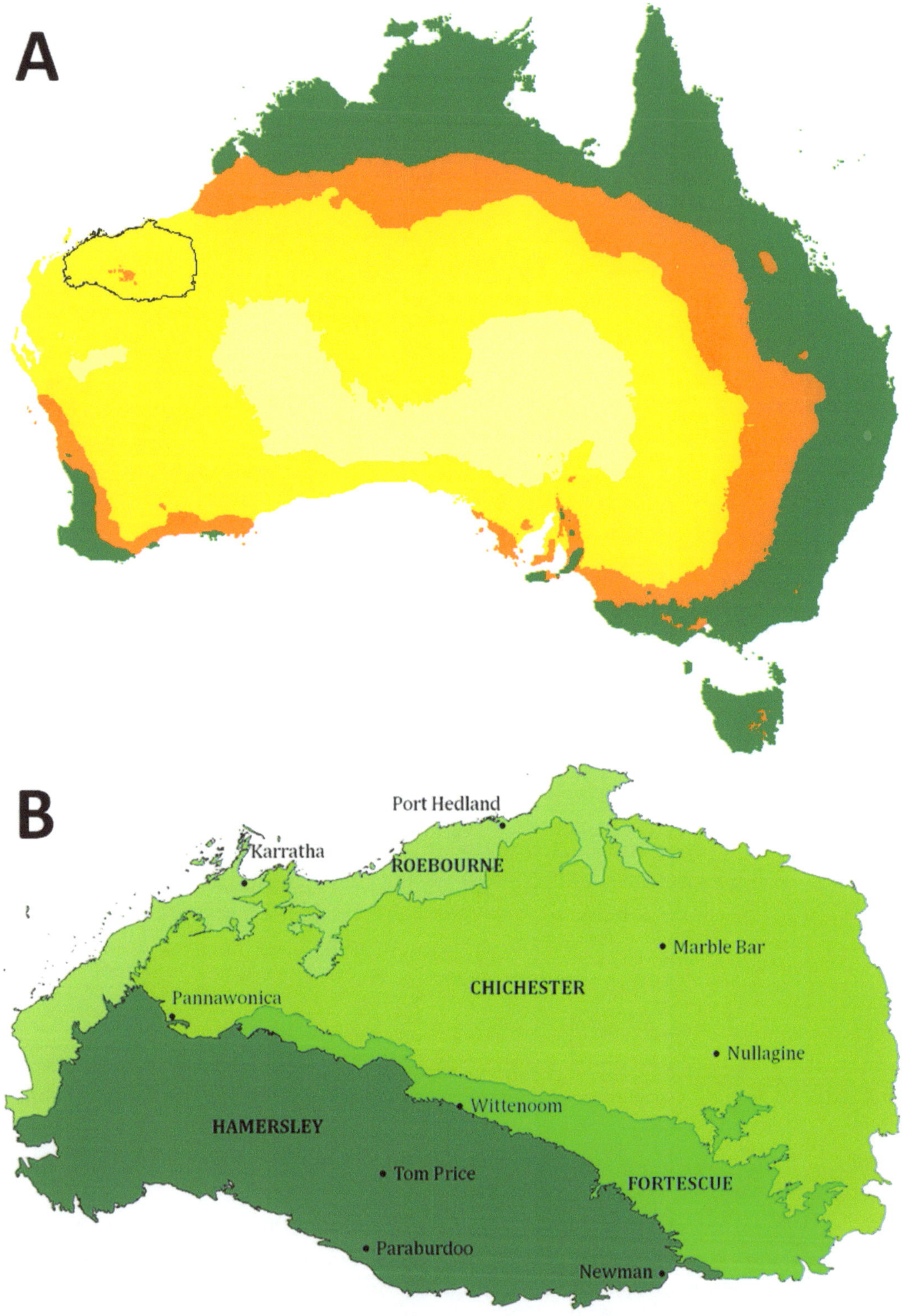

Figure 1.1 Much of inland Australia is considered arid to semi-arid with an average annual rainfall of <350 mm (yellow shading in A). The Pilbara region is situated in northwestern Australia and is known for its unique geology, topography, and vegetation associations (black outline in A). Floristically, the Pilbara is divided into four subregions: the Hamersley, Fortescue, Chichester, and Roebourne (B). Regional towns are scattered throughout the Pilbara and support various mining, pastoral, and tourism industries associated with major national parks.

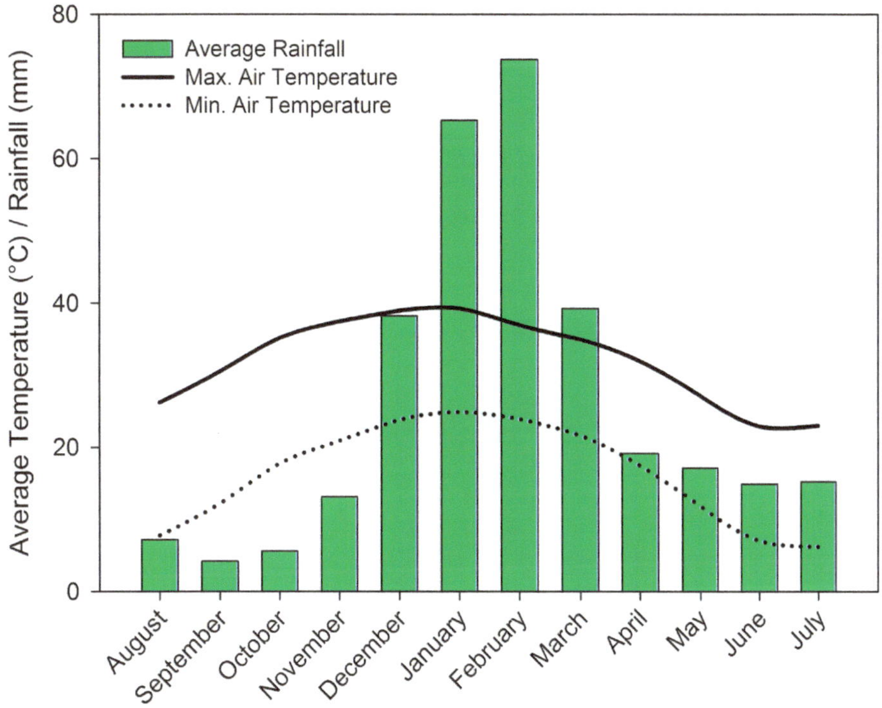

Figure 1.2 Long-term climate data for Newman Airport 1971–2014[2]. Data represents average monthly rainfall and average monthly maximum and minimum temperatures. Peak rainfall and hot summer temperatures in the Pilbara occur between December and March, with rainfall being highly variable from year to year.

Figure 1.3 Escarpments covered in *Triodia* hummock grasslands are characteristic of the Hamersley sub-region. Plant diversity and community structure changes across escarpment ridges (A), mesa cliff faces (B), rocky mid-slopes (C), and drainage flats usually dominated by *Acacia* shrublands (D).

Across the whole Pilbara, total plant diversity is approaching 1,800 species, with approximately 15% of these species endemic to the region[4]. Floristically, hummock grasslands (*Triodia* spp.) dominate the mountain chains and the more skeletal soils of the undulating plains (Figure 1.4). The region is also known as a secondary centre of species richness and endemism for *Acacia* species in Australia[5]. Criss-crossing and bisecting the expansive hummock grasslands are shrublands situated on the deeper soils and drainage flats made up predominately of wattles and grevilleas (*Acacia* and *Grevillea* spp.), scattered mallees and trees (*Eucalyptus* and *Corymbia* spp.), and tussock grasslands (*Aristida* and *Eragrostis* spp.). More striking are annual species that abound after good rains as well as fire ephemerals and herbaceous elements that grow in higher densities immediately after fire and in the buffered corners of the landscapes (e.g. rocky mesa outcrops and permanent waterholes).

Figure 1.4 Hummock grasslands of the Pilbara dominate most landscapes. Currently, 22 different spinifex grasses (*Triodia* species) are recognised in the region[6] (Photo: Alexandra Rouillard).

Species selection for restoration and conservation programs

When considering the differences in plant species and community assemblages, both through space and time, it becomes evident that there is a wide array of vegetation types present in Pilbara ecosystems. If land restoration and conservation projects are to be successful, a wide selection of species from these vegetation types needs to be available. The four most common families in the Pilbara, in terms of species richness, are the Fabaceae, Poaceae, Malvaceae, and Asteraceae (Table 1.1)[4]. The core elements of the vegetation comprise long-lived perennial grasses, shrubs, and trees, and include species of *Triodia*, *Eragrostis*, *Acacia*, *Senna*, *Tephrosia* and *Eucalyptus*. Targeting plants from these framework groups and obtaining sufficient quantities of seeds will ensure that the majority of species are available for restoration and conservation programs.

Table 1.1 The top 10 families and genera by number of species in the Pilbara.

Families	Species	Genera	Species
Fabaceae	320	*Acacia*	122
Poaceae	223	*Eremophila*	60
Malvaceae	137	*Ptilotus*	43
Asteraceae	136	*Euphorbia*	35
Chenopodiaceae	100	*Sida*	33
Amaranthaceae	75	*Senna*	31
Cyperaceae	63	*Goodenia*	29
Goodeniaceae	61	*Tephrosia*	28
Scrophulariaceae	61	*Solanum*	27
Myrtaceae	55	*Cyperus*	26

References

1. McKenzie N., van Leeuwen S., Pinder A. (2009) Introduction to the Pilbara biodiversity survey, 2002–2007. *Records of the Western Australian Museum, Supplement* **78**, 3–89.
2. BoM (2015) *Bureau of Meteorology*, Commonwealth of Australia, URL: http://www.bom.gov.au
3. Pepper M., Doughty P., Keogh J. S. (2013) Geodiversity and endemism in the iconic Australian Pilbara region: a review of landscape evolution and biotic response in an ancient refugium. *Journal of Biogeography* **40**, 1225–1239.
4. DPaW (2007–) *NatureMap: Mapping Western Australia's Biodiversity*, Department of Parks and Wildlife, URL: http://naturemap.dpaw.wa.gov.au/
5. González-Orozco C. E., Laffan S. W., Miller J. T. (2011) Spatial distribution of species richness and endemism of the genus *Acacia* in Australia. *Australian Journal of Botany* **59**, 600–608.
6. Western Australian Herbarium (1998–) *FloraBase – the Western Australian Flora*, Department of Parks and Wildlife, URL: http://florabase.dpaw.wa.gov.au/

Chapter 2
Seed collection, cleaning, and storage procedures

Todd E. Erickson and David J. Merritt

Seed Collection

For many plant species in the Pilbara, outside of those that are dominant and easily collected, knowledge of flowering phenology and seed maturation timing is limited. The phenological information reported in this section is compiled from personal observations of the authors (Chapter 5), the flowering period reported on databases such as FloraBase[1], and, if available, records from commercial seed collections made within the last five years. As further knowledge is gathered, guidelines as to the optimal time for collecting will become more readily available. Nevertheless, it is clear that timing of seed collection in these arid landscapes is heavily influenced by localised rainfall. Seed collection programs need to be carefully planned and communicated to on-ground practitioners well in advance of the collection season to ensure the timely collection of priority species. Importantly, collection activities may need to be organised at short notice to capitalise on sporadic rainfall events that are a characteristic of the Pilbara region.

Two distinct seed collection seasons exist in the Pilbara. Most long-lived tree and shrub species (particularly *Acacia*, *Grevillea*, and *Senna* species) access water from deep within the soil profile and commence flowering between July and September. Seed production and maturation of these species typically occurs between September and December following the onset of hot weather (Spring; Figure 2.1). In contrast, shorter-lived herbs and perennial grasses, such as the dominant *Triodia* species and the common *Cymbopogon*, *Aristida*, and *Enneapogon* species especially, respond to rainfall between January and March. Rapid flowering and seed set then generally occurs 4–6 weeks after sufficient rainfall. Occasionally, above average or out-of-season rainfall can lengthen the flowering and seed maturation season and spread collection times into May–July (Autumn; Figure 2.1).

A number of species in the Pilbara have canopy-stored seeds (serotiny), including species of *Eucalyptus* and *Melaleuca*. However, unlike the serotinous species from more temperate regions of Western Australia, the period of canopy storage appears to be short; ranging between 1 and 3 years, depending on the species. Commonly these *Eucalyptus* and *Melaleuca* species only flower and produce fruit during seasons with above average rainfall and it appears likely that seed production only occurs on a subset of these serotinous species every 3–5 years (Serotinous species; Figure 2.1). With the relatively short periods of canopy storage, timing of collection for serotinous species appears to be species- and rainfall-specific and requires careful observation at local and regional scales.

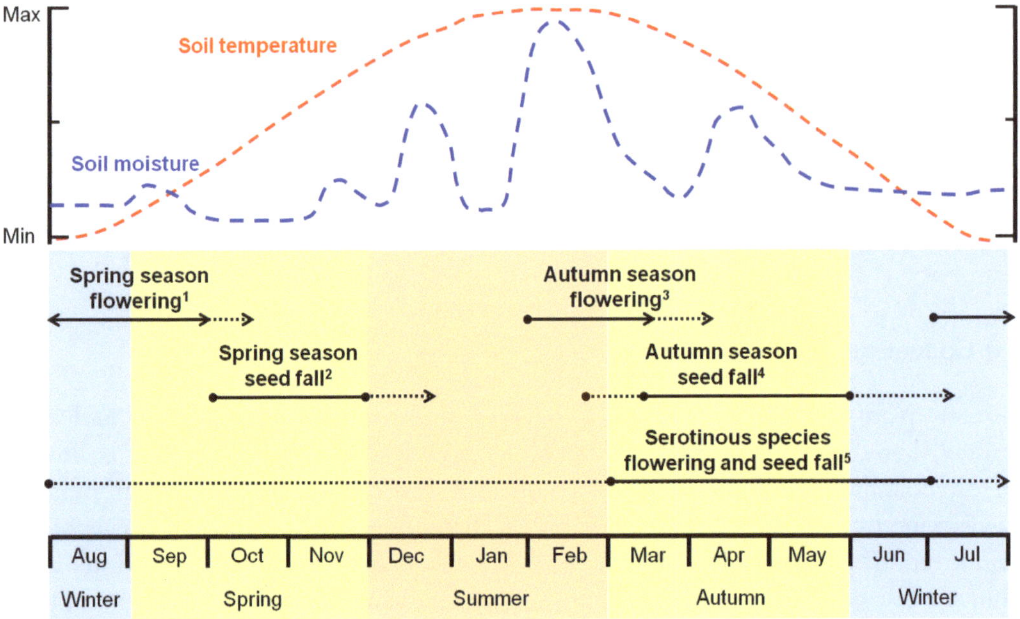

Figure 2.1 A diagrammatic representation of the Pilbara flowering and seed collection seasons.

[1]Spring season species comprise many perennial plants that flower before the onset of the summer rain season and when temperatures are relatively cool to warm (Jul.–Sep.).
[2]Fruit/seed maturation of spring season species occurs in October to November and plants typically dehisce seeds rapidly when the seasonal temperatures become hot and dry.
[3]Autumn season species comprise shorter-lived herbaceous plants and perennial grasses that respond rapidly to elevated soil moisture conditions following sufficient summer rains which initiate new vegetative growth and flowering.
[4]Most herbaceous and grass species shed seed rapidly, within 4–6 weeks of flowering, and there is a short window of time to maximise seed collections.
[5]For serotinous species of the Pilbara, flower and fruit production occurs *en masse* in isolated patches that receive above-average rainfall. Flowering and fruiting events can occur throughout the year, but vary both in space and time. The period of canopy storage of seeds appears to be shorter than the serotinous species of temperate regions of Australia.

Methods for seed collection include hand collection of small herbaceous species (e.g. *Boerhavia coccinea*), the removal of pods and fruits with secateurs (e.g. *Eucalyptus gamophylla*), through to knocking and shaking mature seeds into buckets, canvas bags or ground sheets to obtain larger quantities of seed (e.g. *Acacia inaequilatera*). Commercial seed suppliers also commonly acquire larger collections of grass species (e.g. *Triodia* spp.) with the use of machine-mounted harvesters. More details regarding seed collection methods for specific species can be found in Chapter 5.

Seed Cleaning and Quality Control

When targeting a particular species for seed collection, whether it is for a small-scale landscape and conservation projects or for a mine-site restoration program requiring tonnes of seeds[2], it can be difficult at times to distinguish the seed from the non-seed material, and to determine what proportion of the collection is plant debris and waste. Therefore, it is critical to have a sound understanding of the different types of seeds that exist and the best ways to handle, clean, and maintain quality seeds. Management practices including the timing of seed collection and post-harvest handling procedures, as well as species-specific factors including genetic traits[3] and seasonal conditions during seed maturation[4], can all influence the storage behaviour and longevity of the collected seeds, and their potential to establish and grow into mature plants.

The biology of fruit and seed production and the mechanisms of seed dispersal in Pilbara species differ across the various plant families. This variation can lead to confusion, particularly with respect to the fruit and/or seed type that forms the dispersal unit.

Four distinct 'types' of fruits/seeds are found in Pilbara plants (Figures 2.2 and 2.3):

- Type 1: Individual seeds that are released from dry pods/fruits with minimal excess plant debris (e.g. *Acacia* and *Grevillea* species).
- Type 2: Seeds that are enclosed in an indehiscent floret/'fruit' (e.g. *Triodia* and other grass species; *Ptilotus* species).
- Type 3: Species that release seeds from capsules/fruits after a period of canopy-storage (i.e. serotinous species), mixed with varying levels of plant debris or chaff (e.g. *Eucalyptus* and *Melaleuca* species).
- Type 4: Seeds that are dispersed in an indehiscent woody fruit that can contain one or more seeds (e.g. *Tribulus* and *Eremophila* species).

Collection purity

To assess the purity of a collection the proportion of fruit/floret or seed material to foreign debris must be determined. Depending on the species and collection techniques, foreign and non-vital plant debris can include empty or remnant floral material, chaff, plant or leaf fragments, and soil particles. Effectively, a purity assessment takes into account the seed or fruit type (Figures 2.2 and 2.3) that is dispersed from the parent plant. For species such as *Triodia*, a purity assessment would commonly determine the proportion of fruits/florets to debris. Whereas for species such as *Acacia*, the purity assessment would determine the proportion of seeds to debris. After separating fruit/floret or seed material from debris, fruit/floret/seed batch purity (expressed as a percentage) is defined as the weight of pure fruit/floret/seed material divided by the weight of plant debris (e.g. chaff and pod material). Fruit or seed batch purity in Pilbara species can vary considerably. The commonly applied technique to collect *Acacia* seeds, whereby seeds are shaken and knocked from the pods (Figure 2.2A) into buckets and canvas bags, results in a high seed purity collection, with minimal non-seed material. Collections with low seed purity values are common to species that have large quantities of chaff inside the fruits (e.g. *Eucalyptus* species; Figures 2.3B and 2.4), or a large number of empty dispersal units (e.g. *Triodia* and *Ptilotus* species). Quantifying seed purity is particularly important when it is difficult or impractical to separate the seed from the non-seed material. For example, some species of *Eucalyptus* (Figure 2.4) and *Melaleuca* produce small seeds that are similar in size to the ovulodc (chaff) material. If the seeds cannot be separated from the chaff, data on the seed purity should be gathered to enable the number of seeds in the collection to be accurately calculated. An additional example for assessing purity of collections of *Triodia* seed is provided in Box 2.1.

Figure 2.2 Examples of the different fruit and seed types present in Pilbara species (Type 1 = A and B; Type 2 = C and D). Individual seeds of *Acacia* species are easily separated from dry pods following collection (A), with the resultant individual seeds easy to clean and handle (B). Indehiscent fruits or the 'floret' of *Triodia* species (C) and *Ptilotus* species (D) may or may not contain a viable seed; floret 'fill' must be determined to assess the quality of a collection, as many collections comprise >50% empty florets (see Table 2.1).

Seed 'fill' or viability

Assessing seed 'fill' or seed viability requires the confirmation of whether a seed contains a viable embryo (and endosperm, if present). For an indehiscent fruit, the assessment determines if the fruit contains one (or more) viable seeds. Traditionally, this assessment is carried out by a 'cut test' – the dissection of seed/fruit material under a microscope and inspection of the health and integrity of the embryo/endosperm (Figure 2.5). A firm, moist, and white/cream embryo and endosperm indicates a healthy seed in most species. Some species, such as *Acacia* and *Senna* spp., have a cream/yellow embryo and may or may not have endosperm. More recently, seed fill is often assessed using x-ray analysis, which has the benefit of being non-destructive, accurate and rapid (Figure 2.6). However, x-ray images require training to interpret correctly and their interpretation should always be calibrated by dissecting a sample of the seeds to visually inspect the internal tissues.

Seed Storage

Following collection and cleaning, seeds may be stored for varying periods prior to use in restoration. All seeds have a finite storage life and the rate at which viability declines is dictated by many biological and environmental factors. Effective post-harvest handling

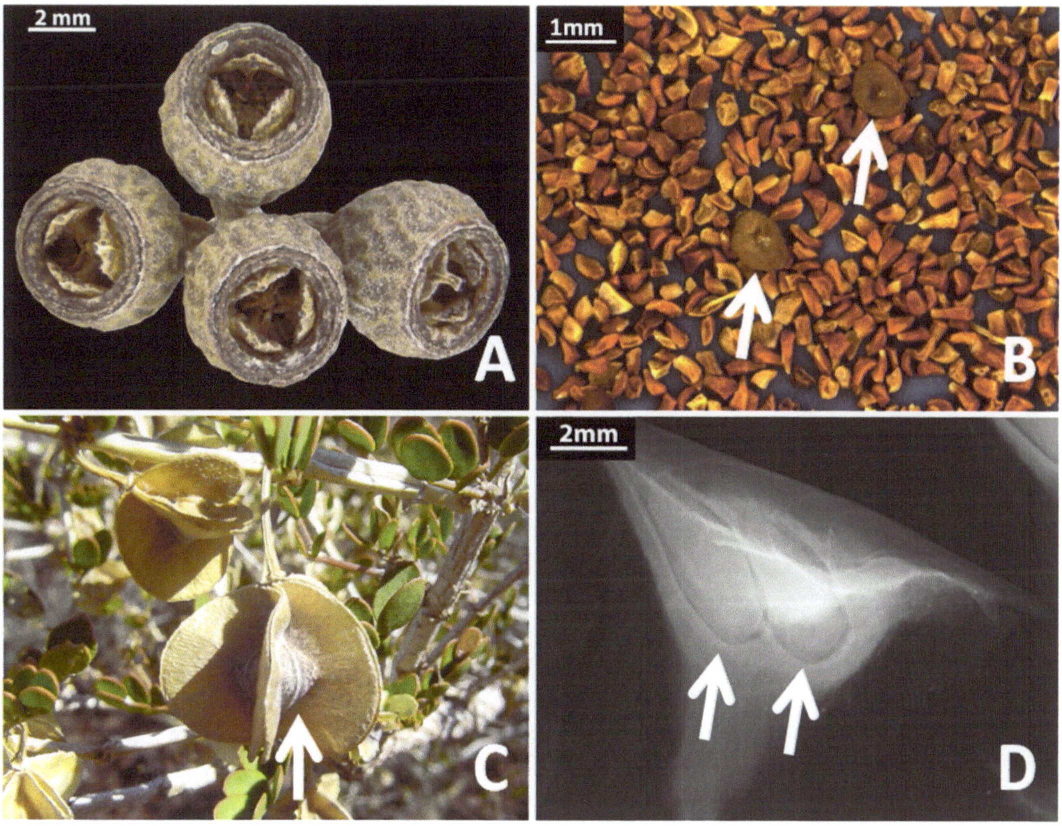

Figure 2.3 Examples of the different fruit and seed types present in Pilbara species (Type 3 = A and B; Type 4 = C and D). Woody capsules that retain their seeds within the plant canopy commonly release seeds shortly after collection (A), with cleaning required for some species, such as *Eucalyptus*, to separate seeds (arrows in B) from the high amounts of ovulode (chaff) material. Woody indehiscent fruits occur in genera such as *Eremophila*, *Newcastelia*, and *Tribulus* (arrow in C), and can contain one or more seeds dispersed whole from the adult plant. For example, the x-ray image of *Tribulus suberosus* shows a fruit with two seeds encased in an indehiscent woody fruit (arrows in D).

and storage procedures are fundamental to ensuring that high quality seeds are available when required. Seeds of most species from arid, temperate, and Mediterranean type climates have 'orthodox' storage behaviour and most, if not all, species in the Pilbara are expected to produce seeds capable of being dried and stored[5]. Orthodox seeds undergo a maturation drying phase during their final stages of development on the parent plant and are therefore naturally tolerant of desiccation. Such seeds can be safely dried to very low moisture contents (to 5% or less of dry weight) and can often be stored at sub-zero temperatures (e.g. in standard freezers at -18°C). Although, some species from other floras can be desiccation tolerant, but susceptible to degradation under cold storage (e.g. some orchid species).

Seeds in storage age at different rates. Different seed batches of thc same species can vary in their longevity under identical storage conditions, and even individual seeds within the same seed batch vary in their longevity. Seed longevity is a complex trait that is influenced by inherent genetic tolerances, the maternal environment during seed development, the initial quality and maturity of the seeds at the time of collection, and the storage environment. In collecting seeds from wild plant populations, there is little that can be

Figure 2.4 Collection of seeds of *Eucalyptus camaldulensis* with a high proportion of ovulode (chaff) material. The low seed purity is due to the small proportion of seeds (see arrows) relative to this chaff (i.e. by weight). Unsorted materials consume valuable space in storage and do not represent a high quality collection. If possible, additional cleaning to separate the chaff from the seeds should be undertaken to improve the quality of the collection. If further cleaning is not possible, data on the purity of the collection is important for planning restoration programs.

Figure 2.5 Healthy and viable 'filled' seeds clearly have a firm and moist embryo (Em) and endosperm (En). Species-specific differences exist in the colouration, and the presence/absence of endosperm. For example, seeds from *Senna* species (left) have a yellow embryo that, including the two cotyledons (C), extends the complete length of the seed, and a glassy (semi-vitreous) endosperm (En). In contrast, seeds from *Acacia* species (right) contain an embryo and cotyledons, but do not have an endosperm.

Box 2.1 Explaining fruit/seed purity and floret fill assessments: an example using commercially supplied *Triodia* (spinifex) material

The terminology used for *Triodia* collections is often confused between commercial suppliers and end-users including researchers and industry, leading to misunderstanding of the true quality of the collection. *Triodia* seeds are dispersed in an indehiscent floret (generally termed a 'fruit') (Figure 2.2 C). Commercial collections of *Triodia* are commonly supplied in bags that contain 5–10 kg of material. The first quality assessment that should be carried out is fruit/floret purity (measured as the proportion of floret to non-floret material). As commercial collections generally contain only florets, and no other plant material, the purity should be close to 100%.

However, not all florets will contain a seed and to assess seed purity, a known quantity of florets (e.g. 100 g) must be passed through an appropriately sized sieve to isolate or remove the seeds from the surrounding floret material. After sieving, the seed and non-seed material (i.e. floret debris) then needs to be separated either by hand or by using vacuum separation. Seed purity is then calculated as the proportion of pure seeds relative to the floret debris. This is determined by weighing the seed and non-seed fractions separately. For example, out of 100 g of starting material, a weight of 30 g of pure seeds, and 70 g of floret debris, would equal a seed purity of 30%.

A floret fill assessment relates to the proportion of florets that contain a seed, determined via x-ray analysis or by conducting a cut test, where the floret is dissected and assessed for the presence or absence of a seed. For example, out of 100 florets, if 40 florets contained a seed, and 60 florets were empty, the floret fill rate would be 40%. Floret fill rates for *Triodia* are highly variable and can range between 10 and 60% (Table 2.1). Therefore, each collection needs to be assessed independently.

In a restoration context, floret purity (or seed purity), floret fill rates, and seed quantities can provide sufficient information to calculate the true number of potential plants (or pure live seeds) that are being sown per hectare. Optimum seeding ratios (kg/hectare) can then be determined to ensure the target number of adult plants is available for restoration programs.

Table 2.1 Example of the variation in floret fill across multiple collections and species of *Triodia* (spinifex). Collections of *Triodia* species often have a low percentage of filled florets, and variation in floret fill between collections of individual species is also common.

Species	Floret Fill % (via x-ray)
T. epactia	10.8
T. epactia	17.6
T. wiseana	20.8
T. lanigera	27.2
T. pungens	32.8
T. basedowii	39.6
T. sp. Shovelanna Hill	42.4
T. lanigera	60.8
AVERAGE	**31.5**

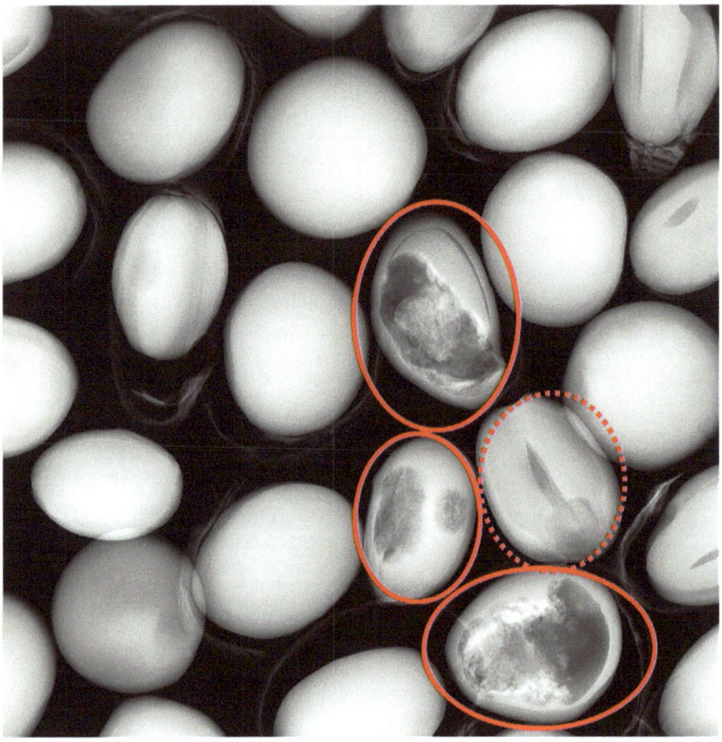

Figure 2.6 An example of an x-ray seed 'fill' assessment of *Acacia inaequilatera* seeds showing healthy, viable seeds (without circle highlights) and non-viable seeds seen as containing shrivelled/empty/predated embryos (red circles). The viability of the seed within the dashed red circle is questionable and requires dissection to clarify the quality.

done to control genetic factors of individual species or the environmental conditions experience by the parent plants. However, the initial quality of the seeds can be managed in part through careful observations of the patterns of seed maturation and dispersal within and amongst different species. Seed maturation is important because seeds become tolerant to desiccation and achieve their maximum potential longevity only during the very final stages of development[4]. This means that wherever possible seeds should be collected at the point of natural dispersal (see information for each species in Chapter 5). In practice this is not always achievable as the onset of flowering and seed maturation can vary between different plants, and even within a single plant. Also, waiting until the point of dispersal to collect seeds obviously risks losing many to the surrounding landscape. Couple these factors with the long distances and remote locations common to many wild seed collection expeditions, and it is clear that compromises must be reached and though not satisfactory, seeds are often collected at varying stages of maturity.

Once seeds are collected the conditions under which they are dried and stored greatly influence storage life. Seed moisture content and storage temperature are the two most important factors governing the longevity of orthodox seeds. Generally, seed longevity is doubled for every 5°C reduction in temperature, and for every 1% reduction in seed moisture content. Seed moisture content is related to the surrounding environment, such that reducing the relative humidity of the surrounding atmosphere reduces the moisture content of the seed. A 1% reduction in seed moisture content is approximately equivalent to a 10% reduction in relative humidity. Orthodox seeds are ideally dried in a cool environment and genebanks storing seeds for long-term conservation generally dry seeds at 15°C and 15–20% relative humidity; these conditions are considered optimal for seeds that are to be stored at -18°C (Table 2.2). Certainly, drying seeds under warmer temperatures (e.g. above 30°C) is best avoided as freshly collected, moist seeds can lose viability rapidly under these conditions. Techniques for drying seeds vary depending on the sophistication of the facilities and the purpose of storage, but include the use of desiccants (e.g. silica gel) whereby seeds are sealed in a container containing the desiccant, or the placement of seeds in porous bags or open trays in air-conditioned rooms, cabinet dehumidifiers, or purpose-built drying rooms. In the Pilbara, drying seeds under ambient conditions is not recommended due to the fluctuating, and potentially high, humidity conditions often experienced. Seeds should be dried to 5–10% moisture content, which should be achieved after 4–6 weeks in an appropriate drying environment. Seeds can then be packaged into an airtight container for storage.

Table 2.2 Recommended environmental conditions for drying and storing seeds based on the intended length of storage[6].

Storage duration	Drying Conditions	Storage Temperature
Short (up to 5 years)	Air-conditioned room or ambient humidity (only if <50%)	Air-conditioned room (23°C) Refrigerator (5–10°C)
Medium (5–10 years)	Relative humidity 15–20% Temperature 15–20°C	Refrigerator (5–10°C)
Long (more than 10 years)	Relative humidity 15–20% Temperature 15–20°C	Freezer (minus 18°C)

References

1. Western Australian Herbarium (1998–) FloraBase – the Western Australian Flora, Department of Parks and Wildlife, URL: http://florabase.dpaw.wa.gov.au/
2. Merritt D. J., Dixon K. W. (2011) Restoration seed banks – a matter of scale. *Science* **332**, 424–425.
3. Probert R., Adams J., Coneybeer J., Crawford A., Hay F. (2007) Seed quality for conservation is critically affected by pre-storage factors. *Australian Journal of Botany* **55**, 326–335.
4. Hay F. R., Probert R. J. (2013) Advances in seed conservation of wild plant species: a review of recent research. *Conservation Physiology* **1**, 1–11.
5. Merritt D. J., Martyn A. J., Ainsley P., Young R. E., Seed L. U., Thorpe M., Hay F. R., Commander L. E., Shackelford N., Offord C. A., Dixon K. W., Probert R. J. (2014) A continental-scale study of seed lifespan in experimental storage examining seed, plant, and environmental traits associated with longevity. *Biodiversity and Conservation* **23**, 1081–1104.
6. Offord C.A., Meagher P.F. (eds) (2009) *Plant germplasm conservation in Australia: strategies and guidelines for developing, managing and utilising ex situ collections.* Fully Revised Edition. Australian Network for Plant Conservation Inc., Canberra.

Chapter 3
Seed dormancy and germination of arid zone species

Todd E. Erickson, David J. Merritt and Shane R. Turner

Seed dormancy and germination

Dormancy and the natural environment

Seed germination and emergence are one of the most critical phases in the lifecycle of plants. As a consequence, seeds have evolved various mechanisms and environmental signal detectors to time their germination to particular periods that are most favourable for seedling establishment and survival[1]. Critical to the regulation of the germination process is seed dormancy, which is present in about 70% of all flowering plant species[2]. Seed dormancy is an adaptive trait that acts to prevent germination when environmental conditions conducive to germination occur at a time when the chance of seedling survival is low (i.e. when suitable germination conditions happen to arise outside of the normal growing season). Dormancy is a property of the seed that defines the environmental conditions that must be met for germination to proceed. In the harsh and highly variable conditions that are found in the Pilbara, seed dormancy prevents germination from occurring during any rainfall events that occur outside of the growing season (i.e. April–November) or following sporadic early rainfall in November and December, when conditions might be temporarily suitable for germination, but seedling survival and establishment is less likely due to short-lived moisture availability (Figures 1.2 & 3.1).

Physiological and/or morphological changes within the embryo, and/or physical and anatomical changes in the fruit or seed coat, act to regulate dormancy loss and germination[3]. Seeds respond to daily or seasonal changes in temperature, water availability, light, and other environmental cues induced by occasional events, such as fire (e.g. smoke and heat)[4]. These environmental signals are utilised by seeds to ensure that once germinated, seedlings can access sufficient resources to establish and survive.

The various classes of seed dormancy

Different classes of dormancy are recognised based on the physical, physiological, and morphological characteristics of a seed[5]. Understanding these seed traits provides insight into the natural germination ecology of different species and allows for the development of effective dormancy alleviating treatments for use in restoration and conservation programs. Of the five classes of seed dormancy recognised, physiological dormancy, and physical dormancy, are the two most common forms observed in seeds from arid climates[2,6,7]. Together, these dormancy classes account for approximately two-thirds of all dormant seeds in the Pilbara[7]. Seeds with physiological dormancy possess an internal physiological 'block' that suppresses embryo growth and prevents the embryo from penetrating the surrounding tissues (i.e. the endosperm, or the seed or fruit coat). This kind of dormancy is controlled principally through the balance of plant hormones within the embryo that act to inhibit (abscisic acid) or promote (gibberellic acid) germination[8]. Seeds with physical dormancy have a water impermeable fruit or seed coat that prevents water uptake. Physical dormancy is largely controlled by a microscopic, specialised structure within the fruit/seed coat called the 'water gap', and germination is prevented until the water gap is ruptured and rendered permeable. Rupture of the water gap occurs naturally via alternating day/night temperatures, or by means such as extreme summer temperatures or fire, but seeds can be made permeable artificially through exposure to hot water, or scarification of the fruit/seed coat with sandpaper or sulphuric acid[9].

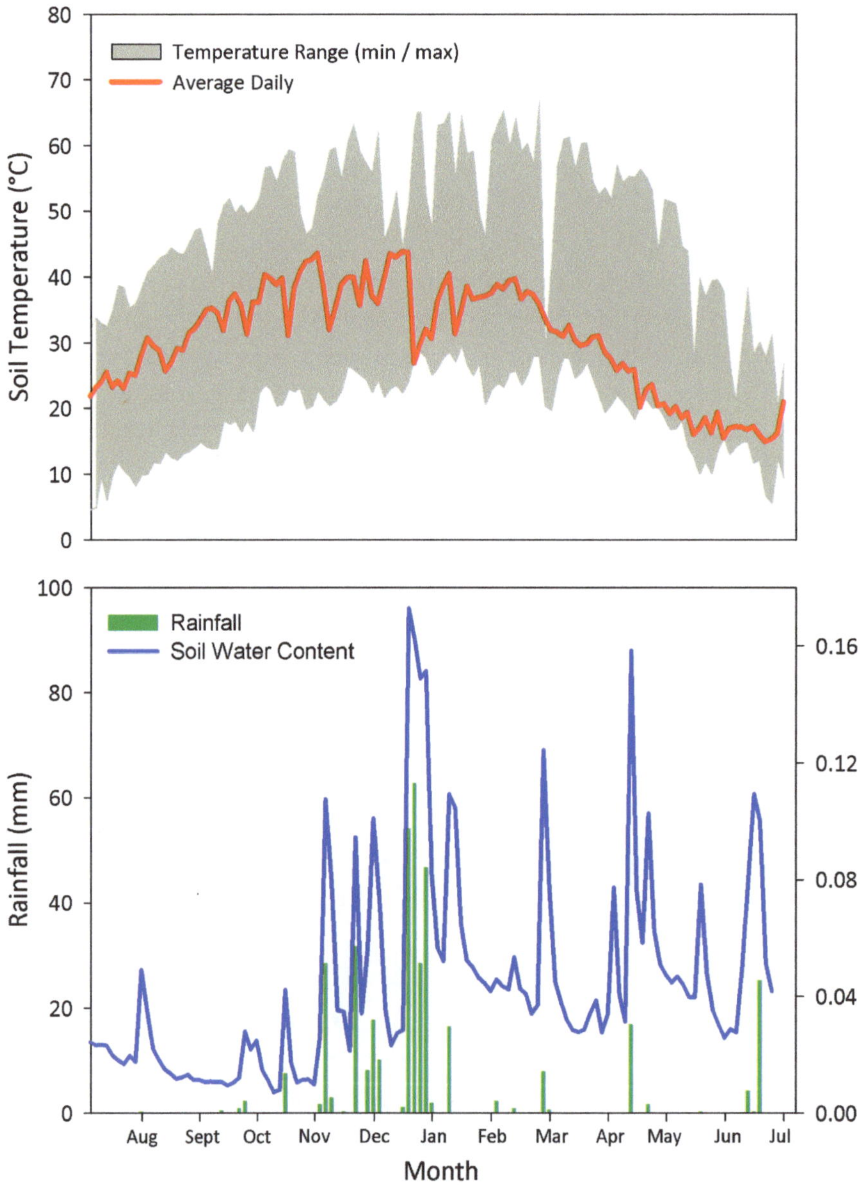

Figure 3.1 In the Pilbara, annual soil temperature (top) and moisture (bottom) conditions vary throughout the year (as demonstrated by these data recorded during 2013–14). Seed dormancy regulates the timing of germination so this critical life-stage transition aligns to more reliable rainfall events that generally occur from December to March.

Three other classes of seed dormancy are also recognised, namely morphological dormancy, morphophysiological dormancy, and combinational dormancy[6]; although these are far less common in seeds from arid regions. The main characteristic for seeds with morphological and morphophysiological dormancy is the presence of a tiny (under-developed) embryo which requires a period of development and/or growth inside the seed prior to radicle emergence (the 'morphological' component). An additional physiological block to germination is also present in seeds with morphophysiological dormancy. These seeds typically require extended periods of moisture availability and are largely excluded from arid environments, such as the Pilbara, due to the unpredictable rainfall and relatively short growing season. However, some examples of seeds with morphophysiological dormancy have been documented for plants that grow in particular environments of the Pilbara (e.g. *Astrotricha hamptonii* and *Hibbertia glaberrima* that grow on south-facing rocky mesas)[7]. Seeds with combinational dormancy possess both physical and physiological dormancy, meaning that they possess a water impermeable fruit/seed coat, and an internal physiological block to embryo growth. While uncommon, a number of plant species from the Pilbara have been recorded as producing seeds with combinational dormancy[7].

In the soil seed bank (see Chapter 4), a population of seeds of any particular species may comprise seeds that were dispersed with differing depths of dormancy, seeds of differing ages, and seeds that have experienced different environmental conditions (e.g. soil temperature and moisture) due to their position in the soil (e.g. depth). As a result, individual seeds within a population can release dormancy at different rates, meaning that only a proportion of the soil seed bank may have the ability to germinate when suitable conditions arise[1]. Alternatively, synchronous dormancy release and germination may occur in response to disturbance, such as fire, resulting in large-scale seedling recruitment (Figure 3.2). In an undisturbed ecosystem (e.g. prior to mining), the above-ground plant community seasonally sheds fresh seeds into the soil seed bank. These regular additions to the soil seed bank through time ensure that there is a continual supply of fresh propagules for the maintenance and persistence of the plant populations that make up the ecosystem.

In a restoration context, whereby practitioners are often faced with a physically or biologically (weedy) disturbed site, one of the most effective ways to return plant communities is through the use of correctly salvaged, well-stored and re-spread topsoil (see Chapter 4)[10,11]. However, there is often a lack of available topsoil and restoration depends upon the sowing of seeds directly into the disturbed landscape. To utilise native seeds for restoration purposes, sufficient quantities of seed must first be collected and stored for a period of time[13], and then spread in sufficient densities to kick-start the restoration process. Yet, if the majority of seeds are dormant, then germination and emergence may not occur following seed sowing, and in some cases, it may be several years before these seeds are released from dormancy and have the capacity to germinate. In essence, seeds that are wild-collected and maintained under ex-situ storage conditions are not exposed to the same environmental conditions that regulate dormancy loss in the natural soil seed bank. Consequently, seed collections that have been stored ex-situ may require artificial cueing (i.e. seed pre-treatments) before germination is possible[14].

Figure 3.2 Following a fire in the Pilbara, seedling recruitment of a wide range of species occurs en masse when there is sufficient rainfall to promote germination (A). Synchronous dormancy release, germination, and seedling emergence results in hundreds of individuals occupying a few square metres. An alternative strategy for plants in fire-prone ecosystems is the ability to regenerate through re-sprouting (B), and this trait is evident in genera such as *Triodia*[12], *Eucalyptus*, and *Eremophila*. However, regeneration both from seeds and via re-sprouting will not occur in severely altered sites, such as reconstructed landforms after mining, as the seed bank and parental plants have often been removed during the mining process.

The step-by-step process of classifying seed dormancy

Classifying the type of dormancy seeds of a particular species possess provides important insight into the type of pre-treatments required to elicit germination. The five classes of seed dormancy recognised are based upon many decades of research on the germination biology of thousands of species from across the world's biomes[3,5,16]. Seed dormancy type can be classified by assessing a variety of morphological, physical, and physiological seed characteristics[5]. This includes an examination of (1) the developmental state and size of the embryo; (2) whether the fruit/seed coat prevents water uptake, and; (3) the germination characteristics of the seed batch across a range of conditions (Box 3.1). Once the type of seed dormancy is known, informed decisions can be made regarding seed pre-treatments that are more likely to be effective at releasing dormancy.

Box 3.1 Seed dormancy classification of *Acacia* seeds

An example of the dormancy classification system applied to seeds of a common Pilbara wattle, *Acacia ancistrocarpa*, is presented in Figure 3.3. *A. ancistrocarpa* seeds have a fully developed, 'investing' embryo type that takes up the whole inner space of the seed coat – this rules out the presence of morphological, or morphophysiological dormancy. Freshly collected seeds do not imbibe water unless treated in such a manner that ruptures the water impermeable barrier in the outer seed coat – this indicates the presence of physical dormancy. Once rendered permeable seeds germinate quickly and readily over a wide temperature range (Figure 3.3) – this excludes physiological dormancy and combinational dormancy. Therefore, the seeds of *A. ancistrocarpa* are determined to possess physical dormancy. Based on this understanding of dormancy type, *A. ancistrocarpa* seeds are now routinely scarified to allow water uptake with the knowledge that they will germinate rapidly over a wide range of temperatures once sufficient moisture is present.

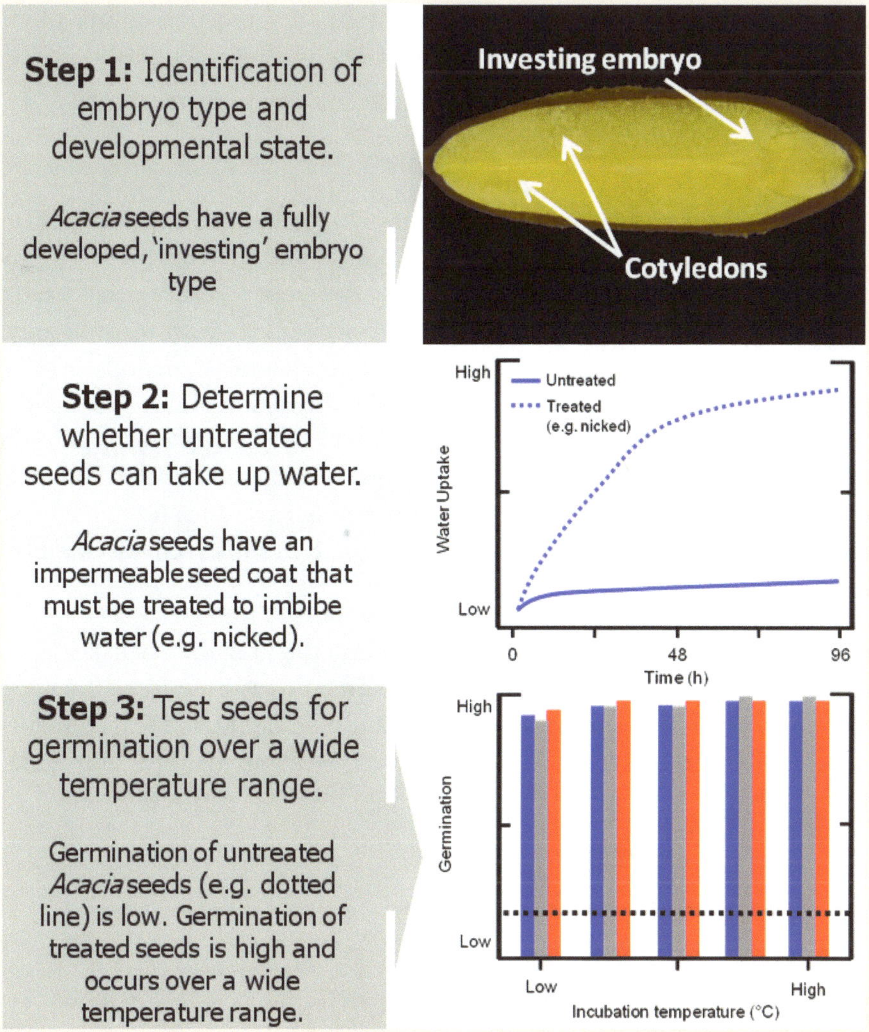

Figure 3.3 The seed dormancy classification process takes into account a variety of seed traits. In this example, seeds of the common Pilbara species *Acacia ancistrocarpa* are deemed physically dormant because: (1) The seeds have a fully developed, investing embryo type. (2) When left untreated, the seeds will not take up water due to a physical barrier present in the outer seed coat. (3) Germination is rapid and occurs over a wide range of temperatures once the seeds have been rendered permeable to water via brief exposure to hot water. Different colours on the bar chart (Step 3) shows germination of seeds in water only (blue), or following treatment with the germination stimulants gibberellic acid (grey) or the smoke-derived chemical karrikinolide (red) (Chapter 5).

This dormancy classification process has been completed for each species presented in Chapter 5. Some form of seed dormancy was confirmed in at least 70% of all Pilbara species studied[7], which is consistent with patterns of seed dormancy seen in arid zone species across the globe. Seeds exhibiting physiological (34% of species) and physical (31% of species) dormancy are the most common. Seeds that are non-dormant are also common and widespread (27%). Relatively few species with morphological, morphophysiological, and combinational (physical + physiological) seed dormancy have been observed in the Pilbara flora (all classes <5%) (Figure 3.4)[7].

In the sections below, the techniques used to assess key morphological, physical, and physiological seed traits are elaborated in greater detail using examples of species found in the Pilbara (also see Chapter 5).

Step 1: Classifying the type and developmental state of embryos

To determine both the embryo type and the embryo to seed length ratio (or E:S ratio), seeds need to be dissected and both the internal seed length and the embryo length measured. The morphology (e.g. position, size, and shape) of the embryo varies considerably among species. There are 13 different embryo types recognised[17,18], with morphology ranging from small, differentiated embryos with minute organs and cotyledons (e.g. *Astrotricha hamptonii* with rudimentary embryos; Figure 3.5 A), through to fully developed embryos that either circumnavigate the inner seed coat wall (e.g. *Gomphrena cunninghamii* with peripheral embryos, or *Dodonaea coriacea* with linear-coiled fully developed embryos; Figure 3.5 D and F, respectively), or fill up the entire internal space (e.g. *Abutilon otocarpum* with folded embryos; Figure 3.5 I). So far, 11 of the currently recognised embryo types have been found in Pilbara species (Figure 3.5).

To determine the E:S ratio, the seed is dissected along the longitudinal axis and the length of the embryo and the length of the seed is measured. Measuring from the internal seed coat wall, the seed length is determined as the longest straight line measurement possible (Figure 3.6). The embryo length is determined in a similar manner, but some embryo

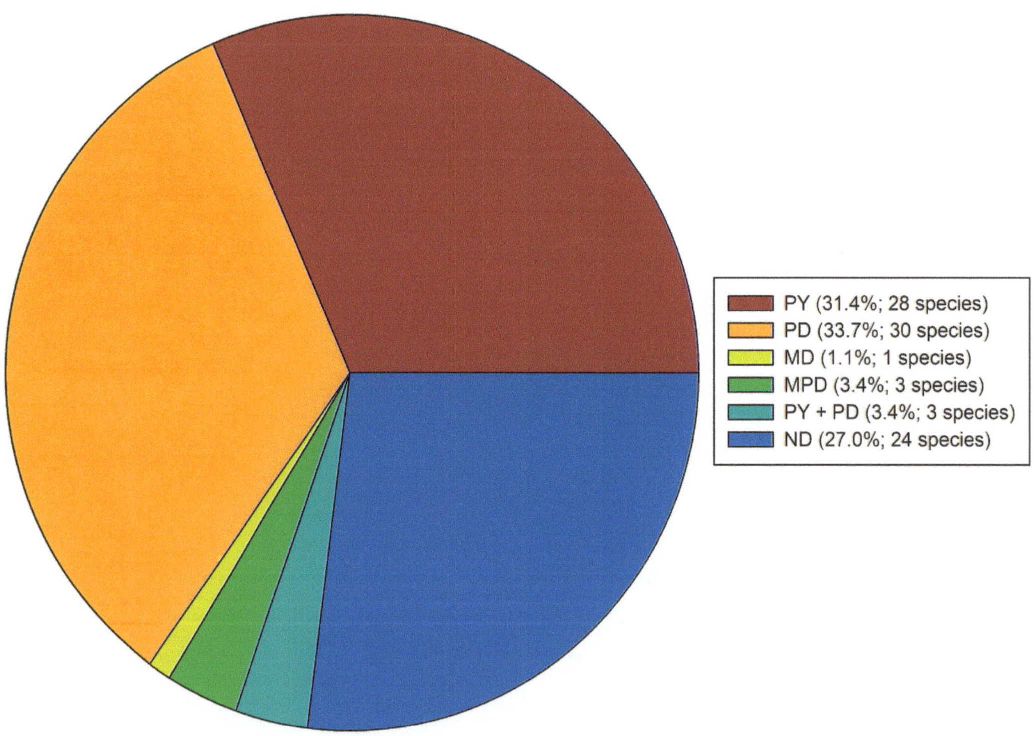

Figure 3.4 Proportions of seed dormancy types present in 89 species collected from across the Pilbara. PY = physical dormancy, PD = physiological dormancy, MD = morphological dormancy, MPD = morphophysiological dormancy, PY + PD = combinational dormancy, and ND = non-dormant species.

types exceed the internal length of the seed coat due to the embryo bending, curving, coiling, or folding back on itself (Figure 3.5 D–F, H, I). In these instances, the embryo length can be estimated by taking a series of measurements and adding together (solid red arrows, Figure 3.6). The E:S ratio can range from very small (e.g. <0.1 in species such as *Astrotricha hamptonii*; Figure 3.6) through to large (e.g. 2.98 in species such as *Dodonaea coriacea*; Figure 3.6).

The developmental state of the embryo is based principally on whether the embryo needs to grow inside the seed prior to radicle emergence, or whether no growth is required. To determine if an embryo is fully developed (no growth required) or underdeveloped (growth is required), the embryo length is measured when seeds are first imbibed (within 24 hours) on a moistened substrate, and again just prior to radicle emergence (completion

Figure 3.5 Examples of different embryo types found in seeds of Pilbara species. Embryo position, size, and shape within the fruit or seed vary across species. In longitudinal section, rudimentary (A), capitate (B), and lateral (C) embryos are positioned in the basal end of the seed and typically have an E:S ratio <0.5. Peripheral (D), linear[-curved] fully developed (E), and linear[-coiled] fully developed (F) embryos are typically large, elongated, and circumnavigate the inner seed coat wall. The E:S ratio in these species is >1 as embryo length exceeds the length of the seed. Spatulate fully developed (G), bent (H), and folded (I) embryos are also large, and are either equal in length (G), or exceed the length (H and I) of the inside of the seed, but have slightly varying shapes and orientations (e.g. the folded embryo type has cotyledons that fold to form an 'm' shape in the upper half of the seed coat). The undifferentiated, broad, spatulate underdeveloped, linear[-straight] underdeveloped, linear[-straight] fully developed, and investing embryos types are not shown here (but see Figures 2.5 and 3.3 to see an example of the investing embryo type). Note: Linear fully developed embryos can contain three sub-categories: linear-straight, -curved, and -coiled. Scale bars = 1mm.

of germination)[19]. If the length of the embryo increases prior to the point of radicle emergence, then the embryo is underdeveloped (this also means that seeds have either morphological or morphophysiological dormancy). Underdeveloped embryos have been confirmed in just three plant species from the Pilbara, *Hibbertia glaberrima*, *Stylidium desertorum*, and *Wahlenbergia tumidifructa* (Table 3.1 and Figure 3.7). However, species of *Astrotricha* and *Trachymene* may also contain underdeveloped embryos, as species that are found in the southwest of Western Australia from the Araliaceae are known to contain these types of embryos. Seeds with underdeveloped embryos usually germinate more slowly, as it may take up to 4–6 weeks for the embryo to grow inside the seed sufficiently for the radicle to emerge.

Table 3.1 Examples of embryo growth prior to germination for seeds of the Pilbara species *Stylidium desertorum* and *Wahlenbergia tumidifructa*. The data show an increase in E:S ratio (and embryo size) from the time when seeds were first imbibed (Initial E:S ratio), to the point in time immediately prior to radicle emergence (Final E:S ratio). The embryo growth is also evident through calculating the percent increase in embryo size.

Species	Initial E:S ratio	Final E:S ratio	Increase in embryo size (%)
Stylidiaceae			
Stylidium desertorum	0.256 ± 0.02 (78 μm ± 5.5)*	0.803 ± 0.05 (286 μm ± 33.1)**	266
Campanulaceae			
Wahlenbergia tumidifructa	0.539 ± 0.01 (256 μm ± 6.7)*	0.687 ± 0.02 (328 μm ± 9.8)**	28

*Embryo size 1 day after incubation. **Embryo size 3–9 days after incubation. Seeds were incubated for up to 9 days at 25°C on water (*Wahlenbergia*) or on the smoke-derived germination stimulant KAR_1 (*Stylidium*).

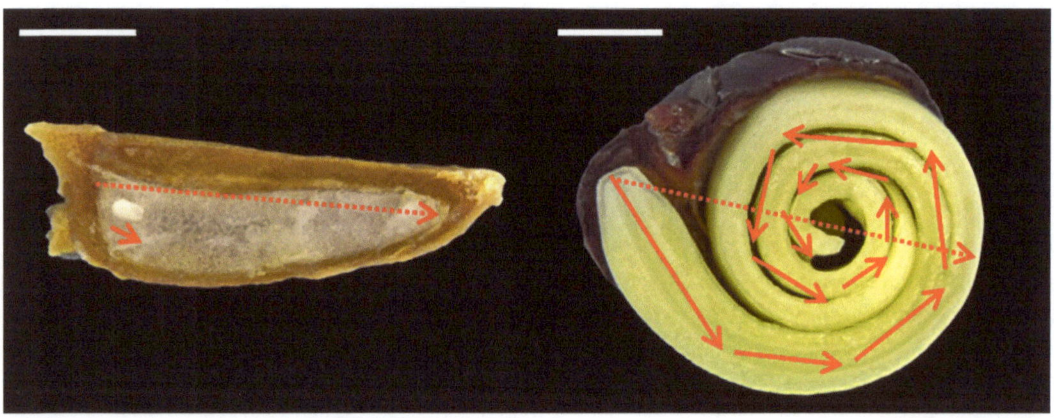

Figure 3.6 The embryo to seed length (E:S) ratio is determined by measuring both the length of the seed and the embryo. The length of the seed is determined by dissecting the seed longitudinally and measuring the longest straight line possible within the inner seed coat wall (red dashed arrows). The length of the embryo is measured in a similar manner, noting that some embryos exceed the length of the seed (solid red arrows). In the example above, the seed of *Astrotricha hamptonii* (left) has a E:S ratio of 0.08, whereas the seed of *Dodonaea coriacea* (right) has a E:S ratio of 2.98. Scale bars = 1 mm.

Step 2: Determining whether fruits/seeds can to take up water using an imbibition test

An imbibition test is used to determine the presence or absence of physical dormancy. The test is straightforward and is conducted by placing untreated fruits/seeds on a moistened substrate and periodically measuring the fruit/seed weight (generally at regular intervals over a 48–96 hour period) (Figure 3.8). If there is no significant increase in weight during the imbibition test, then it can be concluded that the fruit or seed coat has a water impermeable layer. For instance, untreated seeds of many *Acacia* species will not take up water, but once treated (e.g. nicked), seeds readily take up water and the increase in weight can be as much as 120% (Figure 3.8). In contrast, for *Triodia* species, both the florets (the natural dispersal unit) and the seeds (that have been removed from the florets) increase in weight by around 40% without any form of treatment[7]. The similarity in water uptake between the florets and the seeds leads to the conclusion that the florets do not restrict water uptake into the seeds of *Triodia* (Figure 3.8).

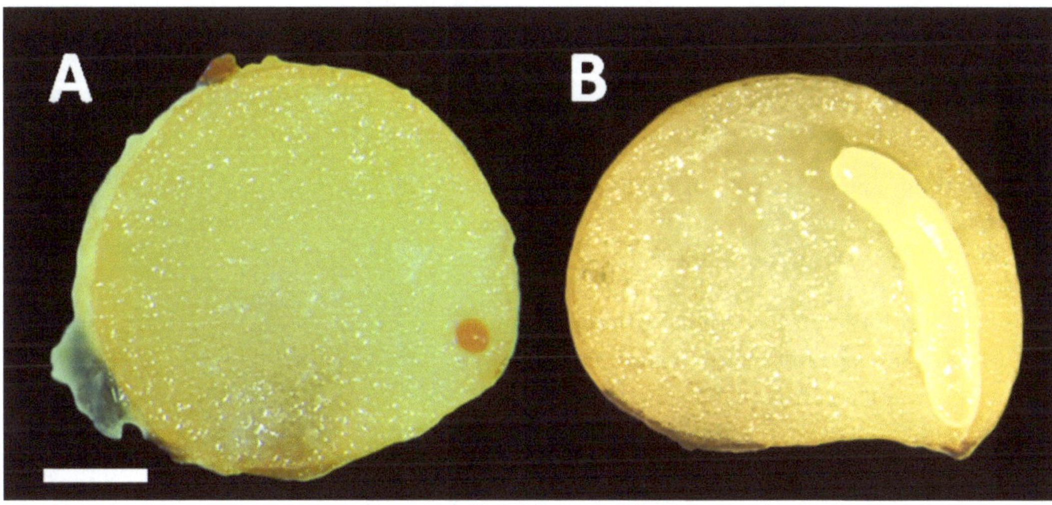

Figure 3.7 An example of embryo growth occurring prior to germination of *Hibbertia glaberrima* seeds. The initial embryo size of *Hibbertia* seeds is minuscule (<250 μm) and seeds have an E:S ratio of 0.11 (embryo stained red; A). Once treated to alleviated dormancy, the embryo grows within the seed prior to radicle emergence (B). At the point of radicle emergence, the embryo is at least 100 times larger. Scale bar = 500 μm.

Step 3: Determining the germination capacity of fresh seeds

The final step in the seed dormancy classification process is to determine the germination capacity of the seed batch over a wide range of conditions that are suitable for germination of non-dormant seeds. In this context, temperature is the most important factor to consider, but the exposure of seeds to light, or darkness, is an additional consideration. Optimal temperatures for germination of non-dormant seeds are generally aligned with those that occur during the period of reliable rainfall in the natural environment[20,21] (Figure 1.2). Species from the northern arid zone of Australia, including the Pilbara, commonly germinate best at warmer temperatures (between 25–35°C), reflecting the elevated soil temperatures during the summer rainfall season (Figure 3.1)[7,20]. Nevertheless, the temperature range over which germination is possible is species-specific, and can be narrow (as little as 10°C), or broad (up to 25°C or greater). Testing germination across

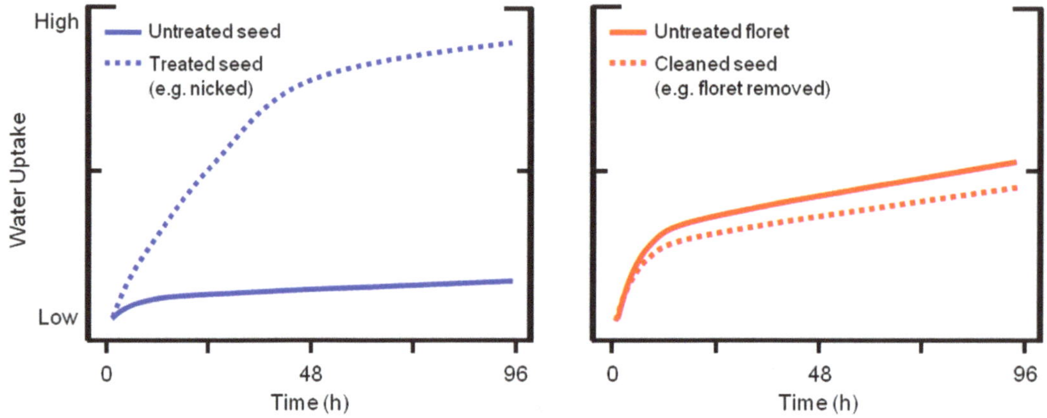

Figure 3.8 Examples of imbibition curves that can be generated from data collected during an imbibition test. Water uptake is measured as the percent increase in seed weight during imbibition. Untreated seeds of *Acacia* (left) possess physical dormancy and do not take up water; therefore, no increase in seed weight is evident. However, once treated (e.g. nicked), seeds readily imbibe and can increase in weight by as much as 120%. Florets are the natural dispersal unit for species of *Triodia* (right), and increase in weight by 40–50% when imbibed. The water uptake of cleaned seeds (i.e. florets removed) is similar, indicating that the florets do not impede water uptake, and ruling out the presence of physical dormancy.

a broad temperature range, for example between 10°C and 35°C, provides detailed information regarding the dormancy status of a seed batch, and is the first step to building a thorough understanding of the germination ecology of a species (Box 3.2).

Testing the response of seeds to chemicals known to promote germination is another useful step for characterising germination behaviour. One of the most ubiquitous chemicals used in germination testing is gibberellic acid (GA), a naturally occurring plant growth regulator. First discovered as a product of a pathogenic fungus in the 1930s, gibberellic acid is known to regulate many developmental events in plants, including seed dormancy, cell elongation, and flowering. There are many types of gibberellins (chemical derivatives of gibberellic acid), but GA_3 is the most commonly used for germination testing. Gibberellic acid can overcome physiological dormancy in many species and is widely used as a dormancy-breaking tool. Chemicals found in smoke generated from burning plant material are also important germination stimulants. Smoke has long been known to promote the germination of seeds of diverse plant species, from many different parts of the world and from many habitats. The primary chemical agent in smoke responsible for germination is karrikinolide (KAR_1) and it is estimated that at least 1,200 species respond to karrikinolide[22]. Like gibberellins, there are numerous types of karrikins that promote germination, but KAR_1 is the most active in most species. Karrikins are most active on seeds with physiological dormancy, but unlike the response of seeds to GA_3, the response to karrikinolide is strongly regulated by seed dormancy status. Some seeds respond immediately and consistently to KAR_1, but others must be cued to respond to KAR_1 through after-ripening, or by other means.

Pre-treatments for common types of seed dormancy

The majority of seeds from arid regions are likely to possess either physiological dormancy, or physical dormancy. Physiological dormancy can be complex and there are

Box 3.2 Species-specific germination responses for seeds of Pilbara species

For seeds of plant species of the Pilbara, a number of patterns of germination response to temperature, GA_3, and KAR_1 have been observed (Figure 3.9)[7]. For instance, freshly collected seeds that have been cleaned out of the florets of most grass species (Poaceae), such as *Eragrostis eriopoda* and *Neurachne muelleri*, have very low germination when tested on water, regardless of incubation temperature (Figure 3.9 A, B). Seeds of both of these species possess physiological dormancy, but have contrasting responses to GA_3; seeds of *N. muelleri* germinate at temperatures above 15°C, but GA_3 is not effective on seeds of *E. eriopoda.* Freshly collected seeds of *Triodia pungens* also have physiological dormancy and relatively low germination, but provided temperatures are greater than 15°C, treatment with GA_3 improves germination, and treatment with KAR_1 greatly improves germination (Figure 3.9 C). It is clear that seeds of these three grasses germinate best when temperatures are between 20°C and 35°C, and this is the most common germination response to temperature for summer-growing species of the Pilbara, but other patterns are observed. For example, a distinct and narrow temperature range for germination exists for some species, such as *Aristida latifolia* (Figure 3.9 D), *Calandrinia schistorhiza*, and *Chrysocephalum apiculatum.* These species usually occur in particular positions within the landscape where water is retained for longer periods (e.g. creek lines, claypans, gullies, and south-facing aspects). The narrow temperature window may, therefore, be acting as a filter to restrict seedling recruitment to specific positions within the landscape that align with the habitat requirements of the adult plant. Notably, the temperature window for germination of *Aristida latifolia* seeds is widened when seeds are first treated with GA_3; this is a common effect of GA_3 or KAR_1. At the other end of the spectrum are seeds that always germinate over a wide range of temperatures. Seeds with physical dormancy, such as those of *Acacia inaequilatera* (Figure 3.9 E), germinate readily across a broad temperature range once rendered permeable to water (e.g. via scarification), and chemical stimulants do not affect germination percentage. Seeds that are non-dormant, such as *Eucalyptus leucophloia* (Figure 3.9 F), also commonly germinate over a wide temperature range without pre-treatment.

These examples highlight different responses of seeds to temperature, GA_3 and KAR_1 and demonstrate the value of assessing germination promoting compounds for their use as seed pre-treatments prior to restoration[7].

many possible treatments (and combinations of treatments) that may need investigation before an optimised method of germination is established. In most cases physical dormancy is relatively simple to alleviate using well-established techniques; however, the optimal approach (e.g. temperature and duration of thermal exposure) requires careful assessment as not all physically dormant species require the same treatment regime.

Box 3.2 continued

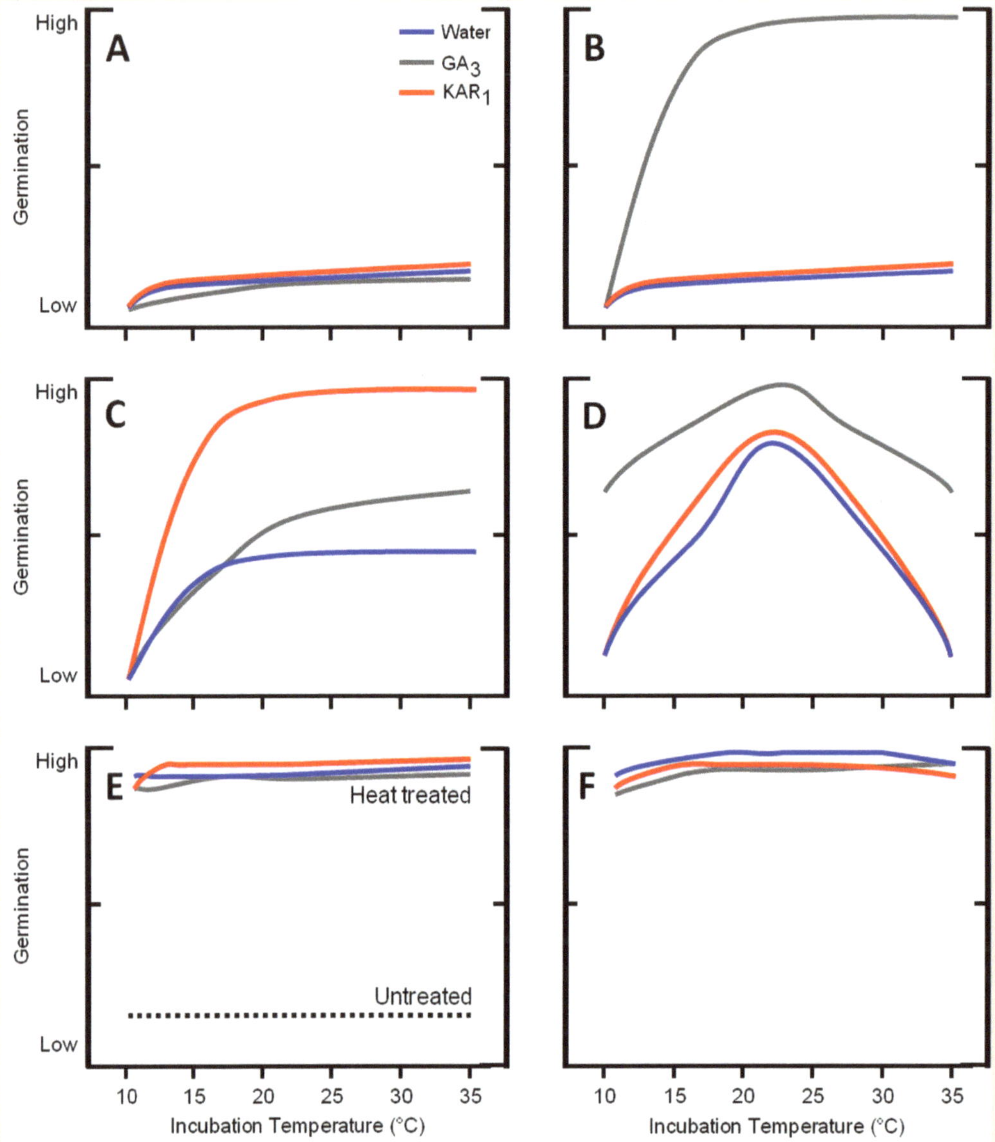

Figure 3.9 Examples of the effect of temperature on germination including the influence of germination stimulants, gibberellic acid (GA_3) and karrikinolide (KAR_1). The charts represent the range of typical germination patterns observed in seeds from the Pilbara. Deeply dormant species do not germinate at any incubation temperature, or after exposure to the germination stimulants, and require an additional form of treatment to alleviate dormancy (*Eragrostis eriopoda* – A). Many seeds with physiological dormancy respond to the germination stimulants GA_3 (*Neurachne muelleri* – B) and/or KAR_1 (*Triodia pungens* – C) and these chemical agents can be effective pre-treatments to promote germination for restoration purposes. Other seeds with physiological dormancy have a narrow temperature range over which germination is possible, but germination stimulants may increase this temperature window (*Aristida latifolia* – D). Seeds with physical dormancy germinate readily over a wide temperature range once dormancy is broken (e.g. via heat-treatment) (*Acacia inaequilatera* – E), as do non-dormant seeds (*Eucalyptus leucophloia* – F) and there is no additional benefit of GA_3 or KAR_1.

Dry after-ripening for the alleviation of physiological dormancy

Dry after-ripening is the process of dormancy loss in dry seeds. Ideally, this pre-treatment involves the storage of fruits or seeds in a controlled environment at a constant temperature and relative humidity. This mimics the conditions that fruits/seeds experience during the dry season in the Pilbara (i.e. when the soil profile is relatively dry and warm in May–October; Figure 3.1 and Box 3.3). Depending on the species and the depth of physiological dormancy, after-ripening may occur within a few weeks, require months or even years[23], or may involve cyclical dormancy. The effectiveness of dry after-ripening is dependent in part on the storage temperature – warmer temperatures increase the rate of dry after-ripening, but if temperatures are too high seeds lose viability in a short period. Usually, a temperature of between 25°C and 35°C is used to after-ripen seeds. The effect of seed water content (as controlled by relative humidity) on dry after-ripening is more complex, but storing seeds at a relative humidity of 20–50% is generally most effective. The sensitivity of seeds to germination stimulants such as GA_3 or KAR_1 often increases as dry after-ripening proceeds, and some seeds will only respond to KAR_1 following a period of dry after-ripening[4]. The removal of covering fruit/floret structures can also work in combination with dry after-ripening to improving germination percentages.

Determining the optimum conditions for dry after-ripening will ensure that dormancy alleviation can occur not only in a timely and efficient manner, but also in a way that is not detrimental to the viability of the seed batch. Evidence of viability loss, for example, can be seen in the reduced germination capacity of *Triodia* seeds after-ripened for greater than 12 months (Figure 3.10). Dry after-ripening of seeds is known to be effective for species of Poaceae and Asteraceae in particular, but examples also exist in other families that have Pilbara representatives including Goodeniaceae, Solanaceae, and Stylidiaceae[15,23,24]. After-ripening is also effective at alleviating the physiological component of dormancy in seeds with combinational dormancy (a seed dormancy type suspected in a number of genera of Malvaceae, see below).

For some seeds with physiological dormancy, interspersing the after-ripening treatment with short periods of hydration can also be effective at breaking dormancy, in some cases reducing the time required for seeds to after-ripen[25]. For these wet/dry cycling treatments, seeds are imbibed for short periods (1–2 days) in water at regular intervals (e.g. once a fortnight, or once a month). Seeds of species of Dilleniaceae[26], Goodeniaceae[15], and Gyrostemonaceae[25] are known to respond to wet/dry cycling. Also, exposure of seeds to short durations (30 mins to three hours) of high temperatures (80–100°C) can also negate the need for after-ripening, and this treatment is particularly effective for seeds of some species of Poaceae and Stylidiaceae[4,23].

Heat treatment for the alleviation of physical dormancy

The presence of a water impermeable fruit or seed coat is the primary reason for the lack of germination in seeds with physical dormancy (Figure 3.3 and 3.8)[9]. Once the fruit/seed coat is rendered permeable to water, germination in these 'hard-seeded' species is generally rapid and occurs over a wide temperature range (e.g. see Figure 3.9 E). In the Pilbara, the presence of physical dormancy has been confirmed in seeds belonging

Box 3.3 Dry after-ripening responses for seeds of *Triodia* species

Some examples of the response of seeds of *Triodia* species to dry after-ripening conditions[7] are presented in Figure 3.10. In these examples, florets of *T. wiseana* and *T.* sp. Shovelanna Hill exhibit very low germination when first placed in the dry after-ripening conditions. As storage time increases, *T. wiseana* florets after-ripen gradually, reaching maximum germination potential after approximately 12 months. Additional treatment of after-ripened florets with KAR_1 further increases germination. Conversely, *T.* sp. Shovelanna Hill florets do not after-ripen within a reasonable time period; germination is still limited after 24 months, with or without KAR_1 treatment. Thus, even though both species of *Triodia* produce seeds with physiological dormancy, they vary in the depth of dormancy, and the dormancy breaking requirements. These differences are also reflected in the patterns of response to dry after-ripening conditions once the covering floret structures are removed from the seeds. Cleaning to pure seeds provides an immediate benefit to germination for *T. wiseana*, and as seeds after-ripen over a 6–12 month period, germination increases steadily. Notably, however, treatment of cleaned seeds of *T. wiseana* with KAR_1 elicits the maximum germination response and removes any requirement for dry after-ripening. For the more dormant seeds of *T.* sp. Shovelanna Hill, there is no option but to after-ripen seeds. While cleaning to pure seeds and treating with KAR_1 increases germination, seeds must still be after-ripened for 6–12 months before the maximum germination response to KAR_1 is achieved.

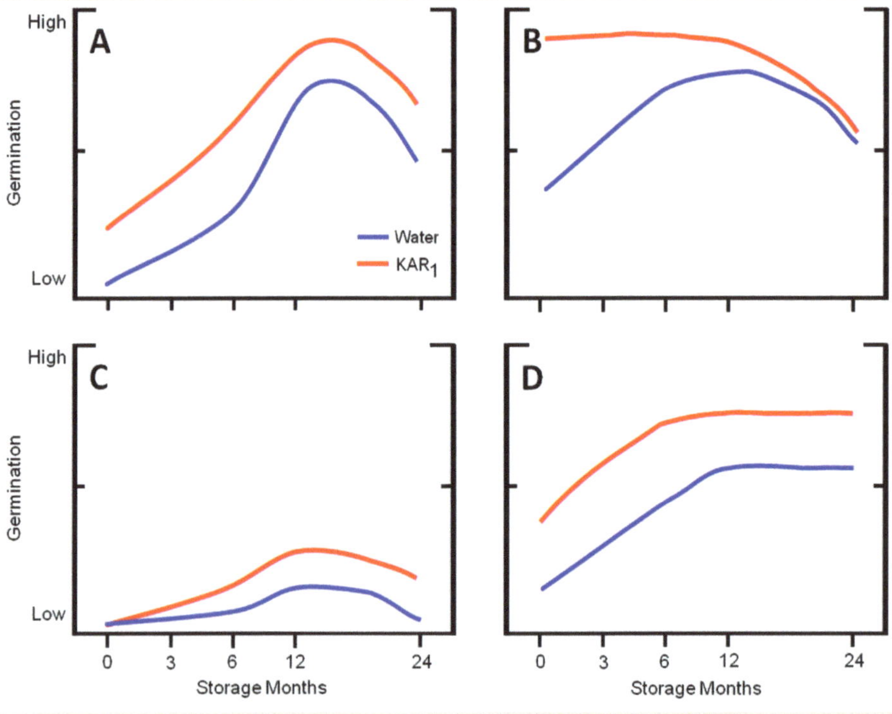

Figure 3.10 Examples of the effect of dry after-ripening (storage at 30°C and 50% relative humidity) on the alleviation of physiological dormancy. Increases in seed germination of two *Triodia* species (*T. wiseana* florets – A and *T.* sp. Shovelanna Hill florets – C) vary depending on the length of dry after-ripening (0–24 months) and whether the florets are exposed to water or KAR_1 (blue and red lines, respectively). Additionally, cleaning down to pure seeds (*T. wiseana* seeds – B and *T.* sp. Shovelanna Hill seeds – D) can improve germination when combined with the same dry after-ripening conditions and KAR_1 treatment.

to five families (Convolvulaceae, Fabaceae, Malvaceae, Sapindaceae, and Surianaceae)[7]. Occasionally, physical dormancy can be coupled with physiological dormancy (termed 'combinational dormancy')[27]. Although combinational dormancy has only been confirmed in a few Pilbara species, there is some evidence that a considerable number of species have this form of dormancy, including species of *Abutilon*, *Gossypium*, *Hibiscus* and *Sida*[7].

Physically rupturing the fruit/seed coat through precision seed nicking (e.g. puncturing the outer coat with a sharp blade) or manual scarification (e.g. abrading the outer coat with sandpaper) will remove the water-impermeable barrier and allow water to enter the seed. These approaches can be time consuming and technically challenging when large numbers of fruits/seeds require treatment, although commercial seed scarifiers are available. Other treatments to overcome physical dormancy include immersion in near boiling water, exposure to dry convection heat, and even soaking in concentrated sulphuric acid (Box 3.4). Trade-offs exist between these various methods in terms of the ability to effectively treat large batches of seeds in a safe and efficient manner without injuring or killing the seeds in the process.

Box 3.4 Heat treatment responses for seeds of the Fabaceae and Malvaceae

Research demonstrates that for many Pilbara species wet-heat outperforms dry-heat in alleviating physical dormancy[7]. The immersion of seeds in water at temperatures of 80–95°C (for up to two minutes) typically results in complete alleviation of physical dormancy (Figure 3.11). The heat treatment breaks dormancy by rupturing a specialised, microscopic structure within the seed or fruit coat, termed the water gap. Once the water gap is ruptured, seeds germinate readily (Figure 3.11). The majority of *Acacia* species respond well to two minutes of exposure to water at 80–95°C. However, there are subtle differences in the optimum temperature and duration of exposure for seeds of other genera. For instance, seeds of *Senna* species are killed by two minutes exposure to water at 90°C or greater (Figure 3.12). Further, species from the Malvaceae (such as *Abutilon*, *Androcalva*, *Corchorus*, *Gossypium*, *Hibiscus* and *Sida* species) produce seeds that vary considerably in their response to wet-heat. As an example, seeds of *Androcalva* and *Corchorus* species follow a similar trend as for seeds of *Acacia* and *Senna* species whereby the seeds lose dormancy at 70–90°C (e.g. blue lines, Figure 3.12). However, others do not respond at all, or only a portion of the seed batch responds (Figure 3.12). The cause of such variation requires further investigation but may be due to: (1) the applied treatment being insufficient to rupture the water gap; (2) the applied treatment alleviating physical dormancy, but injuring or killing the seeds in the process, or; (3) the seeds also possess physiological dormancy (e.g. combinational dormancy) and require an additional treatment to allow germination to proceed (e.g. dry after-ripening).

Box 3.4 continued

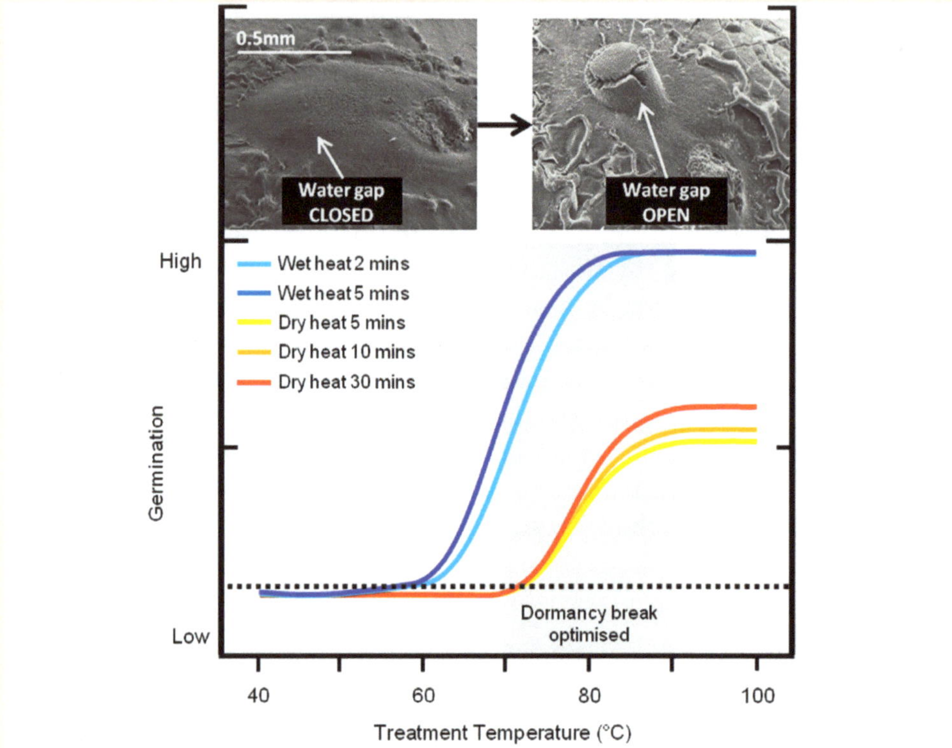

Figure 3.11 An example of the efficacy of wet- or dry-heat treatment in breaking physical dormancy in seeds of *Acacia inaequilatera*. Without treatment (dashed line) the water gap remains closed, preventing water uptake and germination in the majority of seeds. Exposure to wet-heat treatment at >70°C for 2–5 mins (blue lines) promotes the rupture of the water gap, allowing water uptake and germination to occur. Dry-heat (red, orange, and yellow lines) is less effective at rupturing the water gap and germination is limited.

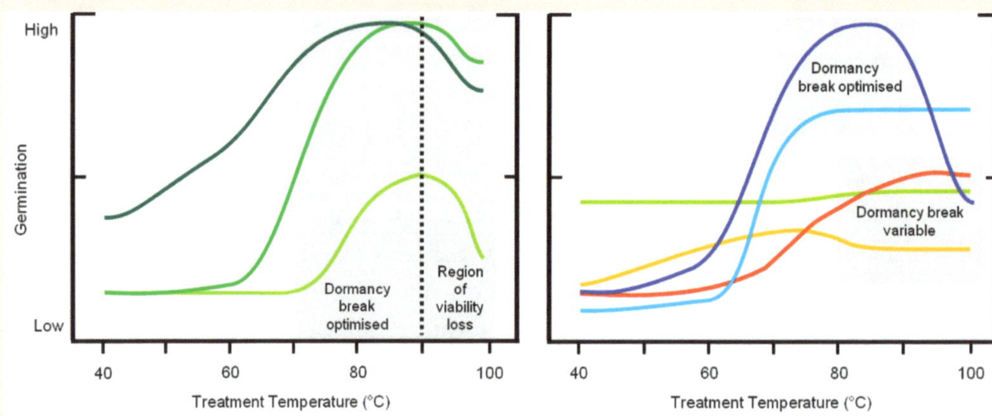

Figure 3.12 The response of seeds with physical dormancy to hot-water treatment can vary substantially between species. In these examples, seeds were immersed for two minutes in water at temperatures ranging from 40–100°C. Seeds of three species of *Senna* (left) require a water temperature of at least 70°C for dormancy-break, but are sensitive to temperatures >90°C. Seeds of five species of the Malvaceae (right) vary widely in their response and further investigation regarding their dormancy type is required. For a few species, two mintutes of wet heat at 70–90°C alleviates physical dormancy (e.g. dark-blue and light-blue lines), but many others either do not respond (e.g. orange and green lines) or respond only marginally (e.g. red line).

References

1. Long R. L., Gorecki M. J., Renton M., Scott J. K., Colville L., Goggin D. E., Commander L. E., Westcott D. A., Cherry H., Finch-Savage W. E. (2014) The ecophysiology of seed persistence: a mechanistic view of the journey to germination or demise. *Biological Reviews* **90**, 31–59.
2. Baskin C. C., Baskin J. M. (2014) *Seeds: ecology, biogeography, and evolution of dormancy and germination.* 2nd edition, Academic Press, San Diego, California.
3. Baskin J. M., Baskin C. C. (2004) A classification system for seed dormancy. *Seed Science Research* **14**, 1–16.
4. Merritt D. J., Turner S. R., Clarke S., Dixon K. W. (2007) Seed dormancy and germination stimulation syndromes for Australian temperate species. *Australian Journal of Botany* **55**, 336–344.
5. Willis C. G., Baskin C. C., Baskin J. M., Auld J. R., Venable D. L., Cavender-Bares J., Donohue K., de Casas R. R., The N. G. W. G. (2014) The evolution of seed dormancy: environmental cues, evolutionary hubs, and diversification of the seed plants. *New Phytologist* **203**, 300–309.
6. Kos M., Baskin C. C., Baskin J. M. (2012) Relationship of kinds of seed dormancy with habitat and life history in the Southern Kalahari flora. *Journal of Vegetation Science* **23**, 869–879.
7. Erickson T. E. (2014) Seed dormancy and germination traits of 89 arid zone species targeted for mine-site restoration in the Pilbara region of Western Australia. PhD Thesis, University of Western Australia.
8. Bewley J. D., Bradford K. J., Hilhorst H. W. M., Nonogaki H. (2013) *Seeds: physiology of development, germination and dormancy.* 3rd edition, Springer, New York, Heidelberg, Dordrecht.
9. Baskin J. M., Baskin C. C., Li X. (2000) Taxonomy, anatomy and evolution of physical dormancy in seeds. *Plant Species Biology* **15**, 139–152.
10. Golos P. J., Dixon K. W. (2014) Waterproofing topsoil stockpiles minimizes viability decline in the soil seed bank in an arid environment. *Restoration Ecology* **22**, 495–501.
11. Golos P. J. (2013) Restoring vegetation on waste rock dumps at the Telfer mine site in Australia's Great Sandy Desert: topsoil management and plant establishment. PhD Thesis, University of Western Australia.
12. Rice B., Westoby M. (1999) Regeneration after fire in *Triodia* R. Br. *Australian Journal of Ecology* **24**, 563–572.
13. Merritt D. J., Dixon K. W. (2011) Restoration seed banks – a matter of scale. *Science* **332**, 424–425.
14. Turner S. R., Steadman K. J., Vlahos S., Koch J. M., Dixon K. W. (2013) Seed treatment optimizes benefits of seed bank storage for restoration-ready seeds: the feasibility of prestorage dormancy alleviation for mine-site revegetation. *Restoration Ecology* **21**, 186–192.
15. Hoyle G. L., Daws M. I., Steadman K. J., Adkins S. W. (2008) Mimicking a semi-arid tropical environment achieves dormancy alleviation for seeds of Australian native Goodeniaceae and Asteraceae. *Annals of Botany* **101**, 701–708.
16. Baskin C. C., Baskin J. M. (2004) Germinating seeds of wildflowers, an ecological perspective. *HortTechnology* **14**, 467–473.
17. Baskin C. C., Baskin J. M. (2007) A revision of Martin's seed classification system, with particular reference to his dwarf-seed type. *Seed Science Research* **17**, 11–20.
18. Martin A. C. (1946) The comparative internal morphology of seeds. *American Midland Naturalist* **36**, 513–660.
19. Baskin C. C., Baskin J. M. (2005) Underdeveloped embryos in dwarf seeds and implications for assignment to dormancy class. *Seed Science Research* **15**, 357–360.
20. Jurado E., Westoby M. (1992) Germination biology of selected central Australian plants. *Austral Ecology* **17**, 341–348.
21. Commander L. E., Merritt D. J., Rokich D. P., Dixon K. W. (2009) Seed biology of Australian arid zone species: germination of 18 species used for rehabilitation. *Journal of Arid Environments* **73**, 617–625.
22. Dixon K. W., Merritt D. J., Flematti G. R., Ghisalberti E. L. (2009) Karrikinolide – a phytoreactive compound derived from smoke with applications in horticulture, ecological restoration and agriculture. *Acta Horticulturae* **813**, 155–170.
23. Turner S. R., Merritt D. J., Renton M. S., Dixon K. W. (2009) Seed moisture content affects afterripening and smoke responsiveness in three sympatric Australian native species from fire-prone environments. *Austral Ecology* **34**, 866–877.

24. Commander L. E., Merritt D. J., Rokich D. P., Dixon K. W. (2009) The role of after-ripening in promoting germination of arid zone seeds: a study on six Australian species. *Botanical Journal of the Linnean Society* **161**, 411–421.
25. Baker K. S., Steadman K. J., Plummer J. A., Merritt D. J., Dixon K. W. (2005) The changing window of conditions that promotes germination of two fire ephemerals, *Actinotus leucocephalus* (Apiaceae) and *Tersonia cyathiflora* (Gyrostemonaceae). *Annals of Botany* **96**, 1225–1236.
26. Hidayati S. N., Walck J. L., Merritt D. J., Turner S. R., Turner D. W., Dixon K. W. (2012) Sympatric species of *Hibbertia* (Dilleniaceae) vary in dormancy break and germination requirements: implications for classifying morphophysiological dormancy in Mediterranean biomes. *Annals of Botany* **109**, 1111–1123.
27. Turner S. R., Merritt D. J., Baskin J. M., Baskin C. C., Dixon K. W. (2006) Combinational dormancy in seeds of the Western Australian endemic species *Diplopeltis huegelii* (Sapindaceae). *Australian Journal of Botany* **54**, 565–570.

Chapter 4
A systematic approach to seed management for restoration

David J. Merritt, Peter J. Golos and Todd E. Erickson

Conservation and restoration programs can take on many forms. In the Pilbara, these range from small-scale landscaping projects that provide locally sourced plant material for community gardens and mine-site village plantings (Figure 4.1)[1], through to the large-scale application of high diversity seed mixes, or topsoil, to sites that have been disturbed by mining operations[2–4]. While the goals of these programs are similar (to reinstate an array of species and/or plant communities through the deliberate re-introduction of seeds or seedlings[5]), the scale is different, and different approaches are required. For instance, local knowledge exists within the Pilbara regarding the growing of native plants in nurseries for managed gardens and landscaping[1], with successful establishment of trees and shrubs in town and mine-site villages in the region. However, the large-scale application of these techniques to restoration sites is limited by the costs of producing plants in a nursery, the need for a continual supply of water at the landscaping site via reticulation, and the limited numbers of plants and species available.

Figure 4.1 Nursery propagation of Pilbara plant species at a commercial nursery in Marble Bar. The plants pictured here are grown for specific landscaping projects such as mine-site village plantings.

In an era of larger-scale restoration programs that aim to return biodiverse plant communities across 10's to 1000's of hectares of degraded land, the broad-scale application of seeds through seed broadcasting, mechanised direct seeding, or the re-spreading of stockpiled topsoil is fundamental to cost-effective plant establishment[2,6]. It is well known that topsoil contains seeds and soil microbes from the local area, and if harvested, stored, and re-spread correctly, topsoil can be an effective means of rapidly reinstating plant diversity (Box 4.1). Unfortunately, there is often insufficient topsoil available to cover the areas requiring restoration and seed broadcasting becomes the only option.

Using knowledge of seed quality to inform seeding operations

Seed broadcasting or direct seeding commonly entails the addition of seed mixes comprising a few, or many, diverse species in anticipation of the reinstatement of a plant community that reflects at least some level of the biodiversity present prior to disturbance[5,7,8]. In calculating the quantity of seeds required it is important to incorporate biological information pertaining to each seed batch. Ideally, seeding rates are based on known parameters and data including seed size (i.e. number of seeds per gram), viability, germination, and establishment[9]. These features of seed quality are captured in the concept of 'pure live seed' (a measure of the purity, viability and germination capacity of a seed batch), which forms part of an accreditation system for seeds produced in wild seed farming enterprises in the United States and some parts of Europe[10]. Without this information it is not possible to monitor and document seedling emergence to determine the proportion of seeds that emerge to evaluate the success (or otherwise) of seeding to guide future practices (Box 4.2).

Box 4.1 Utilising topsoil in disturbed arid environments: benefits and management options

Topsoil provides a valuable resource for reinstating a high level of species diversity to disturbed sites, as topsoil contains seeds and soil micro-organisms for a large proportion of species indigenous to the local area[4,6]. Topsoil can include seeds that are difficult to germinate and/or seeds that are not currently available in the commercial market. The use of topsoil can, therefore, be particularly advantageous for the return of plant species that produce seeds with complex or poorly understood dormancy-break and germination requirements[3] (see Chapter 3). Further, the reinstatement of soil microorganisms is known to create positive plant–soil relationships (e.g. rhizobia of *Acacia* species) that increase the ability of plants to establish and promote ecosystem processes such as nutrient cycling.

The methods used for the harvesting, handling, and storage of topsoil are vital to protecting its value for restoration[4,6]. Stripping of topsoil should be limited to the depth at which the majority of the soil seed bank resides (i.e. in the upper 10 cm), as stripping topsoil too deeply can result in dilution of the soil seed bank[4]. Stripping of topsoil must also be conducted under dry soil conditions because the physical disturbance, coupled with higher moisture levels, can lead to the stimulation of the soil seed bank prematurely (this concept includes avoiding the wetting-up of topsoil to minimise dust creation during stripping activities). The original landform and site characteristics need to be considered during stripping as the terrain and soil profile (e.g. shallow, rocky soils) may make it difficult to harvest topsoil at the required depth. Lastly, land management history, such as time since fire and the presence of weeds, needs to be taken into account as this can affect the quality and size of the soil seed bank (e.g. there is a depleted soil seed bank immediately after fire).

Due to operational and logistical constraints, it is common to stockpile topsoil before undertaking restoration. While the decline in the soil seed bank can be rapid in wetter climates, the decline can be slower in arid environments if the topsoil is managed correctly. In the Great Sandy Desert (east of the Pilbara) seeds of *Triodia* species have been found to be particularly prone to viability decline in stockpiled topsoil stored in the open (i.e. exposed to seasonal rainfall), with considerable losses documented during two years of storage (e.g. a 70% decline in seedling emergence from the re-spread topsoil)[6]. Notably, this viability decline is not as pronounced in seeds of *Acacia* species that exhibit physical dormancy[6] demonstrating that storage practices can also alter the composition of the viable seed bank. To slow the decline in the soil seed bank, simple techniques such as protecting stockpiles from rainfall by covering with industrial-grade tarpaulins have been shown to be effective[6].

When formulating seed mixes for broadcast seeding it is important to consider the potential contribution of the soil seed bank to plant recruitment. There can be large differences in plant establishment from topsoil depending on the source location, storage history, handling practices, and seed persistence of individual species. Good record keeping is essential to topsoil management, and audits of the topsoil are useful to determine the germinable seed fraction. With a greater knowledge of the potential plant return from topsoil (i.e. species richness and seedling density), practitioners can avoid over- or under-seeding restoration sites.

Equally, the dormancy status of the seed batch, and the requirements for dormancy-break, are important considerations (Box 4.3). When a seed harvest is undertaken, the seed collector or the purchaser of the seeds holds the seeds in a storage facility for a period of time (usually at least 3–6 months, or longer) prior to use. Common practice in the restoration industry is to then remove the seeds from storage and sow without pre-treatment (i.e. regardless of dormancy state) and allow nature to take its course. However, seeds can take a long period of time to lose dormancy even if conditions are favourable. In arid systems where rainfall is patchy and unpredictable, suitable rain events that provide sufficient moisture for germination may only occur sporadically. Seed dormancy can, therefore, impede seedling recruitment as there are limited 'germination windows' in each season, and if seeds are dormant during this window, germination will not proceed. Further, if the appropriate cues for dormancy loss and germination are absent (e.g. where there is a requirement for smoke to promote germination), then seedling recruitment may never occur.

Box 4.2 The impact of seed quality on restoration planning

With suitable rains and a well executed collection program, large quantities (i.e. hundreds of kilograms) of *Triodia* seeds can be available in the restoration market in any particular year. Since 2011, over 12 tonnes of *Triodia* florets (that contain the seeds) have been collected by the commercial seed sector in the Pilbara. However, the quality of each individual collection is highly variable and this variation must be taken into account when planning a restoration program. For example, taking the data presented in Table 2.1, it is evident that two different batches of *Triodia lanigera* seeds vary widely in their quality – either 27% or 61% of florets contain a seed, depending upon the batch. With an estimated 400,000 florets per kilogram, there exists 108,000 viable seeds per kilogram in the first batch, and 244,000 viable seeds per kilogram in the second batch. This difference in quality therefore equates to a two-fold difference in the amount of seeds being sown per hectare and could ultimately result in vastly different numbers of plants establishing. Without an understanding of this variation in seed quality between the two batches it is not possible to accurately calculate seeding rates required to achieve a desired or predictable level of plant establishment, and it is also not possible to account for differences in plant establishment that might be evident between different restoration sites.

The data presented in Chapter 5 can be used as an indication of the seed quality that might be expected for a collection of any particular species and as a guide to seed purchasing, but the seed quality of each individual collection should always be determined prior to use in restoration.

Box 4.3 Developing mixes of seeds with common dormancy-break requirements

As techniques to overcome seed dormancy for individual species are developed, this knowledge can be applied to improve the implementation of restoration programs. For instance, for species that produce seeds with the same type of dormancy and similar requirements for dormancy-break, diverse seed mixes can be easily created and then pre-treated in bulk. It is more efficient to mix seeds of a number of *Acacia* species and treat the entire seed mix immediately prior to seeding (e.g. through immersing the seeds in hot water at 90°C for two minutes) than to treat each seed batch individually, and then mix for seeding at a later date.

The data presented in Chapter 5 and Appendix A can be used to group species based on their seed pre-treatment requirements to assist in creating these types of diverse seed mixes.

Even with the best information, current seeding practices and technologies frequently result in poor seedling establishment. In arid regions it is common that 90% or more of the broadcast seeds fail to produce a seedling[2,5,11,12]. The loss of time, money, and seeds under these scenarios is considerable. Some of the most common factors contributing to poor seedling establishment include: the failure to assess, manage and account for seed quality during the collection, storage, and seeding phases; the failure to consider and alleviate seed dormancy prior to seeding; and hindrances to germination and seedling emergence created by environmental conditions at the restoration sites such as insufficient soil moisture, modified and unnatural soil conditions, seed predators, and wind erosion[2,7,12,13].

To improve the effectiveness of seeding programs, information regarding the biological properties of the seeds needs to be integrated with that of the environmental and physical characteristics of the restoration site[14]. The success of seeding hinges on optimising handling practices from the point at which seeds are collected, through to the point at which they are delivered to site. Each of these steps should be viewed not in isolation, but rather as a continual journey, with the effectiveness of the previous steps impacting on the success of the next (Figure 4.2)[2,12,15]. Once seeds enter the soil, their interactions with the surrounding micro-environment shape the chances of seedling survival. One of the most critical phases in plant establishment is the transition from the germinated seed to the emerged seedling. Research in arid regions (including the Pilbara) has demonstrated that the number of seeds that germinate underground can be much higher than the number of germinated seeds that subsequently emerge from the soil as seedlings[11,12,16].

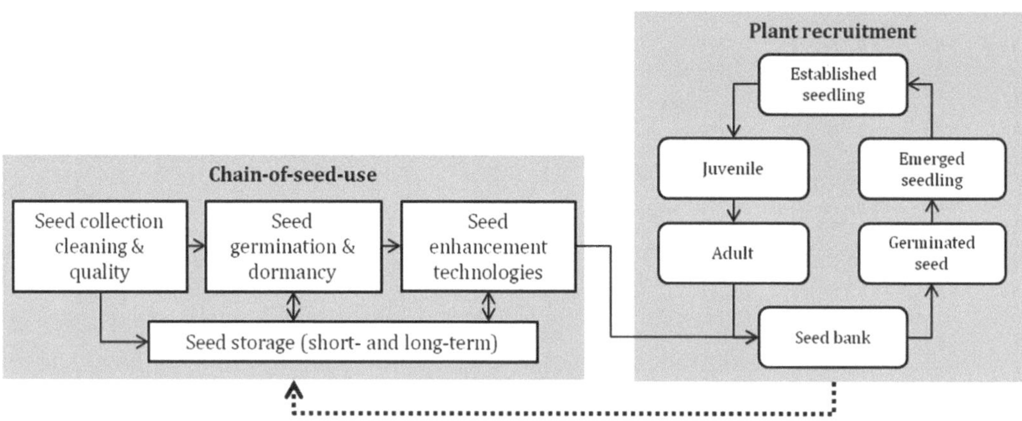

Figure 4.2 The 'chain-of-seed-use' concept links each step in the seed handling process to maximise the chances of seedling establishment in conservation and restoration programs (solid arrows). If aspects such as the seed quality (e.g. viability) and the storage conditions are not considered, then practitioners may be seeding sites with seeds of low and varied quality, or, even worse, seeding sites with seed batches that contain no seeds. Failing to consider seed dormancy and the influence this has on the germination potential of a seed batch is another example of how seedling establishment can be impeded. Combined with various site-specific conditions, such as the species contribution from re-spread topsoil, or localised landscape/ environmental conditions during early seedling establishment, the chain-of-seed-use forms a vital seed management strategy. Continual monitoring and evaluation of restoration performance ensures successes and failures of plant establishment are identified and modified in future seeding attempts (e.g. the feedback pathway indicated by the dashed arrow).

As a result, much research effort is now directed at modifying the soil–seed environment (the 'growing niche') through enhancing the ability of seeds to germinate and emerge, and through creating a soil substrate that has physical, chemical, and biological properties more conducive to growth.

As restoration programs aspire to ever-greater levels of complexity and scale, seed handling practices must be developed that are reliable, repeatable, and applicable to large quantities of seeds from a diverse range of species. Seeds are complex biological organisms and there is still much to be learnt regarding the science and practice of restoration. However, recognising that there are many stages in handling seeds, each of which can impact on the success or failure of a restoration project, is the first step to developing an integrated or holistic approach to seed management for restoration.

Seed enhancement technologies and the future of seed management

Seed enhancement refers to a suite of techniques and technologies that may be applied to seeds to improve seed germination performance and the tolerance of seedlings to environmental stress. These technologies include seed priming, coating and pelleting. While seed enhancement is widely adopted in agriculture and horticulture, its adoption to seeds of wild species is in its infancy. However, seed priming (the controlled hydration of seeds prior to sowing, followed by re-drying for ease of handling) has been demonstrated to increase seedling emergence of native grass species in the deserts of the United States[17] and is showing promise for native seeds in Western Australia. Seed coating involves covering seeds with thin layers of soil or mineral products (e.g. filler materials such as calcium carbonate or talc) and polymer binding agents. These are applied with repeated applications of a binding agent and a filler material using a rotary seed coater. Through adjusting the seed coating approach, seeds of different species can be 're-sized' to a uniform size to aid in mechanised delivery to site. These coats can also incorporate beneficial germination stimulants (such as KAR_1), herbicides, and soil surfactants[18] to aid germination and establishment. Similarly, seed pellets consist of a blend of soil and mineral products. These are mixed together with seeds into a dough-like material and extruded to make pellets that can contain up to five seeds[19]. Only recently tested on *Triodia* florets and seeds, field-trials have demonstrated that seedling establishment from coated florets and/or cleaned seeds can be at least two-fold higher than the <10% emergence commonly reported[2,12] (Figure 4.3). Coating of seeds with artificial polymers has also been demonstrated to increase seedling emergence of *Banksia* woodland species in southwest Western Australia, as well as acting to decrease losses of seeds through predation and wind erosion[20].

In the future, seed coating and extruded pellet techniques may be useful as secondary pre-treatments (i.e. following dormancy-break treatments) to address the biotic and abiotic constraints to seedling recruitment that are often observed in degraded or altered soil environments[18–21]. The development of these integrated seed-based technologies for restoration will, over time, allow practitioners to streamline seed batch handling and strengthen their ability to reinstate complex and biodiverse plant communities in arid land restoration programs.

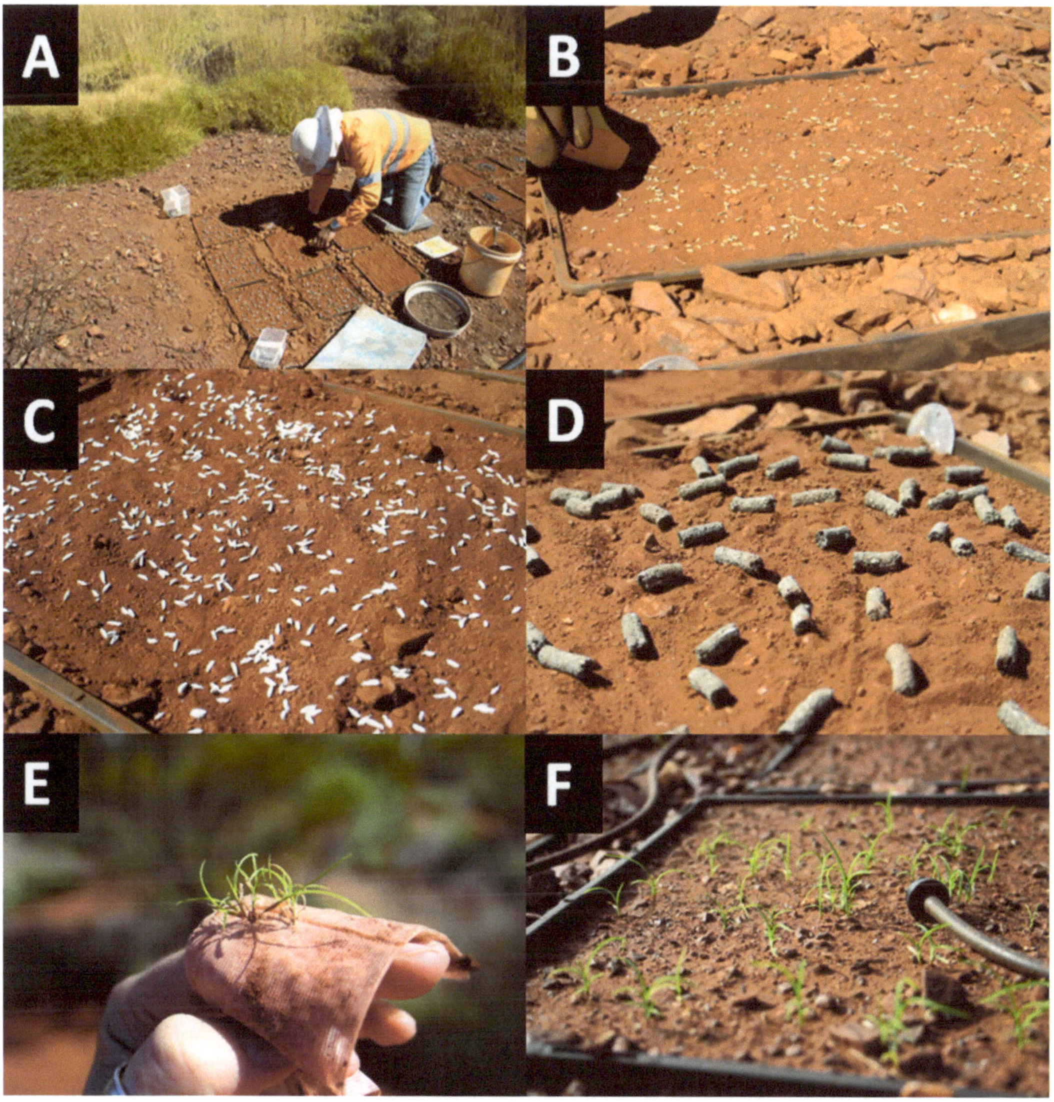

Figure 4.3 Seed enhancement technologies include seed coatings and extruded seed pellets using artificial polymers, soil, and/or mineral products. These types of technologies are currently being evaluated in the Pilbara (A). The coating or pelleting of seeds in beneficial soil and mineral products, along with the incorporation of germination stimulants, herbicides, nutrients, and/or soil surfactants, is showing promise in seed-based restoration programs. Trials are comparing cleaned *Triodia* florets and seeds (B), with polymer-coated florets (C) and pelleted florets and seeds (D), to determine the effects on germination (checked by recovering nylon sachets that contain seeds – E) and seedling emergence (F).

References

1. Smith N., Clark M. (2014) *Pilbara native plants for gardens and landscapes.* Greening Australia (WA) Ltd., Shenton Park, Western Australia.
2. Merritt D. J., Dixon K. W. (2011) Restoration seed banks – a matter of scale. *Science* **332**, 424–425.
3. Erickson T. E. (2014) Seed dormancy and germination traits of 89 arid zone species targeted for mine-site restoration in the Pilbara region of Western Australia. PhD Thesis, University of Western Australia.
4. Golos P. J. (2013) Restoring vegetation on waste rock dumps at the Telfer mine site in Australia's Great Sandy Desert : topsoil management and plant establishment. PhD Thesis, University of Western Australia.

5. Commander L. E., Rokich D. P., Renton M., Dixon K. W., Merritt D. J. (2013) Optimising seed broadcasting and greenstock planting for restoration in the Australian arid zone. *Journal of Arid Environments* **88**, 226–235.
6. Golos P. J., Dixon K. W. (2014) Waterproofing topsoil stockpiles minimizes viability decline in the soil seed bank in an arid environment. *Restoration Ecology* **22**, 495–501.
7. De Falco L. A., Esque T. C., Nicklas M. B., Kane J. M. (2012) Supplementing seed banks to rehabilitate disturbed Mojave Desert shrublands: where do all the seeds go? *Restoration Ecology* **20**, 85–94.
8. Williams M. I., Schuman G. E., Hild A. L., Vicklund L. E. (2002) Wyoming big sagebrush density: effects of seeding rates and grass competition. *Restoration Ecology* **10**, 385–391.
9. Gibson-Roy P., Moore G., Delpratt J., Gardner J. (2010) Expanding horizons for herbaceous ecosystem restoration: the Grassy Groundcover Restoration Project. *Ecological Managment and Restoration* **11**, 176–186.
10. Jones T. A., Young S. A. (2005) Native seeds in commerce: more frequently asked questions. *Native Plants Journal* **6**, 286–293.
11. James J. J., Svejcar T. J., Rinella M. J. (2011) Demographic processes limiting seedling recruitment in arid grassland restoration. *Journal of Applied Ecology* **48**, 961–969.
12. James J. J., Sheley R. L., Erickson T., Rollins K. S., Taylor M. H., Dixon K. W. (2013) A systems approach to restoring degraded drylands. *Journal of Applied Ecology* **50**, 730–739.
13. Carrick P. J., Krüger R. (2007) Restoring degraded landscapes in lowland Namaqualand: lessons from the mining experience and from regional ecological dynamics. *Journal of Arid Environments* **70**, 767–781.
14. Carrick P. J., Erickson T. E., Becker C. H., Mayence C. E., Bourne A. R. (2015) Comparing ecological restoration in South Africa and Western Australia: the benefits of a 'travelling workshop'. *Ecological Management and Restoration* **16**, 86–94.
15. Perring M. P., Standish R. J., Price J. N., Craig M. D., Erickson T. E., Ruthrof K. X., Whiteley A. S., Valentine L. E., Hobbs R. J. (2015) Advances in restoration ecology: rising to the challenges of the coming decades. *Ecosphere* **6**, art131. doi: http://dx.doi.org/10.1890/ES15-00121.1
16. Larson J. E., Sheley R. L., Hardegree S. P., Doescher P. S., James J. J. (2014) Seed and seedling traits affecting critical life stage transitions and recruitment outcomes in dryland grasses. *Journal of Applied Ecology* **52**, 199–209.
17. Hardegree S. P., Van Vactor S. S. (2000) Germination and emergence of primed grass seeds under field and simulated-field temperature regimes. *Annals of Botany* **85**, 379–390.
18. Madsen M. D., Kostka S. J., Inouye A. L., Zvirzdin D. L. (2012) Postfire restoration of soil hydrology and wildland vegetation using surfactant seed coating technology. *Rangeland Ecology and Management* **65**, 253–259.
19. Madsen M. D., Davies K. W., Mummey D. L., Svejcar T. J. (2014) Improving restoration of exotic annual grass-invaded rangelands through activated carbon seed enhancement technologies. *Rangeland Ecology and Management* **67**, 61–67.
20. Turner S. R., Pearce B., Rokich D. P., Dunn R. R., Merritt D. J., Majer J. D., Dixon K. W. (2006) Influence of polymer seed coatings, soil raking, and time of sowing on seedling performance in post-mining restoration. *Restoration Ecology* **14**, 267–277.
21. Madsen M. D., Davies K. W., Williams C. J., Svejcar T. J. (2012) Agglomerating seeds to enhance native seedling emergence and growth. *Journal of Applied Ecology* **49**, 431–438.

Chapter 5
An atlas to the plants and seeds of the Pilbara region

Todd E. Erickson, Russell L. Barrett, David R. Symons, Shane R. Turner and David J. Merritt

Introduction

The Pilbara region, in the far northwest of Australia's arid zone (Figure 1.1), has never had a comprehensive guide to its diverse flora. Despite significant interest in the region driven by mining, pastoralism and tourism, there is still much to learn about the plant life of the Pilbara[1]. What is known regarding the diversity of the flora and fauna of the Pilbara is captured in the Pilbara Region Biological Survey[1] – a project that spanned more than 10 years from 2002 through to 2013. Along with the vegetation surveys routinely undertaken for mineral exploration and mine development, this survey has led to a much-improved knowledge base for understanding plant diversity and vegetation distribution[1]. Yet still to this day, discoveries of new species occur on a regular basis, from large and showy species such as *Cochlospermum macnamarae*[2] and *Grevillea saxicola*[3], to the more diminutive *Tephrosia oxalidea*[4] and *Eragrostis surreyana*[5], through to species that are located in isolated and unusual habitats (e.g. *Synostemon hamersleyensis*[6]). Currently, there are 1,794 plant taxa known from the Pilbara (Table 5.1; Dillon, Appendix C). Of those taxa, 1,674 are native, and 120 are introduced. In addition, 178 taxa (10.6% of the total) are listed as having conservation priority in Western Australia and 147 taxa (8.8%) are as yet not named and require formal taxonomic descriptions.

Table 5.1 Number of plant taxa recorded for the Pilbara Region.

	Families	Genera	Species	Subspecies	Varieties	Forma	Hybrids	Total Taxa
Native	103	418	1559	56	41	3	15	1674
Weeds	33	98	118	1	0	0	1	120

Very little information is available regarding the ecology of the seeds of the unique and diverse plants of the Pilbara. As the cumulative impacts of resource development and human activity increase throughout the region, addressing the gaps in our knowledge of seed biology, and developing appropriate seed management practices based on this knowledge, is critical to developing the capacity for biodiverse restoration and for conserving plant species.

This chapter provides detailed taxonomic descriptions and a range of diagnostic photographs for 103 species edited by taxonomist Russell Barrett. Images for each species were taken in the field by Todd Erickson and Russell Barrett. Detailed information is also provided on the seed ecology and biology for each species including recommended methods and timing for seed collection, seed cleaning techniques, and data on seed quality and germination requirements. This information has been compiled through direct experimentation by seed ecologists Todd Erickson, Shane Turner, and David Merritt[7]. A guide to seed and/or fruit identification, and key features of the seeds, is also provided through photo-microscope images taken by David Symons. A detailed explanation of the layout of the information is provided on the opposite page.

The methods used to test germination while gathering the data for this chapter included the use of karrikinolide (KAR_1)[8,9]. This chemical is a pure, synthetic version of the karrikinolide that is naturally produced during the combustion of plant material, and is the primary agent responsible for promoting the germination of smoke-responsive seeds. While KAR_1 is not yet readily available to the restoration practitioner, there are various well-established methods for producing smoke and for treating seeds with smoke, including the use of smoke water or aerosol smoke. More details regarding the production of smoke products for seed germination can be found in Appendix B.

References

1. McKenzie N., van Leeuwen S., Pinder A. (2009) Introduction to the Pilbara biodiversity survey, 2002–2007. *Records of the Western Australian Museum, Supplement* **78**, 3–89.
2. Hislop M., Thiele K. R., Brassington D. (2013) *Cochlospermum macnamarae* (Bixaceae), a rare, new endemic from the Pilbara bioregion of Western Australia. *Nuytsia* **23**, 89–94.
3. Dillon S. J. (2014) *Grevillea saxicola* (Proteaceae), a new species from the Pilbara of Western Australia. *Nuytsia* **24**, 103–108.
4. Butcher R., Hurter P. J. H. (2012) *Tephrosia oxalidea* (Fabaceae: Millettieae), a new species from the Pilbara and Gascoyne bioregions of Western Australia. *Nuytsia* **22**, 341–349.
5. Shepherd K. A., Trudgen M. E. (2011) *Eragrostis surreyana* (Poaceae) an uncommon, habitat restricted new species from the Pilbara Bioregion of Western Australia. *Telopea* **13**, 143–148.
6. Telford I. R. H., Naaykens J. (2015) *Synostemon hamersleyensis* (Phyllanthaceae), a new species endemic to the Pilbara, Western Australia. *Nuytsia* **25**, 31–37.
7. Erickson T. E. (2014) Seed dormancy and germination traits of 89 arid zone species targeted for mine-site restoration in the Pilbara region of Western Australia. PhD Thesis, University of Western Australia.
8. Flematti G. R., Ghisalberti E. L., Dixon K. W., Trengove R. D. (2005) Synthesis of the seed germination stimulant 3-methyl-2H-furo[2,3-c]pyran-2-one. *Tetrahedron Letters* **46**, 5719–5721.
9. Dixon K. W., Merritt D. J., Flematti G. R., Ghisalberti E. L. (2009) Karrikinolide – a phytoreactive compound derived from smoke with applications in horticulture, ecological restoration and agriculture. *Acta Horticulturae* **813**, 155–170.

Genus and species name

Common names: Common names are listed, though many species have no common name.

Indigenous names: There are relatively few published records of indigenous plant names in the Pilbara and many more names will be in use than are listed here.

Description: The important morphological features are provided, adapted from previously published descriptions (as listed on pages 252–253).

Key features for identification: The most important features for the recognition of the species are described.

Similar species: Closely related or confusing species in the area are distinguished.

Distribution: A general distribution in Western Australia is provided and occurrence in other states is noted.

Habitat: The general habitat in which the species is found is described.

Flowering period: The most common flowering period is noted. Flowering outside these times can be expected in unusual seasons.

The pictorial guide above indicates the main landscape positions in which species grow:

A: Lowland plains.
B: Lowland wetlands.
C: Lowland waterways.
D: Rocky slopes.
E: Upland plateaux.
F: Upland waterways.

Habitats not occupied are greyed out.

A map of the species' distribution in the Pilbara region is provided based on data from the Australian Virtual Herbarium. Records known to be inaccurate have been removed in some cases.

Photos are provided that aim to capture the most useful features for the identification of each species. Usually the habit, leaves, flowers and fruit.

J F M A M J J A S O N D

The time-plot above indicates the most common flowering period (blue) and peak fruiting period (brown). These are guides for average seasons.

Genus species

Fruit/Seed collection

Fruit/Seed maturity: The peak period in which fruit/seed is known to be mature.

Appearance of mature fruit/seed: Characteristics that guide collection of mature fruit/seed.

Cleaning requirements

Cleaning techniques: Methods for the optimal cleaning of fruit/seed that has been collected.

Purity and viability: Expected viability for mature seed and notes on purity issues.

Seeds per gram: Average weight of seeds for estimation of sowing requirements.

Germination requirements

Dormancy type: The class of dormancy recognised for this species (if present).

Embryo type: The type of embryo. **Embryo:seed ratio:** embryo to seed length ratio.

Pre-treatments: Suggested treatments that may be required prior to germination being possible.

Germination: Methods required to attain optimal germination percentages.

Notes: Other notes that may improve seed use outcomes.

Images of fruit/seeds and in some cases the flowering/fruiting parts (e.g. florets of grasses) where relevant to identification or recognition of fruit and seed that is clean and ready for storage or germination treatments.

Gomphrena cunninghamii

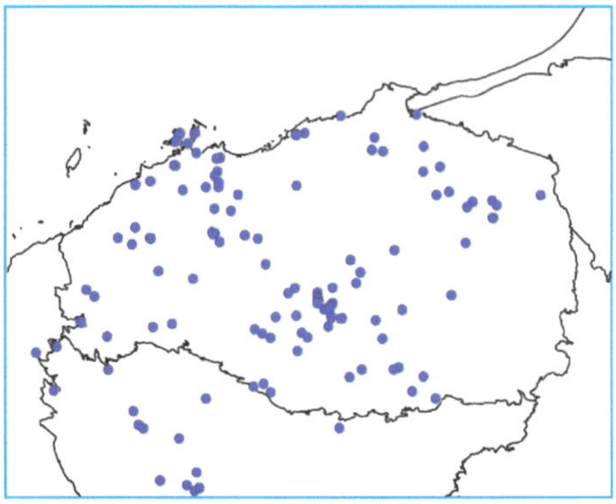

Common name: Cunningham's gomphrena, Silky bachelor's buttons.

Indigenous names: None recorded.

Description: Procumbent to ascending annual herb, 3–30 cm high, to 40 cm diam. Leaves sessile, opposite, lanceolate to broadly ovate, 7–30 mm long, 3–10 mm wide, softly hairy. Inflorescence 9–12 mm long, 12–20 mm wide, not elongating. Bracts 4–9 mm long, 2–4 mm wide, glabrous, translucent throughout or opaque white at apex only. Bracteoles 4.2–8.5 mm long, slightly or distinctly longer than tepals, 2.6–5 mm wide, translucent throughout or opaque white at apex only, midnerve lacking a dorsal crest. Tepals 2.5–6.5 mm long, 0.6–1.5 mm wide, white or pink; apex acute; outer surface with dense, white or brown, crimped or straight hairs covering most of tepal length. Staminal filaments united for 1/2–2/3 of their length in a tube shorter than the fruit, glabrous inside; anthers 0.5–1 mm long; pseudostaminodes absent. Style 0.3–0.9 mm long; stigma 0.1–0.25 mm long. Fruit ellipsoid, 1.9–3.2 mm long, 1.1–1.5 mm wide. Seeds 1.1–1.7 mm long, and 0.8–1.2 mm wide.

Key features for identification: Bracts and bracteoles translucent throughout. Staminal filaments united in a tube shorter than the fruit; no pseudostaminodes.

Similar species: Most similar to *G. leptoclada*. It differs from this species in having bracts and bracteoles that are translucent throughout or with only the apex being opaque white, the bracteoles being 2.6–7 mm wide. *G. leptoclada* has bracts and bracteoles that are opaque white for half their length and bracteoles that are 1.5–3 mm wide.

Distribution: Widespread in the Pilbara, extending to the southern Kimberley and east to the Queensland border.

Habitat: Grows in shallow rocky soils on slopes and in watercourses with *Triodia* spp.

Flowering period: February to August.

Gomphrena cunninghamii

Fruit/Seed collection

Fruit/Seed maturity: May to December.

Appearance of mature fruit/seed: An individual seed is enclosed in each flowering unit (white, spherical floral bracts). When mature, the floral bracts release easily from the plant when touched. Seeds are dark orange in colour and oval-shaped, but flattened longitudinally.

Cleaning requirements

Cleaning techniques: To clean to pure seeds, the floral bracts can be gently passed through a sieve to separate the seeds. Multiple sieve passes are sometimes required. The lighter floral bracts can then be removed via vacuum separation. Care must be taken not to damage the embryo, which protrudes at one end of the seed, during the sieving process.

Purity and viability: Viability is generally high (>85%).

Seeds per gram: 1214.

Germination requirements

Dormancy type: Physiological.

Embryo type: Peripheral. **Embryo:seed ratio:** 2.22.

Pre-treatments: Cleaning to pure seeds by removing bracts and treating seeds with GA_3 or smoke increases germination by >40%.

Germination: Maximum germination achieved via bract removal is 45% at 20°C. Treatment with GA_3 or smoke increases germination to >90%, and vastly improves the germination temperature range (15–35°C).

Notes: Germination of seeds when retained in the floral bracts is likely impeded. Additional dormancy break pre-treatments such as dry after-ripening or wet/dry cycling may assist in overcoming physiological dormancy if bract removal is not achievable.

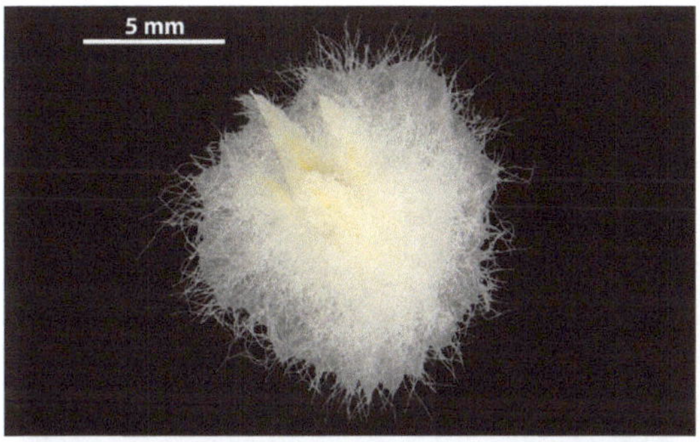

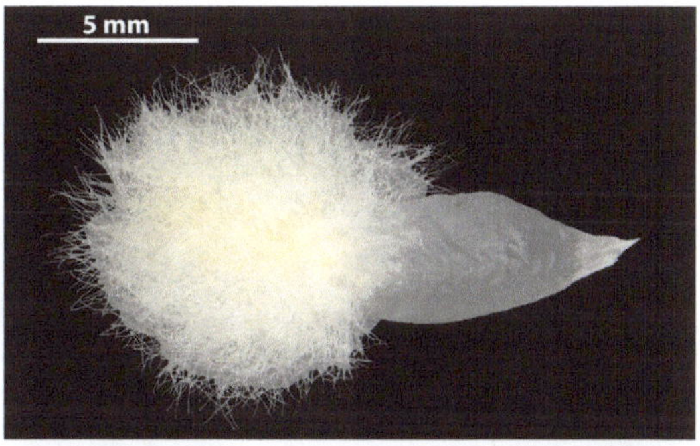

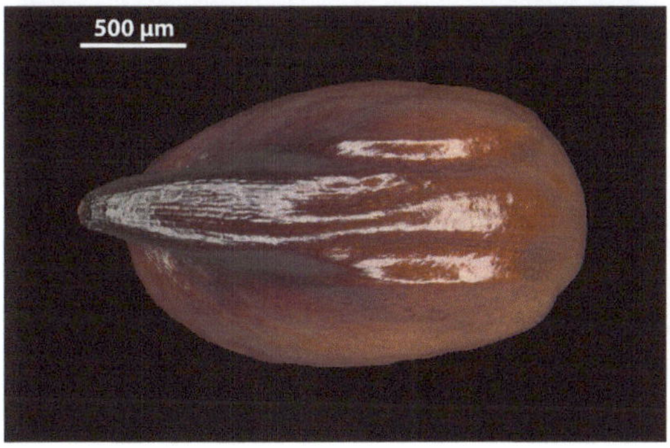

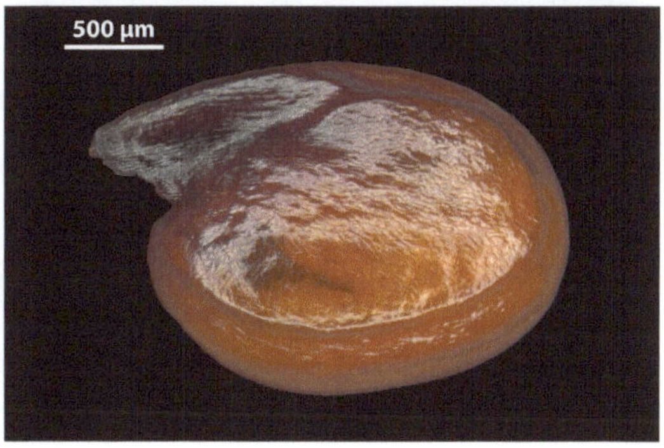

Ptilotus auriculifolius

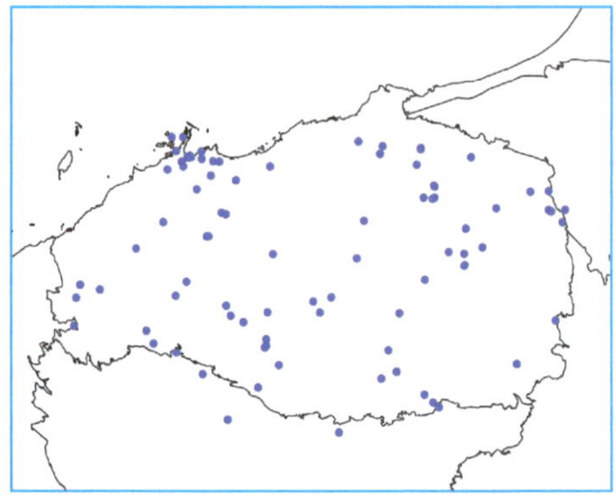

Common names: Nodding mulla mulla.

Indigenous names: None recorded.

Description: Perennial, tufted herb. Basal leaves obovate or obovate-oblong, very obtuse, glabrous, rather thick, with undulate margins, 25–50 mm long, on rather short petioles. Flowering stems erect, 15–30 cm high, nearly glabrous below, clothed upwards with soft hairs, and bearing a few small distant leaves. Spikes nearly globular, 4 or 5 together nearly sessile and crowded in a terminal dense ovate panicle. Bracts broadly ovate, acute, to 5 mm long; bracteoles twice as long, obtuse, very broad and enveloping the flowers, all very thin and transparent. Perianth to nearly 14 mm long, the turbinate tube about 6 mm long; segments narrow, nearly equal, densely plumose with rather long hairs, with short glabrous tips rounded or truncate in the outer ones, narrower in the inner ones, all glabrous inside. Staminal cup shortly free, surrounded by long woolly hairs reaching to the top of the stamens and some of them proceeding from the filaments themselves; anthers often all 5 perfect. Ovary on a long stipe, woolly at the top.

Key features for identification: Broad glabrous leaves; nodding inflorescence, green, densely hairy flowers.

Similar species: Similar to *P. nobilis* which has erect inflorescences with pink flowers in this region.

Distribution: Common and widespread in the Pilbara region.

Habitat: On red sandy soils, rocky hills, or stony undulating plains.

Flowering period: May to October.

Ptilotus auriculifolius

Fruit/Seed collection

Fruit/Seed maturity: December (based upon limited observations).

Appearance of mature fruit/seed: Progressively flowering from the base to the apex of the elongated inflorescence, timing of seed maturation is staggered. The often drooping flower head turns from pale lime-green to off-white when mature. Individual floral bracts are mature when they fall from the inflorescence with gentle pressure. Seeds contained within the floral bracts are dark red to almost black.

Cleaning requirements

Cleaning techniques: Collections will most likely consist of pure floral bracts (which may or may not contain viable seeds) and are commonly of low 'fruit' fill. To clean to pure seeds, collected floral bract material can be passed through a suitably sized sieve using a rubber bung. Lighter floral bract debris can then be removed using a vacuum separator. Ensure seeds are not being damaged during the sieving process as the embryo is located close to the edge of the seed coat. Mechanised cleaning using a thresher is possible and can assist with bulk collections. However, the risk of damaging individual seeds is increased when using machinery.

Purity and viability: Floral bracts containing seeds are prone to insect predation. Therefore, in combination with low 'fruit' fill (typically <10%), bulk collections can have significant quality issues. Once cleaned to seeds, ensuring embryo damage does not occur during cleaning, viability is high (>95%).

Seeds per gram: 785 (seeds). Filled floral bracts not evaluated.

Germination requirements

Dormancy type: Non-dormant (tested on seeds).

Embryo type: Peripheral. **Embryo:seed ratio:** 1.91.

Pre-treatments: After cleaning to seeds, germination is high when tested on water. Germination is further improved after exposure to GA_3 and smoke.

Germination: Maximum germination of fresh seed is >75% on water across a wide temperature range (15–35°C). Once treated with GA_3 or smoke germination increases to 80–100%. Germination at 10°C is reduced (55–75%).

Notes: Fresh seeds that remain in the floral bracts likely possess physiological dormancy due to the mechanical impedance of the covering floral structures (see *Ptilotus* data[2]). Given that germination increases after exposure to GA_3, dry after-ripening may assist in improving germination of pure seeds, and germination of seeds contained within the floral bracts.

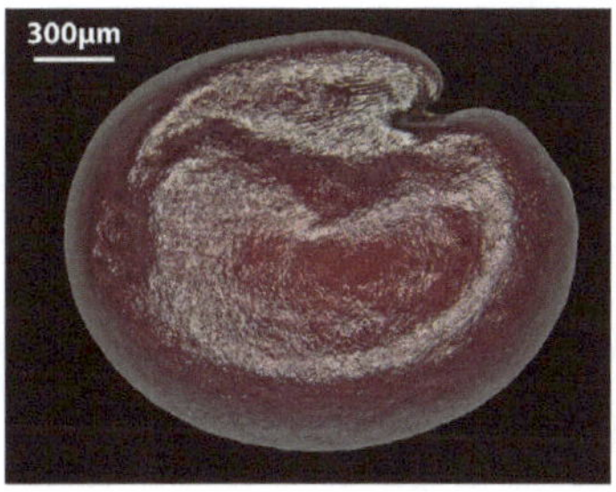

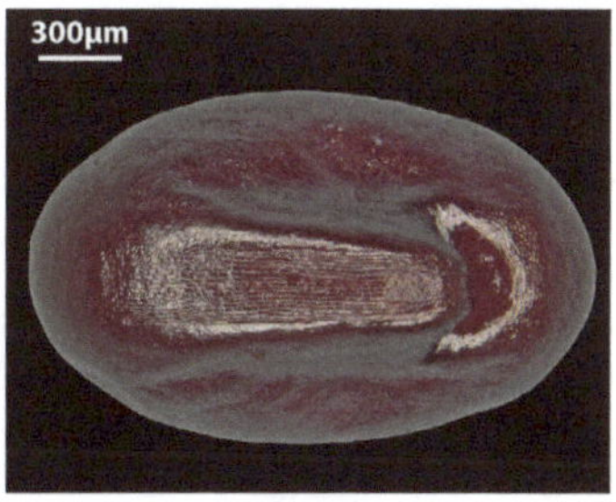

Ptilotus macrocephalus

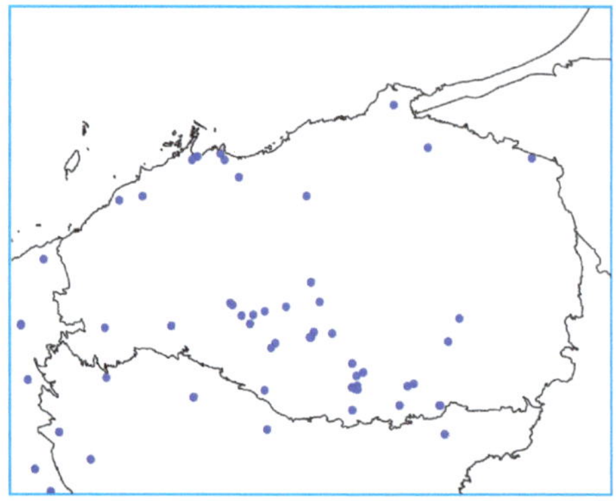

Common names: Featherheads.

Indigenous names: None recorded.

Description: Erect perennial herb to 1 m high, almost glabrous to sparsely hairy apart from curled septate hairs on young shoots and inflorescence. Leaves alternate, subsessile to petiolate, linear to very narrowly elliptic, 40–145 mm long, 2–11 mm wide, undulate, mucronate. Spikes on a thick peduncle, solitary, broadly ovate and *c.* 30 mm long, 30–35 mm wide, but becoming cylindric and 55–105 mm long, 30–40 mm wide. Bracts and bracteoles straw-coloured, transparent, shiny, glabrous or sparsely hairy on the midrib; bracts ovate, 6–8 mm long, acuminate to somewhat aristate; bracteoles broadly elliptic, 5–6 mm long, obtuse and mucronulate. Sepals greenish white, linear, subequal, 13–20 mm long, ribbed, with long hairs which have a whorl of projections at each septum, only glabrous at the obtuse apex. Stamens not all with anthers, unequal; staminal cup adnate to sepals, sometimes somewhat hairy; filaments linear. Ovary glabrous. Style slightly excentric.

Key features for identification: The large, greenish flowers and long, slender leaves.

Similar species: Similar to *P. giganteus* and *P. polystachyus* which have smaller flowers (spikes 15–32 mm wide and bracts 2–7 mm long) and the sepal hairs lack a whorl of projections at each septum.

Distribution: From the southern Kimberley, throughout the Pilbara, south to Carnarvon and inland to the Gibson Desert. Occurs in all mainland states.

Habitat: Occurs on clay soil.

Flowering period: April to July.

Ptilotus macrocephalus

Fruit/Seed collection

Fruit/Seed maturity: May to June.

Appearance of mature fruit/seed: Progressively flowering from the base to the apex of the elongated inflorescence, timing of seed maturation is staggered. The upright flower head turns from green to off-white when mature. Individual floral bracts are mature when they fall from the inflorescence with gentle pressure. Seeds contained within the floral bracts are covered in a papery thin, black outer seed coat which rubs off to expose an orange seed.

Cleaning requirements

Cleaning techniques: Collections will most likely consist of pure floral bracts (which may or may not contain viable seeds) and are commonly of low 'fruit' fill. To clean to pure seeds, collected floral bract material can be passed through a suitably sized sieve using a rubber bung. Lighter floral bract debris can then be removed using a vacuum separator. Ensure seeds are not being damaged during the sieving process as the embryo is located close to the edge of the seed coat. Mechanised cleaning using a thresher is possible and can assist with bulk collections. However, the risk of damaging individual seeds is increased when using machinery.

Purity and viability: Floral bracts containing seeds are prone to insect predation. Therefore, in combination with low 'fruit' fill, (typically <10%) bulk collections can have significant quality issues. Once cleaned to seeds, and ensuring embryo damage does not occur, viability is moderate to high (>80%).

Seeds per gram: 836 (seeds). Filled floral bracts not evaluated.

Germination requirements

Dormancy type: Non-dormant (tested on seeds).

Embryo type: Peripheral. **Embryo:seed ratio:** 2.73.

Pre-treatments: After cleaning to seeds, germination is high when tested on water. Germination is marginally improved at lower temperatures after exposure to GA_3 and smoke.

Germination: Maximum germination of fresh seed is >85% on water across a temperature range of 20–35°C. Once treated with GA_3 or smoke germination increases to >90% at incubation temperatures >15°C. Germination at 10°C with all treatments is variable (60–100%).

Notes: Fresh seeds that remain in the floral bracts likely possess physiological dormancy due to the mechanical impedance of the covering floral structures (see *Ptilotus* data[2]). Given that germination increases after exposure to GA_3, dry after-ripening may assist in improving germination of pure seeds, and germination of seeds contained within the floral bracts.

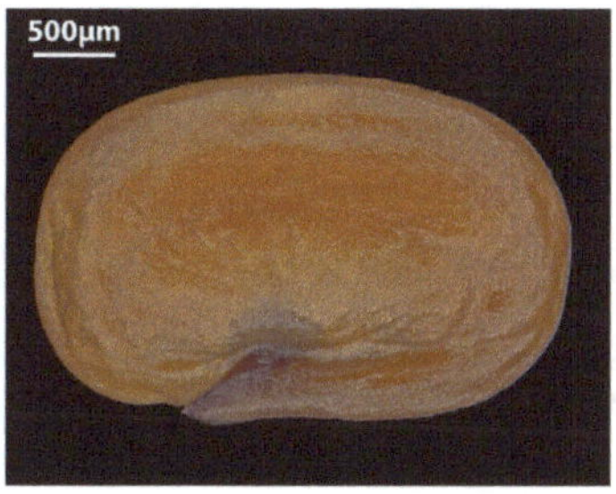

Ptilotus nobilis

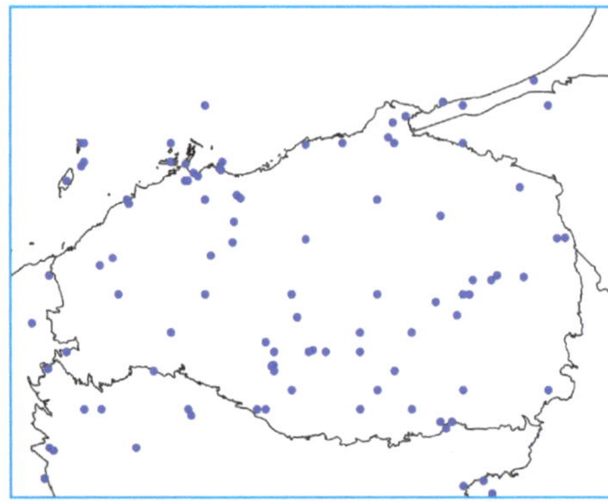

Common names: Regal foxtail, Tall mulla mulla.

Indigenous names: Gumbarli jiruna (Yindjibarndi); Murlumurlu (Ngarluma).

Description: Erect perennial herb to 1 m high, more or less glabrous apart from young shoots and inflorescence. Leaves somewhat succulent, mucronate; lower leaves petiolate, spathulate, to 100 mm long, 25 mm wide; upper leaves alternate, subsessile, elliptic, 18–75 mm long, 6–24 mm wide, margins sometimes undulate. Spikes solitary, on a long peduncle, ovoid to conic but becoming cylindric, 50–130 mm long, 35–45 mm wide. Bracts and bracteoles translucent, straw-coloured but often with a dark midrib, ovate, 6–8 mm long, sparsely hairy, acuminate and mucronate. Sepals pink to mauve, linear, rigid, connate at the base, subequal, 15–21 mm long, ribbed, hairs short-dendritic and also long-septate with a whorl of projections at each septum, apex petal-like and truncate, the inner 3 sepals with woolly margins inside. Stamens 2 or 3 much shorter and without anthers, unequal; staminal cup adnate to the sepals; filaments subulate. Ovary stipitate, sparsely hairy below the style. Style excentric.

Key features for identification: Glabrous leaves; erect inflorescence with bright pink flowers; erect habit.

Similar species: *P. axillaris* is a prostrate herb with hemispheric to globular spikes 17–40 mm long and sepals 8–16 mm long.

Distribution: Occurs in a variety of habitats.

Habitat: Widespread in the Pilbara region and from the north Kimberley south to Coorow and Norseman. Occurs in all mainland states.

Flowering period: April to October.

Ptilotus nobilis

Fruit/Seed collection

Fruit/Seed maturity: July to October.

Appearance of mature fruit/seed: The pink-purple cylindrical inflorescences progressively flower from the base to the apex of the flowering stem. Timing of seed maturation is therefore staggered. The upright flower head turns from pink-purple to pale-pink/white when mature. Individual floral bracts are mature when they fall from the inflorescence with gentle pressure. Seeds contained within the floral bracts are dark red to black.

Cleaning requirements

Cleaning techniques: Collections will most likely consist of pure floral bracts (which may or may not contain viable seeds) and are commonly of low 'fruit' fill. To clean to pure seeds, collected floral bract material can be passed through a suitably sized sieve using a rubber bung. Lighter floral bract debris can then be removed using a vacuum separator. Ensure seeds are not being damaged during the sieving process as the embryo is located close to the edge of the seed coat. Mechanised cleaning using a thresher is possible and can assist with bulk collections. However, the risk of damaging individual seeds is increased when using machinery.

Purity and viability: Floral bracts containing seeds are prone to insect predation. Therefore, in combination with low 'fruit' fill, (typically <10%) bulk collections can have significant quality issues. Once cleaned to seeds, and ensuring embryo damage does not occur, viability is high (>95%).

Seeds per gram: 645 (seeds). Filled floral bracts not evaluated.

Germination requirements

Dormancy type: Non-dormant (tested on seeds).

Embryo type: Peripheral. **Embryo:seed ratio:** 2.18.

Pre-treatments: After cleaning to seeds, germination is high when tested on water. Germination is not improved after exposure to GA_3 and smoke.

Germination: Maximum germination of fresh seed is >95% on water across a wide temperature range (10–35°C). Once treated with GA_3 or smoke germination is 95–100%.

Notes: Fresh seeds that remain in the floral bracts likely possess physiological dormancy due to the mechanical impedance of the covering floral structures (see *Ptilotus* data[2]). Given that germination increases after exposure to GA_3, dry after-ripening may assist in improving germination of pure seeds, and germination of seeds contained within the floral bracts.

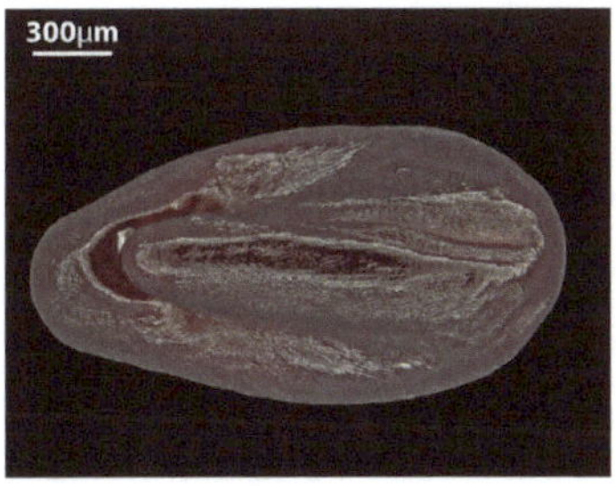

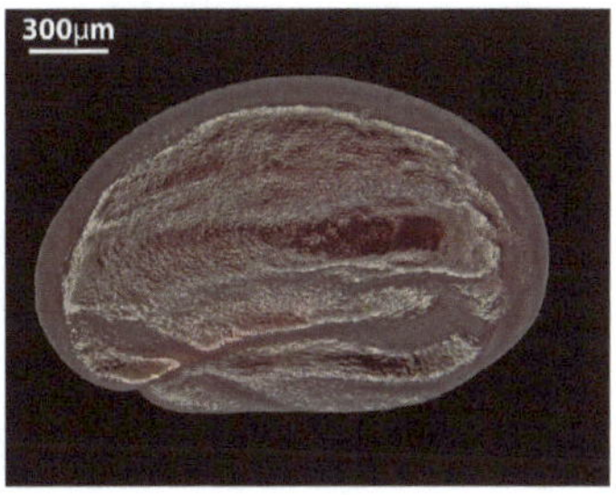

Astrotricha hamptonii

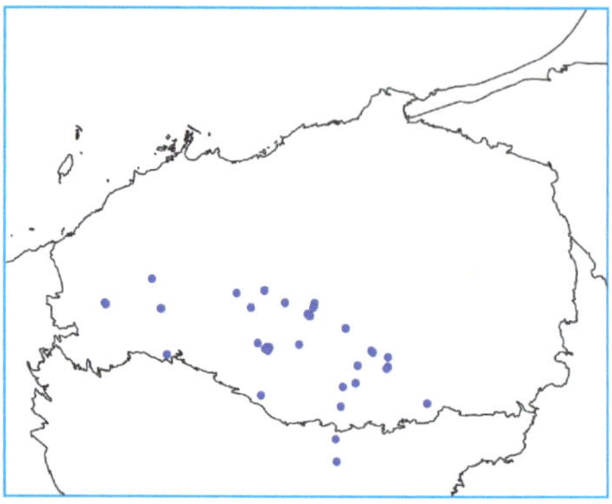

Common names: Ironplant, McCamey's ironplant.

Indigenous names: None recorded.

Description: Shrub, 1–2 m high, stems felted; bark fissured. Stems, petioles, leaves and bracts stellate hairy. Leaves spirally alternate, narrowly lanceolate to narrowly ovate, 75–170 mm long, 17–30 mm wide; densely stellate hairy, dark green above, pale green below; petioles 33–88 mm long. Panicles terminal, 15–39 cm long, moderately stellate-hairy. Unit inflorescence compound-umbellate, 1–4 umbels per lateral branch; 8–26-flowered; pedicels 1.2–2.6 mm long. Flowers 2.0–2.5 mm across. Hypanthium 3.3–4.3 mm long, campanulate; calyx lobes lanceolate to ovate, 0.5–0.8 mm long, green. Petals 5, 1.9–2.8 mm long, 1.0–1.2 mm wide, yellowish green, caducous. Stamens 5, filaments 1.2–2.3 mm long; anthers cream, 1.4–2.1 mm long. Styles 2, fused at the base, 2.6–4.4 mm long. Fruit a schizocarp, 5.5–6.3 mm long, 2.5–2.8 mm wide, with narrow wings. Seeds 3.6 mm long, 0.7 mm wide, cream, not readily released from mericarps.

Key features for identification: Erect, few-branched habit. Habitat on massive ironstone cliffs. Long, hairy leaves. Flowers in umbellate heads.

Similar species: There are no similar species in WA.

Distribution: Restricted to the southern Pilbara, in the Hamersley Range.

Habitat: On rocky hills, growing almost exclusively on ironstone.

Flowering period: Mainly August to November but also December and March to May, so probably all year round under favourable conditions.

Astrotricha hamptonii

Fruit/Seed collection

Fruit/Seed maturity: November to December.

Appearance of mature fruit/seed: Immature fruits are light green in colour and are clustered at the apex of the flowering stems. Once mature, the fruits turn light brown and fall easily from the plant when touched.

Cleaning requirements

Cleaning techniques: Typically, fruits are collected with high purity and minimal plant debris. However, lighter debris can be removed via vacuum separation if present.

Purity and viability: High (>90%).

Seeds per gram: Filled fruits = 494.

Germination requirements

Dormancy type: Morphophysiological.

Embryo type: Rudimentary. **Embryo:seed ratio:** 0.08.

Pre-treatments: Dormancy alleviation treatments unknown.

Germination: Very low germination(0–5%) of seeds retained within the indehiscent fruits has occurred across all incubation temperatures (10–35°C) and after treatment with GA_3 and smoke. Maximum germination occurred at 25°C after treatment with smoke.

Notes: Seeds with morphophysiological dormancy can be difficult to germinate. Combinations of dry after-ripening, wet/dry cycling or warm stratification coupled with exposure to GA_3 or smoke have been shown to improve germination in fruits/seeds with this kind of dormancy[5].

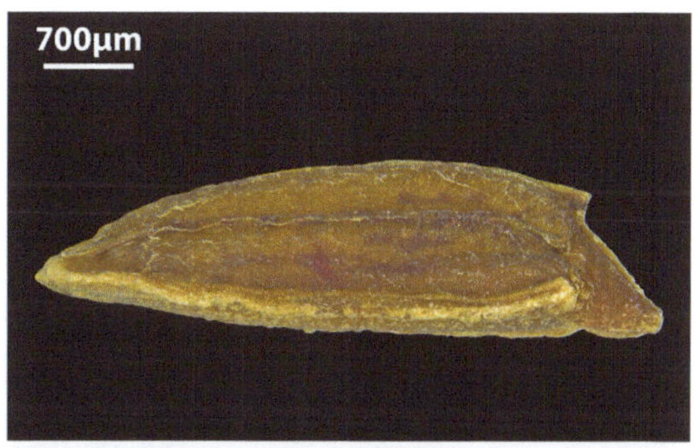

Trachymene oleracea subsp. *oleracea*

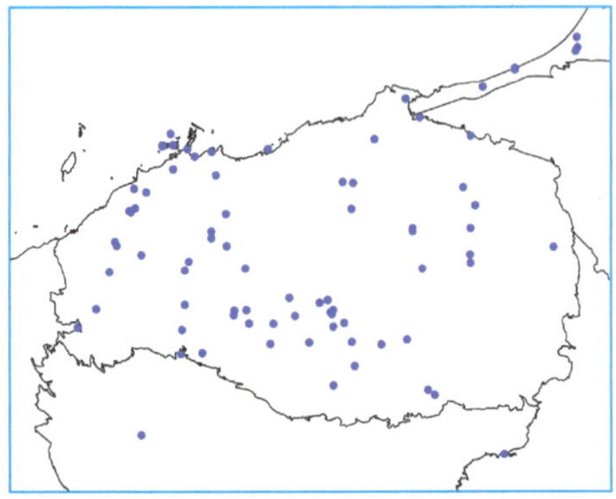

Common names: Fan-leaved lace flower.

Indigenous names: Gujiwanggarn (Ngarluma).

Description: Annual or biennial herb, glabrous or with a sparse to moderately dense indumentum of multicellular glandular hairs. Leaves more or less broadly ovate to circular in outline and at least the lowermost deeply 3–7-lobed, 15–50 mm long, 20–60 mm wide, glabrous or glandular-hairy, base often cordate, the lobes dentate. Inflorescence glabrous or glandular-hairy throughout; bracts broad, obcuneate to semicircular, sessile and amplexicaul, dentate; involucral bracts very narrowly elliptic to subulate, 4–6 mm long, often with sparse marginal glandular hairs, connate towards the base forming a ribbed shallowly obconic base; pedicels 5–6 mm long. Petals white, broadly elliptic, *c.* 1 mm long. Fruiting umbel 15–25 mm across. Fruit usually with 1 mericarp aborted; mericarp 4.5–6.5 mm long, 4–5.5 mm wide, faintly to distinctly papillose-tuberculate, with a smooth scarcely winged margin.

Key features for identification: Tall annual herb with umbellate flowers; leaves broadly ovate to circular in outline and often deeply 3–7-lobed.

Similar species: *T. pilbarensis* and *T. bialata* have the uppermost leaves with slender lobes and/or long petioles.

Distribution: From the southern Kimberley region, south and widespread in the Pilbara Region.

Habitat: Sandplains and low hills.

Flowering period: May–October.

Trachymene oleracea subsp. *oleracea*

Fruit/Seed collection

Fruit/Seed maturity: October to December.

Appearance of mature fruit/seed: After flowering, thin, flat, circular- to ovate-shaped seeds develop at the apex of flowering stems. When mature, the seeds turn from a lime-green yellow to a brown-reddish colour. Mature seeds fall easily from the plant when touched and can be collected in bulk using buckets or calico bags.

Cleaning requirements

Cleaning techniques: Seed collections should be fairly pure. If not, plant material and debris can easily be separated using a suitably sized sieve or by vacuum separation.

Purity and viability: Seed viability is low (55%). Non-viable seeds have successfully been removed using vacuum aspiration.

Seeds per gram: 287 (range 186–287).

Germination requirements

Dormancy type: Morphophysiological.

Embryo type: Rudimentary. **Embryo:seed ratio:** 0.08.

Pre-treatments: Dormancy alleviation treatments unknown.

Germination: Typically 0–5%. Very low germination of seeds occurs across all incubation temperatures (10–35°C) and after treatment with GA_3 and smoke.

Notes: Seeds with morphophysiological dormancy can be difficult to germinate. Combinations of dry after-ripening, wet/dry cycling or warm stratification coupled with exposure to GA_3 or smoke have been shown to improve germination in fruits/seeds with this kind of dormancy[5].

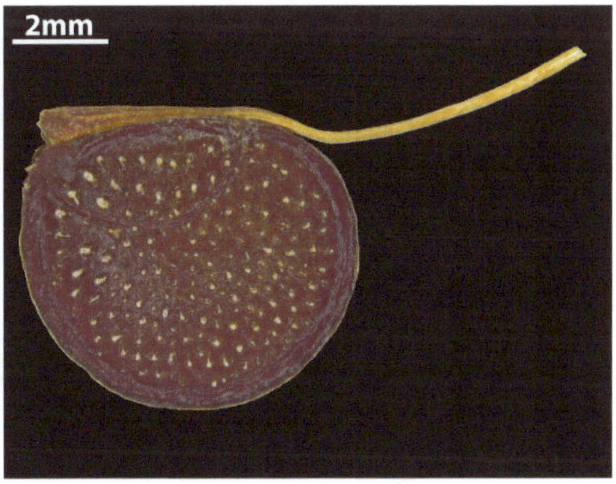

Chrysocephalum apiculatum

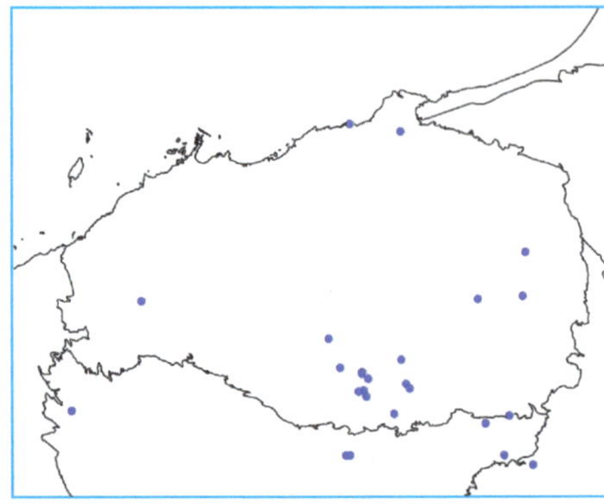

Common names: Yellow buttons, Common everlasting.

Indigenous names: None recorded.

Description: Variable, sometimes bushy, perennial herb 7–60 cm high; branches with woolly hairs and underlying glandular hairs. Leaves linear-lanceolate to spathulate, 10–60 mm long, 10–25 mm wide, apex acuminate to rounded and apiculate, margins flat or recurved, indumentum as on branches. Heads several to numerous in compact clusters, terminal on branches; heads sessile or shortly pedunculate, campanulate to obconical, 5–8 mm long, 7–15 mm diam.; intermediate involucral bracts longest; outer bracts scarious, opaque, golden-yellow or tinged-brown (occasionally white when sometimes tinged pink); intermediate bracts about equal to the florets, with laminas erect. Florets female and bisexual. Achenes narrowly oblong-obovoid, 4-angled, slightly compressed, papillose; pappus bristles finely barbellate and white in the lower half, bright-yellow and subplumose in the upper half, with 6–12 bisexual florets, and 3–8 female florets.

Key features for identification: Bright yellow flower heads; leaves with matted grey hairs; achenes with a pappus, 4-angled, slightly compressed, papillose.

Similar species: No similar species in the Pilbara region.

Distribution: Widespread in W.A. Occurs in all Australian states.

Habitat: Grows in various communities and soil types, usually on open or disturbed sites.

Flowering period: December to May.

Chrysocephalum apiculatum

Fruit/Seed collection

Fruit/Seed maturity: (February to) June.

Appearance of mature fruit/seed: When immature, the involucral bracts of the flowers are green and remain tightly packed around immature seeds. Each flowering unit contains many seeds (10+). At maturity, the bracts dry off and open up revealing silvery seeds with an obvious pappus.

Cleaning requirements

Cleaning techniques: Collections can contain the involucral bracts if collected in bulk. Without due care during collection the proportion of non-seed material, such as the bracts, can increase and prolong cleaning time. The bracts can generally be separated from the seeds by sieving to remove the larger floral bracts.

Purity and viability: Due to floral bracts, purity can be as low as 40%. Once cleaned, seed viability is high (>95%).

Seeds per gram: 18900.

Germination requirements

Dormancy type: Non-dormant

Embryo type: Linear-straight. **Embryo:seed ratio:** 1.00.

Pre-treatments: Seeds are moderately responsive to GA_3, but generally, no pre-treatments required.

Germination: Maximum seed germination achieved with no pre-treatment is 75–80% at 20–30°C. Treatment with GA_3 increases germination to 75–100% over a wide temperature range (10–35°C).

Notes: Additional dormancy break pre-treatments such as dry after-ripening or wet/dry cycling may assist to increase germination to 100%[6].

Pterocaulon sphacelatum

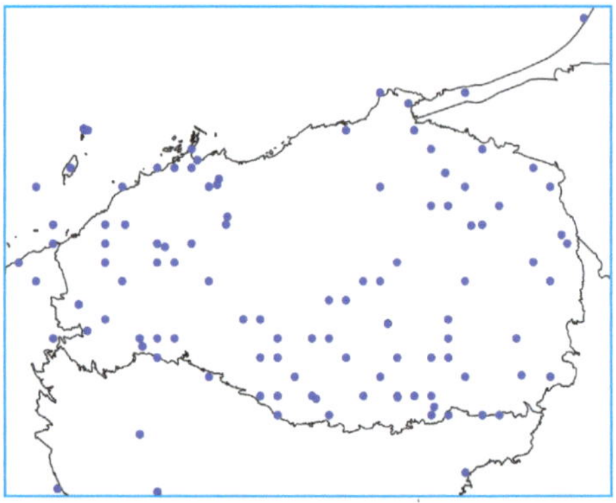

Common names: Apple bush, Fruit salad plant.

Indigenous names: None recorded.

Description: Strongly aromatic shrub 40–100 cm high; stems densely lanate; stem wings entire or sparsely toothed, 0.7–2 mm wide. Leaves oblanceolate or spathulate, 20–66 mm long, 4–15 mm wide, apex obtuse or acute; margins recurved, denticulate, with lanate multicellular hairs only. Upper surface bullate (blistered); hairs erect to spreading, sparse to moderately dense. Inflorescence globose or ellipsoidal, 10–15 mm long, 10–14 mm wide; side branchlets never continuous, at 30–60°. Outer involucral bracts 2.2–3.7 mm long. Inner bracts violet, pink or white on dorsal surface; longest ones linear, 3.5–4.6 mm long, 0.3-0.5 mm wide. Filiform (female) florets 17–29, pink, corolla 2.4–3.2 mm long, styles extending 0.4–0.6 mm beyond corolla. Hermaphrodite floret solitary, corolla pink, 2.5–3.7 mm long; corolla lobes 0.4–0.6 mm long, globose sessile glands present, eglandular hairs absent. Achenes narrowly ellipsoid, 0.7–1 mm long, terete or slightly quadrangular, dark brown, with 25–50 twin hairs; pappus 2.3–3.3 mm long.

Key features for identification: Inflorescence globose; stem wings narrow; leaves with a blistered appearance.

Similar species: *P. xenicum* has stem wings 0.3–0.5 mm wide and inner bracts 4.8–5.6 mm long.

Distribution: Widespread in northern Australia. Also occurs in N.T., Qld., S.A. and N.S.W.

Habitat: In a wide variety of habitats.

Flowering period: April to October.

Pterocaulon sphacelatum

Fruit/Seed collection

Fruit/Seed maturity: October to November.

Appearance of mature fruit/seed: During flowering, the spherical inflorescence contains numerous individual flowers protruding radially from the flower head. When immature, the involucral bracts of the flowers are green-grey and remain tightly packed around immature seeds. Upon maturity, individual flowers and the involucral bracts dry off and release seeds. Each seed contains a pappus to aid in wind dispersal and therefore a small collection window exists.

Cleaning requirements

Cleaning techniques: Most seed collections contain both the seeds (with pappus) and loose involucral bracts. Cleaning to pure seeds is difficult due to the similar size and weight of the bracts versus seeds.

Purity and viability: Involucral bracts can reduce purity values slightly. However, given the very light weight of bulk collections (bracts and seeds) this is not a major concern. Viability of seeds is generally high (>90%).

Seeds per gram: 23100.

Germination requirements

Dormancy type: Physiological.

Embryo type: Linear-straight. **Embryo:seed ratio:** 1.00.

Pre-treatments: Exposure of fresh seeds to GA_3 and smoke significantly increases germination.

Germination: Maximum germination achieved on water is variable, with germination ranging between 5% and 65% across a wide temperature range (10–35°C). Once treated with GA_3 or smoke germination increases to 45–95%.

Notes: Additional dormancy break pre-treatments such as dry after-ripening may assist in overcoming physiological dormancy, promote more uniform germination, and remove the need for treatment with GA_3 or smoke[6].

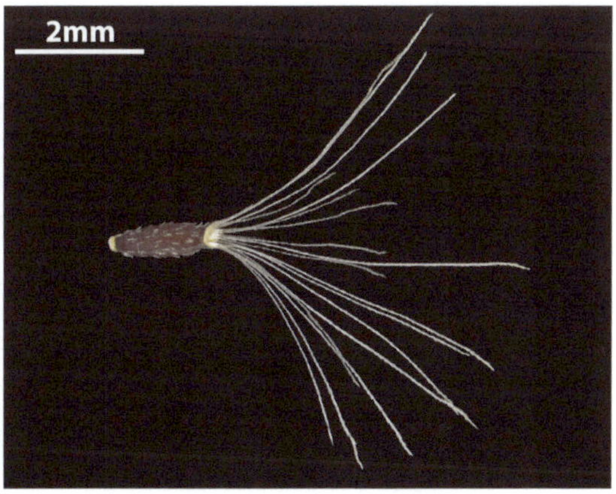

Streptoglossa decurrens

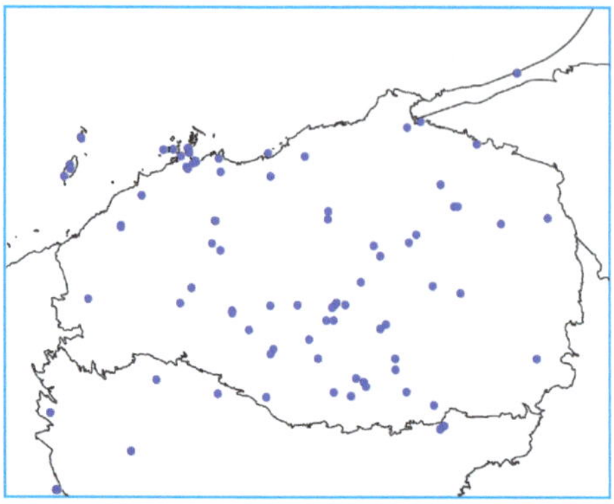

Common names: Wing-leaf Streptoglossa.

Indigenous names: None recorded.

Description: Shrub to 70 cm; strongly aromatic, pilose to villous, densely glandular; glands on long and short stipes. Stem erect, leafy or the older branches leafless. Leaves oblong to elliptic, acute; bases decurrent; leaves of the stem and main branches 20–60 mm long, 8–23 mm wide, leaves of the branchlets 8–20 mm long, 2–6 mm wide; margins entire or serrate. Capitula usually in corymbose clusters on lateral branches; florets 45–80, disc fewer than marginal. Involucre 8–14 mm long. Several of the outer phyllaries with leaf-like tips, outer and median ciliate, pilose, densely glandular, inner glabrous or glabrescent, usually ciliate, glandular or non-glandular. Receptacle areolate, glabrous, non-glandular. Marginal florets ligulate, ligules 1.5–3 mm long, regularly or irregularly 3- or 4-lobed, one lobe often more deeply incised; glabrous, non-glandular. Corollas of disc florets 6–7 mm long, 5-lobed, glabrous, usually non-glandular. Achenes 2–2.5 mm long, densely sericeous, without ribs. Pappus in 2 or 3 series, setae of the outer series, if present very short.

Key features for identification: Leaves decurrent; involucres <15 mm long.

Similar species: *S. odora* has leaves on primary stem rarely more than 10 mm wide; scattered heads with 15–30 florets.

Distribution: Widespread in the arid zone of Western Australia. Also in N.T. and Qld.

Habitat: Rocky hillslopes, creeks, flats and depressions.

Flowering period: July to November.

Streptoglossa decurrens

Fruit/Seed collection

Fruit/Seed maturity: August to October.

Appearance of mature fruit/seed: When immature, the large involucral bracts of the flowers are lime green and remain tightly packed around immature seeds. Each flowering unit contains many seeds (20+). At maturity, the bracts dry off and open up revealing silvery and densely-haired seeds with an obvious pappus.

Cleaning requirements

Cleaning techniques: If collected with care, most collections will contain predominantly seeds (with pappus). Occasional loose involucral bracts will also be present. If the involucral bracts are present, cleaning to pure seeds is possible using a vacuum separator, but this can be difficult at times.

Purity and viability: Purity can be reduced due to floral bracts. However, if collected carefully or once cleaned, seed viability is high (>95%).

Seeds per gram: 1096 (range 1096–1250).

Germination requirements

Dormancy type: Non-dormant.

Embryo type: Linear-straight. **Embryo:seed ratio:** 1.00.

Pre-treatments: Germination on water is high at warmer temperatures (>25°C). Exposure of fresh seeds to GA_3 and smoke significantly increases germination.

Germination: Maximum germination achieved on water is variable with germination ranging between 30–100% at temperatures of 20–35°C. Once treated with GA_3 or smoke germination increases to 100% at 15–35°C.

Notes: Additional dormancy break pre-treatments such as dry after-ripening or wet/dry cycling may assist to increase germination to 100% without the need to treat with GA_3 or smoke[6].

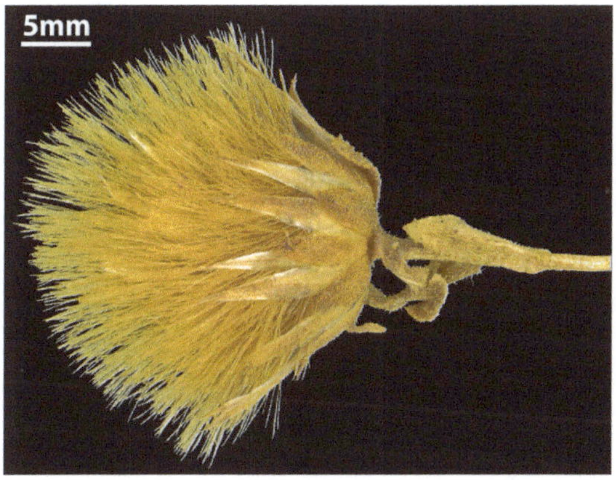

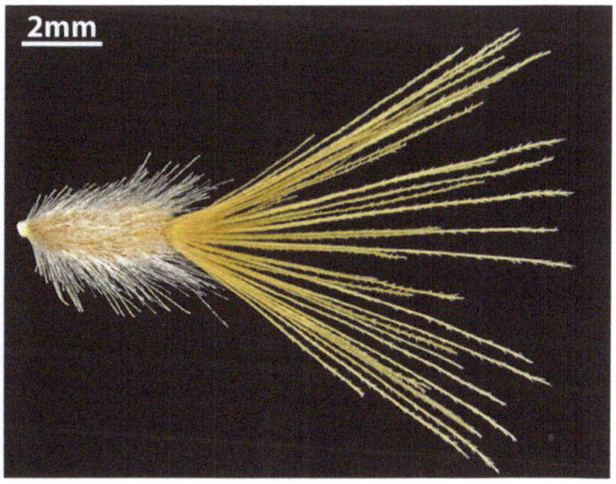

Trichodesma zeylanicum

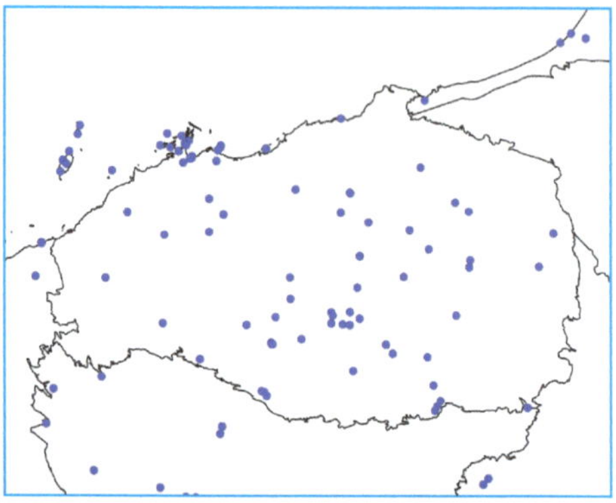

Common names: Northern bluebell, Camel bush, Rough bluebell.

Indigenous names: Bigurrga (Thalanyji); Gumbarli or Bawularri (Yindjibarndi); Jabajaba (Ngarluma).

Description: Coarse annual or perennial, 0.3–2 m high, with a dense indumentum of minute hairs and much longer spreading hairs on the stems, leaves and pedicels, the long hairs usually stiff and often pungent. Leaves opposite on main stems and alternate on flowering stems, sessile or with a petiole up to 9 mm long; blade pale grey-green to medium green, very narrowly ovate to ovate, 17–160 mm long, 5–42 mm wide, entire, acute. Inflorescence a simple raceme; bracts smaller than, but similar to, leaves. Pedicels 12–38 mm long, often S-shaped. Sepals more or less free, narrowly to almost broadly ovate, 7–26 mm long, indumentum similar to the leaves, acuminate or sometimes acute. Petals usually pale blue, rarely white or medium blue, glabrous or with a few hairs outside, shortly connate; lobes broadly or very broadly ovate, 3–11 mm long. Stamens 6–11 mm long, with a very dense indumentum of long, closely arranged hairs in basal 2–4 mm, the free glabrous portion of each appendage spirally twisted. Style not exserted, sometimes visible among the twisted anther appendages. Schizocarp broadly ovoid, 4–5 mm long.

Key features for identification: Blue pendulous flowers, silvery hairy heaves.

Similar species: No similar species in the region.

Distribution: Widespread in the Kimberley and Pilbara, south to near Geraldton and inland to the Warburton area. Also occurs in N.T., S.A., Qld, N.S.W., extends through southeast Asia to India.

Habitat: Occurs along watercourses and on floodplains in a variety of soils, also on sand dunes, often in red soil.

Flowering period: March to October.

Trichodesma zeylanicum

Fruit/Seed collection

Fruit/Seed maturity: October to November.

Appearance of mature fruit/seed: When mature, four individual seeds are arranged in the central axis of the remnant flower (petals dehisced) surrounded by the persistent sepals, and fall from the plant when gently touched. The outer (visible) sides of the seeds are shiny and mottled white-black in colour. The inner side of the seed coat is dull black in colour.

Cleaning requirements

Cleaning techniques: Collections should only contain the remnant flower with sepals and the individual seeds. The plant material can easily be separated using a suitably sized sieve or by vacuum separation.

Purity and viability: Viability high (>95%).

Seeds per gram: 95 (range 60–112).

Germination requirements

Dormancy type: Physiological.

Embryo type: Spatulate. **Embryo:seed ratio:** 1.00.

Pre-treatments: Dormancy alleviation treatments unknown.

Germination: Typically 0–20%. Very low germination of seeds occurs across all incubation temperatures (10–35°C) and after treatment with GA_3 and smoke.

Notes: Dormancy break pre-treatments such as dry after-ripening or wet/dry cycling may assist in overcoming physiological dormancy.

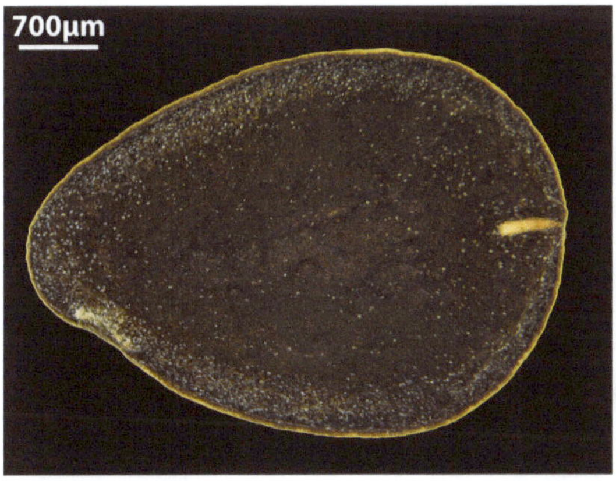

Lepidium catapycnon

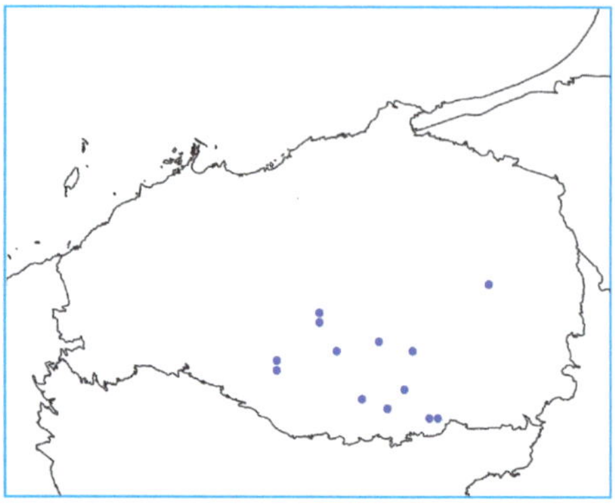

Common name: Hamersley lepidium.

Indigenous names: None recorded.

Description: Plant a papillose, perennial shrub. Stems bent at the nodes in a marked zig-zag manner (flexuose). Leaves papillose, linear, terete, succulent, to 4 cm long and 2 mm wide. Inflorescence a dense terminal raceme. Sepals 4 mm long. Petals somewhat rhomboid, pseudotubular with cucullate base and attenuate tips, 6 mm long. Stamens 6. Pedicels spreading to 8 mm long. Silicula winged and sparsely papillose, the style is markedly exserted.

Key features for identification: Zig-zag stems; papillose, terete, succulent leaves.

Similar species: Close to *L. pedicellosum* but it has very different leaf morphology.

Distribution: Restricted to the southern Pilbara region.

Habitat: Rocky hillsides.

Flowering period: October.

Note: This species is listed as Threatened.

Photographs by Darren Brearley (top; centre left; bottom left & right); Belinda Barnett (centre right; bottom left).

Lepidium catapycnon

Fruit/Seed collection

Fruit/Seed maturity: December (limited observations, but likely to be October to December).

Appearance of mature fruit/seed: The flat, oval-shaped, two-seeded capsules mature progressively from the earlier flowering basal flowers, to the flowers located at the apex of the flowering stem. Therefore, timing of seed maturation is staggered and is dependent on the timing and retention of soil moisture. Once mature, the capsules turn from light green to pale brown. Seeds are orange to light brown in colour.

Cleaning requirements

Cleaning techniques: Capsules release seeds relatively easily when mature and can be passed through a suitably sized sieve. Additionally, the pods can be gently crushed with a bung, passed through a sieve, and separated using vacuum separation.

Purity and viability: Viability high (>90%).

Seeds per gram: 361 (range 360–377).

Germination requirements

Dormancy type: Physiological.

Embryo type: Bent (inferred from other *Lepidium* species). **Embryo:seed ratio:** Not determined.

Pre-treatments: Exposure to GA_3 significantly enhances germination of fresh seeds. Exposure to smoke has not been tested.

Germination: Maximum germination achieved on water is 18%. Once treated with GA_3, germination increased to *c.* 90%. Germination has only been tested at constant 15°C and alternating 25/10°C.

Notes: The seed quality and germination data for this species was provided by Dr Anne Cochrane at the Threatened Flora Seed Centre (Science and Conservation Division, Department of Parks and Wildlife, Perth, Western Australia). Given fresh seeds show a strong germination response to GA_3, it is likely that dormancy break pre-treatments such as dry after-ripening may assist in overcoming physiological dormancy, promote more uniform germination, and remove the need for treatment with GA_3. Further, treating seeds with smoke may provide similar germination results, but this remains to be tested.

Lepidium echinatum

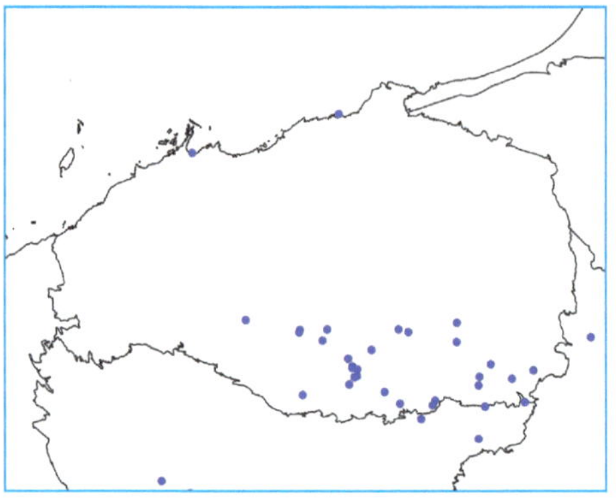

Common names: Prostrate peppercress.

Indigenous names: None recorded.

Description: Plants prostrate, spreading, perennial herbs (perennating from a woody rootstock); stems glabrous, up to 1 m long. Leaves succulent, papillose, obovate to lanceolate to linear, obtuse to acute tipped, basally attenuate, up to 6 cm long and 3 mm wide. Inflorescences very elongate racemes. Sepals 4–5 mm long, often pink to purple pigmented, papillose. Petals white, somewhat lanceolate, tending to be pseudotubular and cucullate basally, 5–7 mm long, 1–1.5 mm wide. Stamens 6. Pedicels papillose, up to 8 mm long, tending to twist to orientate flowers and fruits vertical. Silicula 5–6 mm long, 6–7 mm wide, rotund, inflated, unwinged and without a notch; valves very pubescent with somewhat acicular hairs; style 3–6 mm long. Seeds 3–4 mm long, 2.5–3 mm wide; cotyledons incumbent.

Key features for identification: Glabrous stems; 6 stamens; hairy siliculas and entire leaf margins.

Similar species: *L. pedicellosum* has glabrous siliculas.

Distribution: Restricted to the southern Pilbara, Fortescue and Ashburton districts.

Habitat: Red sand and clay soils, often with gravel.

Flowering period: April to October.

Lepidium echinatum

Fruit/Seed collection

Fruit/Seed maturity: October to November.

Appearance of mature fruit/seed: Throughout the growing season the flat, circle-shaped, two-seeded capsules mature progressively, from the oldest to the youngest flowers along the prostrate flowering stems. Therefore, timing of seed maturation is staggered. Once mature, the capsules turn from light green to pale brown. Seeds are orange to light brown in colour.

Cleaning requirements

Cleaning techniques: Capsules release seeds relatively easily when mature and can be passed through a suitably sized sieve. Additionally, the pods can be gently crushed with a bung, passed through a sieve, and separated using vacuum separation.

Purity and viability: Viability high (>90%).

Seeds per gram: 149.

Germination requirements

Dormancy type: Non-dormant.

Embryo type: Bent. **Embryo:seed ratio:** 1.99.

Pre-treatments: No pre-treatment requirements.

Germination: Unimpeded (85–100%) across a wide temperature range (10–30°C). Germination is reduced slightly at 35°C (75% on water, but >85% using GA_3 or smoke).

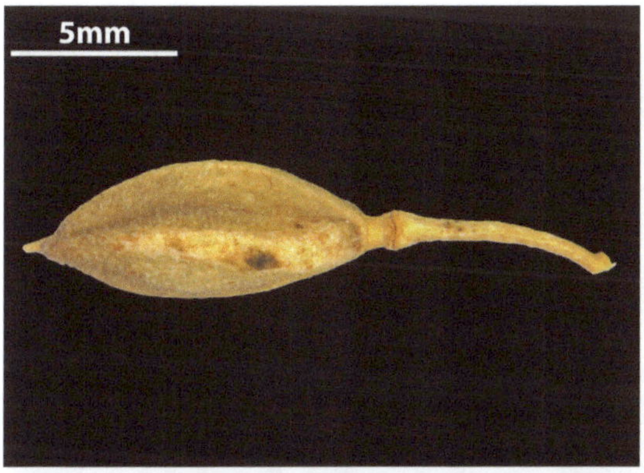

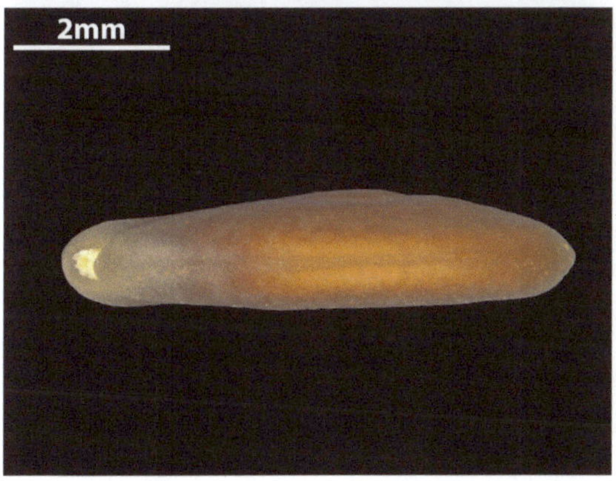

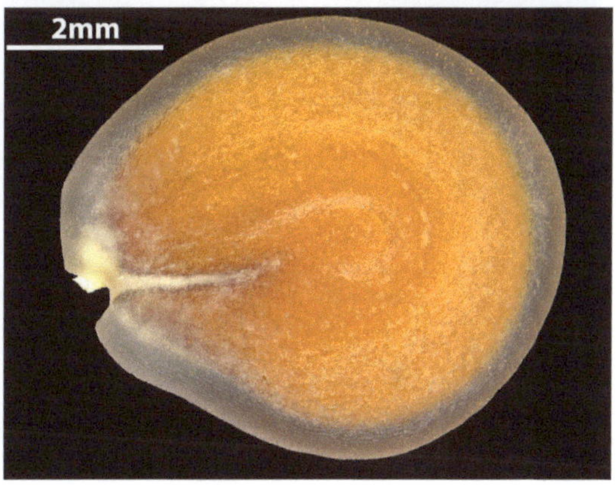

Lepidium pedicellosum

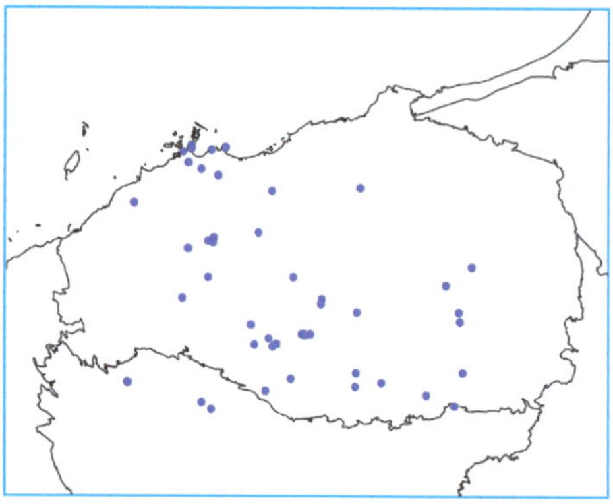

Common names: Fluted peppercress.

Indigenous names: None recorded.

Description: Plants erect, perennial shrubs, up to 50 cm tall, glabrous or pubescent with short acicular hairs. Stems tend to be anfractuose. Leaves entire, leathery, obtuse to acuminate, basally attenuate to petiolate, large, 20–45 mm long, 10–20 mm wide. Inflorescences dense terminal racemes. Sepals 4–6 mm long. Petals white, somewhat rhomboid, pseudotubular with cucullate base and attenuate tips, 5–7 mm long, 3 mm wide. Stamens 6. Strongly nectariferous. Pedicels spreading, 5–8 mm long. Silicula 6–9 mm long, 4–6 mm wide, elliptical, winged; wings approximately 1/8 of the length of the silicula, somewhat acute, forming an open notch; style 3–4 mm long, markedly exserted; (when the plant is pubescent, the valves and style are also beset with similar acicular hairs). Seeds 2.7 mm long, 1.5 mm wide; cotyledons incumbent.

Key features for identification: Stems and leaves glabrous; leaves >3 mm wide; flowers conspicuous; petals more than 3 mm long; 6 stamens; silicula glabrous.

Similar species: *L. echinatum* has hairy siliculas.

Distribution: Widespread in the Pilbara, south to the Carnegie district.

Habitat: Stony soils on rocky hillsides and along creeks.

Flowering period: May to October.

Photographs by Eddy Wajon.

Lepidium pedicellosum

Fruit/Seed collection

Fruit/Seed maturity: October to December.

Appearance of mature fruit/seed: The flat, oval-shaped, two-seeded capsules mature progressively from the earlier flowering basal flowers, to the flowers located at the apex of the flowering stem. Therefore, timing of seed maturation is staggered. Once mature, the capsules turn from light green to pale brown. Seeds are orange to light brown in colour.

Cleaning requirements

Cleaning techniques: Capsules release seeds relatively easily when mature and can be passed through a suitably sized sieve. Additionally, the pods can be gently crushed with a bung, passed through a sieve, and separated using vacuum separation.

Purity and viability: Viability high (100%).

Seeds per gram: 316.

Germination requirements

Dormancy type: Physiological.

Embryo type: Bent. **Embryo:seed ratio:** 1.86.

Pre-treatments: Exposure to GA_3 or smoke significantly enhances germination of fresh seeds.

Germination: Maximum germination achieved on water is variable with germination ranging between 5% and 50% across a wide temperature range (10–30°C). Once treated with GA_3 or smoke germination increases to 45–90%. Germination is absent at 35°C.

Notes: Additional dormancy break pre-treatments such as dry after-ripening or wet/dry cycling may assist in overcoming physiological dormancy, removing the need for treatment with GA_3 or smoke.

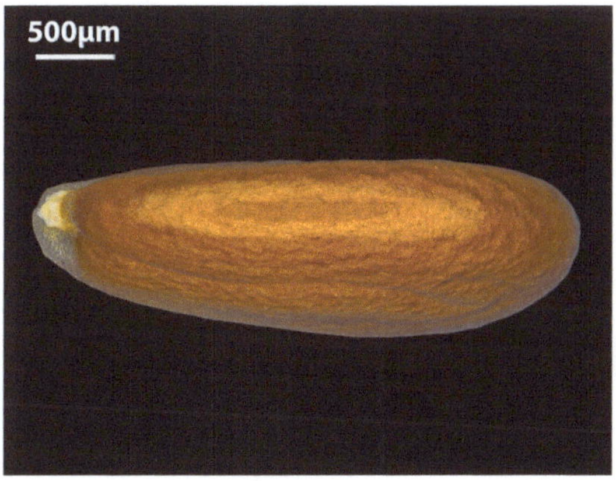

Wahlenbergia tumidifructa

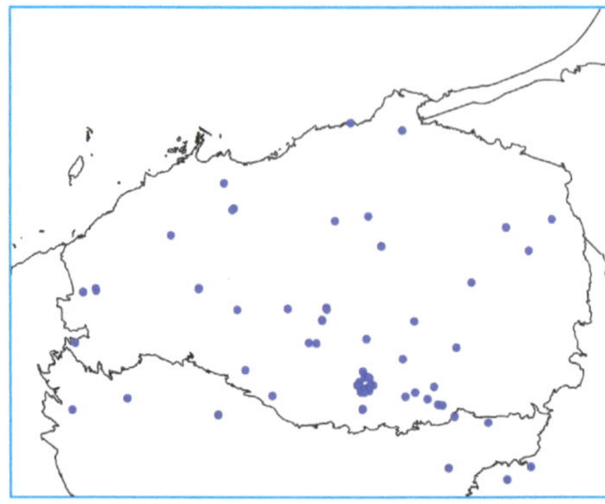

Common names: Mallee annual-bluebell.

Indigenous names: None recorded.

Description: Perennial or annual tufted herb, 4–90 cm high, much-branched below inflorescence, upper parts usually glabrous, lower parts hirsute. Leaves alternate or sometimes the lowermost opposite, obovate or oblanceolate, becoming linear up the stem, 2–60 mm long, 0.2–12 mm wide, apex obtuse to acute, margins flat or undulate, entire or with small callus teeth, lower leaves hirsute, upper leaves glabrous. Flowers in cymes; pedicels 20–110 mm long, glabrous or sometimes sparsely hirsute; bracteoles linear, 1.5–6 mm long, glabrous or sometimes sparsely hirsute. Hypanthium globose, cylindric or elongated-obconic, glabrous. Sepals 1–4 mm long, glabrous. Corolla deeply campanulate, usually blue, sometimes white or pink, puberulous inside at base, otherwise glabrous; tube 1–5 mm long; lobes 2–7 mm long, 1–3.5 mm wide. Stamens 5; filaments white, 0.7–1.5 mm long; anthers 1–3 mm long. Ovary 3-locular. Style white, 2.5–6 mm long, 3-fid, usually not constricted; stigmatic lobes 0.5–1 mm long. Capsule cylindric or globose, often appearing swollen, 2.5–9 mm long, 2–4.5 mm wide, glabrous. Seeds *c.* 0.4 mm long.

Key features for identification: Lowermost leaves alternate; corolla tube 1–5 mm long; lobes 2–7 mm long.

Similar species: *W. queenslandica* has a longer corolla tube 4–11 mm long, distinctly longer than the ovary and corolla lobes 6–20 mm long.

Distribution: Widespread in the Pilbara and in all mainland states.

Habitat: Grows in open sites on red sand, stony soils, sandy loam, along drainage lines and in claypans.

Flowering period: March to December.

Wahlenbergia tumidifructa

Fruit/Seed collection

Fruit/Seed maturity: December (limited observations).

Appearance of mature fruit/seed: Capsules containing numerous seeds are located at the apex of many flowering stems. When mature, the capsules will turn brown and split open. Seeds can be collected by upturning mature capsules. Whole plants can be upturned to collect bulk seeds if the parent plant has senesced for the season.

Cleaning requirements

Cleaning techniques: If collected carefully by upturning old flower stems, near pure seeds can be collected as seeds easily fall from the open capsules.

Purity and viability: Viability is very difficult to assess in this species as it has very small seeds. X-ray analysis indicates viability may be as low as 20–30%. Under the microscope, seeds appear aborted, with very few healthy seeds observed. However, the non-viable fraction of collections have successfully been removed using vacuum separation. Repeat collections are required to determine if low viability is common.

Seeds per gram: 77300.

Germination requirements

Dormancy type: Morphological.

Embryo type: Linear-straight. **Embryo:seed ratio:** 0.54.

Pre-treatments: No pre-treatment requirements.

Germination: Unimpeded (70–95%) across a wide temperature range (10–30°C). Germination is reduced at 35°C to <10%.

Notes: Embryo growth has been observed in seeds of this species confirming that seeds have morphological dormancy.

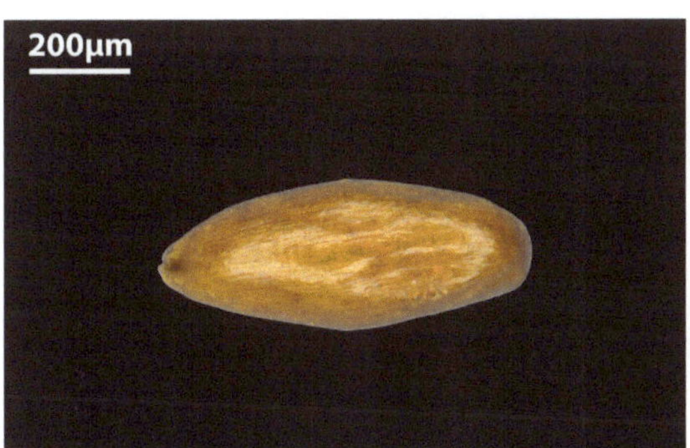

Polycarpaea corymbosa

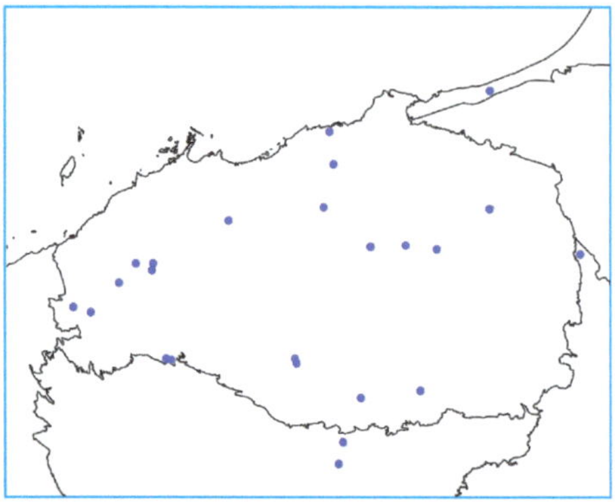

Common names: Slender polycarpaea.

Indigenous names: None recorded.

Description: Erect or ascending herb, 0.1–0.3 m high. Stems with curled hairs. Stipules white, narrowly triangular to narrowly ovate, 2–6 mm long, narrowly acute to aristate; glands between the stipules usually conspicuous. Basal leaves spathulate; cauline leaves sessile, linear, 5–22 mm long, *c.* 0.5 mm, glabrous or sparsely hairy, aristate. Inflorescence of terminal cymes arranged in loose or dense corymbs without floral leaves; bracts white, ovate, 2–4 mm long, divided into 2 narrowly acute segments. Sepals white, narrowly ovate, 2–3 mm long, without a conspicuous midrib, acute; conspicuous red glands sometimes present between the bases of the sepals. Petals oblong, 0.5–1 mm long, more or less free, obtuse. Stamens 5, 0.5–1 mm long; staminodes absent. Capsule ovoid, *c.* 1.5 mm long. Seeds several, minute.

Key features for identification: Annual; inflorescence without involucre of leaves; sepals 2–3 mm long; petals free, 0.5–1 mm long; sepals 2–3 mm long.

Similar species: *P. holtzei* and *P. involucrata* both have the inflorescence subtended by an involucre of leaves.

Distribution: Widespread in the Kimberley Region south to near Meekatharra and inland to the Southesk Tablelands. Also occurs in N.T., S.A., Qld and N.S.W.

Habitat: Occupying a variety of soils and habitats.

Flowering period: February to September.

Polycarpaea corymbosa

Fruit/Seed collection

Fruit/Seed maturity: December (limited observations).

Appearance of mature fruit/seed: The thin, elongated capsules are located at the apex of flowering stems and covered by the remnant papery flower petals. When mature, seeds will readily drop from the capsule when upturned. Each capsule contains numerous seeds (10–20). Whole plants can be upturned to collect bulk seeds if the parent plant has senesced.

Cleaning requirements

Cleaning techniques: If collecting seeds by upturning the stems, the collection should contain near pure seeds. Plant debris and capsules can be separated by vacuum separation if required.

Purity and viability: Viability high (>95%).

Seeds per gram: 7630.

Germination requirements

Dormancy type: Non-dormant.

Embryo type: Peripheral. **Embryo:seed ratio:** 2.29.

Pre-treatments: No pre-treatment requirements.

Germination: Unimpeded (80–100%) across a wide temperature range (15–35°C). Germination is reduced to <10% at 10°C. Exposure to GA_3 or smoke does not enhance germination.

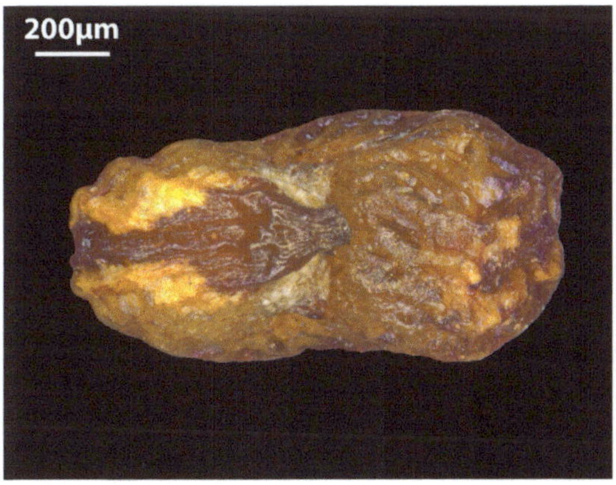

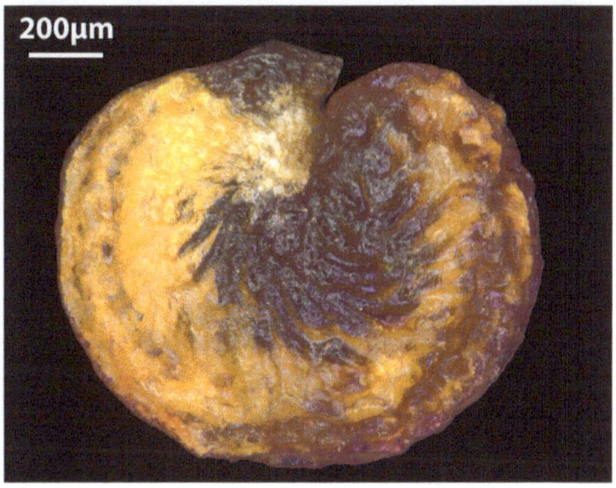

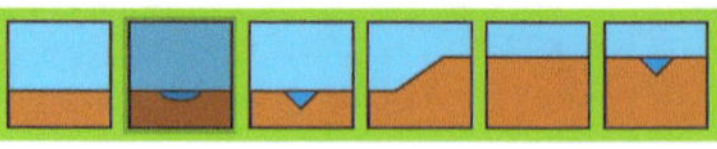

Dysphania rhadinostachya var. *rhadinostachya*

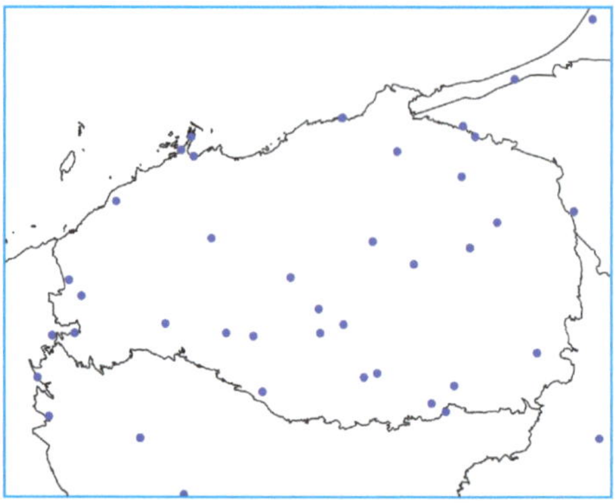

Common names: Green crumbweed.

Indigenous names: None recorded.

Description: Erect annual to 30 cm high, with a single main stem, copiously branched above the base, pilose all over with short glandular and larger simple hairs, aromatic; leaves elliptic, undulate or obtusely lobed; lamina 10–20 mm long; petiole somewhat shorter. Inflorescence of paniculately arranged spikes composed of compact 3–7-flowered clusters; flowers bisexual and female; perianth-segments 4, erect, 1 mm long, hooded, united below, sometimes horizontally keeled, glabrous to hirsute, inflated in fruit; stamen 0 or 1; styles 2; pericarp dull, somewhat adherent; seed subglobular, horizontal to erect, 0.5 mm diam. Infructescence breaking up at the apex of the pedicels with the seed enclosed in the perianth.

Key features for identification: Plant erect with one main stem; flowers in spike-like, paniculate inflorescences; tepals 4; seed dull, almost globular.

Similar species: *D. kalpari* is a much smaller, prostrate or decumbent species with erect, spicate inflorescences and glossy seeds.

Distribution: Widespread in the Pilbara and arid zone of W.A. Also in N.T., Qld, N.S.W. and S.A.

Habitat: Usually found on rocky slopes in skeletal soil.

Flowering period: August to September.

Dysphania rhadinostachya var. *rhadinostachya*

Fruit/Seed collection

Fruit/Seed maturity: October to December.

Appearance of mature fruit/seed: Densely packed flower spikes radiate off the main stems. Immature spikes containing very small fruits/capsules are dull green in colour. Once mature, the whole plant turns yellow. Bulk collections can be carried out by taking whole branches, although this leads to lower seed purity. Hand collected capsules can be gathered by running hands up individual stems.

Cleaning requirements

Cleaning techniques: Larger stem debris can be removed by passing material through sieves. Separation of very fine capsule debris can be achieved using vacuum separation. The seeds are black and very small in size (diam. <1 mm).

Purity and viability: Viability high (>95%).

Seeds per gram: 6013 (range 6013–8333).

Germination requirements

Dormancy type: Physiological.

Embryo type: Peripheral. **Embryo:seed ratio:** 1.94.

Pre-treatments: Dormancy alleviation treatments unknown.

Germination: Maximum seed germination achieved with no pre-treatment is 30% at 15°C. Treatment with GA_3 increases germination to 20–35% at 15–35°C. Dry after-ripening for up to 24 months has not improved germination.

Notes: Additional dormancy break pre-treatments such as wet/dry cycling may assist in improving germination.

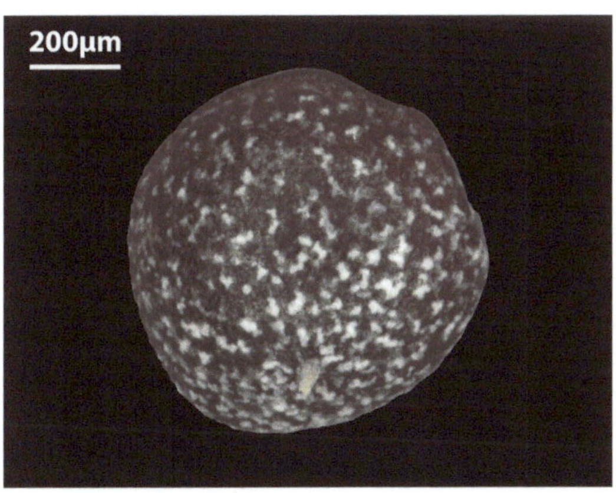

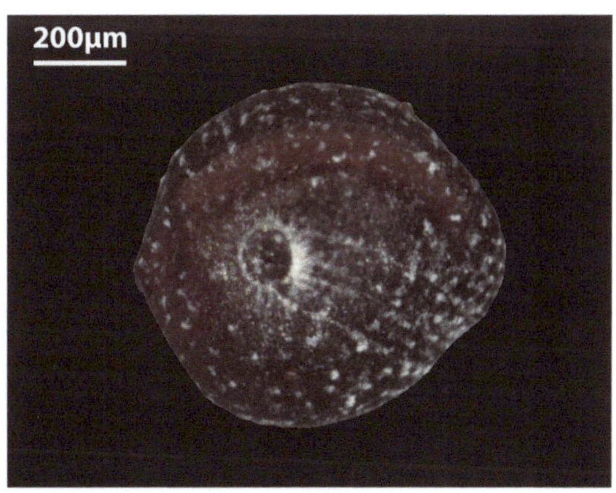

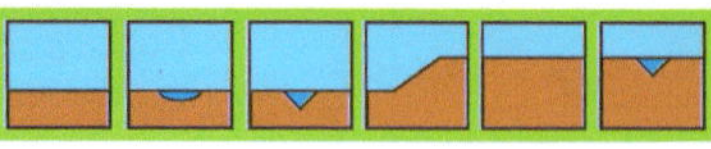

Cleome viscosa

Common names: Tick weed.

Indigenous names: Jirnduwatha (Yindjibarndi & Ngarluma).

Description: Annual or perennial herb usually to 1.5 m high, glandular-hairy with rather harsh hairs. Leaves cauline, petiolate, 3–5-foliolate; leaflets subsessile to petiolulate, elliptic to obovate, 4–25 mm long, 1.5–10.5 mm wide, acute to obtuse. Flowers in leafy racemes, more or less actinomorphic. Sepals narrowly elliptic, 4–7 mm long, acute. Petals yellow, 8–13 mm long, scarcely clawed, limb elliptic to obovate. Stamens 10–26, free; anthers curved to revolute. Ovary sessile, densely glandular-hairy. Style short. Siliqua very narrowly cylindric, 35–98 mm long, 2–3.5 mm wide, more or less straight, longitudinally striate. Seeds transversely rugose, not arillate.

Key features for identification: Erect annual herb, very sticky glandular-hairy.

Similar species: *Cleome uncifera* is a low-growing perennial with the petals all held erect.

Distribution: Widespread in the Pilbara and across northern and inland Australia. Also in N.T., Qld, S.A. and N.S.W. Extends in tropical regions from Africa to Australia, introduced in central America.

Habitat: Often on rocky ridges, on a variety of soils. Often seen on roadsides and other disturbed areas.

Flowering period: February to September.

Note: A variable species requiring further study.

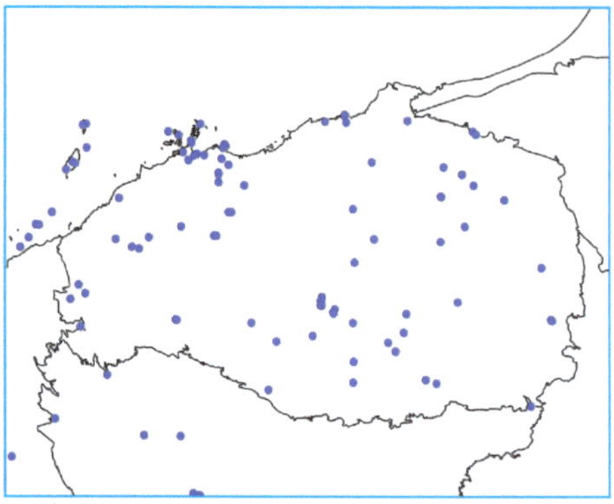

Cleome viscosa

Fruit/Seed collection

Fruit/Seed maturity: November to May.

Appearance of mature fruit/seed: Pods tend to mature progressively throughout the growing season, turning from green to light brown. Pods exude a sticky resin. When mature, the pods 'harden' off and split with gentle pressure. Numerous seeds are contained within each pod.

Cleaning requirements

Cleaning techniques: Passing seeds through a small aperture sieve easily separates pods from seeds. Use of a rubber bung assists in splitting unopened pods.

Purity and viability: Viability high (100%).

Seeds per gram: 1847 (range highly variable, 210–1847).

Germination requirements

Dormancy type: Physiological.

Embryo type: Linear-curved. **Embryo:seed ratio:** 1.75.

Pre-treatments: Exposure of fresh seeds to GA_3 and smoke increases germination slightly. Dry after-ripening for 12–24 months increases germination.

Germination: Maximum seed germination achieved with no pre-treatment is 15% (at 20°C). Treatment with GA_3 or smoke increases germination to 15–40% at 20–35°C. Maximum germination achieved after 12 months of dry after-ripening is 55% (in water) and 75% (following smoke treatment).

Notes: Additional dormancy break pre-treatments such as wet/dry cycling may assist in maximising germination in shorter treatment periods.

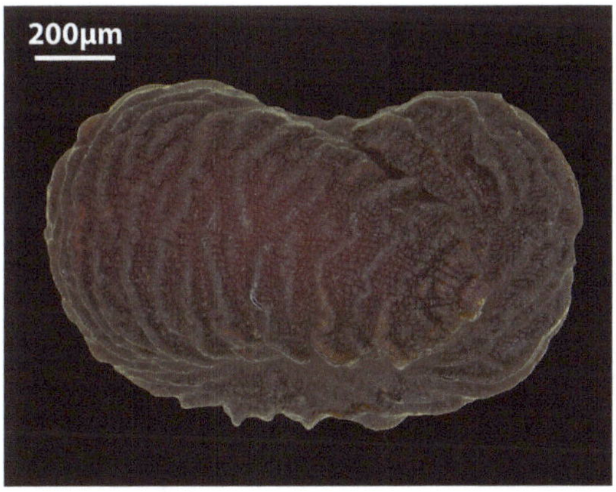

Convolvulus clementii

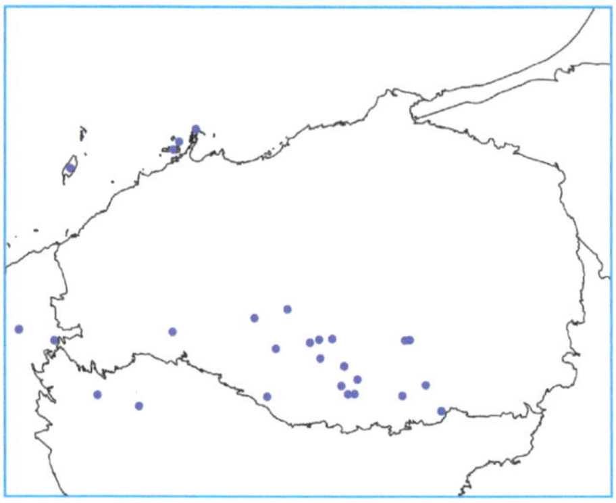

Common names: Desert bindweed.

Indigenous names: None recorded.

Description: Stems trailing and twining; leaves very variable in shape, upper cauline leaves ovate in outline, 10–60 mm long, 5–15 mm wide, obtuse to rounded, emarginate, base cordate, deeply lobed, with a prominent pair of basal lobes, ascending lobes linear to narrowly oblong; terminal lobe linear to linearly obovate, margins entire to shallowly lobed, moderately to sparsely hairy with crisped-appressed to ascending hairs, becoming glabrous. Inflorescence cymose 1- or 2-, rarely 3-flowered; peduncle 5–50 mm long; pedicel 3–15 mm long; outer sepals obovate, orbicular to broadly elliptic, barely acute to rounded with a recurved apiculum, 4–5.5 mm long, moderately to sparsely hairy with crisped-appressed to erect hairs, inner sepals similar; corolla white to pink, rarely purplish, 5–8 mm long; filaments adnate at base for 1.5–2.5 mm, free for 1.5–3 mm; style 2–3 mm long. Capsule globular, 4–6 mm diam.; seeds 2.5–3.2 mm long with irregular short ridges and tubercles made of fused hairs and a serrated or continuous wing.

Key features for identification: Leaves deeply lobed; sepals moderately to sparsely hairy; seeds tuberculate, winged.

Similar species: *C. remotus* has less divided leaves; corollas >10 mm long; and ridged seeds.

Distribution: In the southern Pilbara, south to Wiluna. Also occurs in S.A., N.T., Qld, N.S.W. and Vic.

Habitat: Grows mainly in grassy woodlands on plains, often in flooded and swampy situations; also on claypans between sand dunes.

Flowering period: Throughout the year following rainfall.

Convolvulus clementii

Fruit/Seed collection

Fruit/Seed maturity: June.

Appearance of mature fruit/seed: When ripe, semi-spherical, 4-valved capsules appear off-white to light brown. The thin-walled capsules split when ripe and contain 4 black seeds. Can be collected by hand.

Cleaning requirements

Cleaning techniques: The thin-walled capsules can be easily broken down by rubbing by hand. Capsule debris can be separated easily through a sieve, with the debris falling through the sieve. Vacuum separation is an alternative method to separate the capsule debris.

Purity and viability: Viability generally high (>90%).

Seeds per gram: 112.

Germination requirements

Dormancy type: Physical.

Embryo type: Folded. **Embryo:seed ratio:** 1.95.

Pre-treatments: Wet heat for 1 min at 90–95°C or immersion in sulphuric acid (98%) for 1 hour.

Germination: >85%. Unimpeded across a wide temperature range (15–35°C) once treated to alleviate physical dormancy.

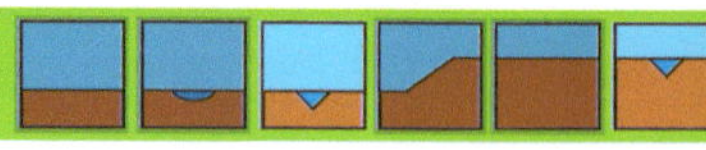

Cyperus ixiocarpus

Common names: Viscid cyperus.

Indigenous names: None recorded.

Description: Robust perennial, tufted, deep-rooted, 50–80 cm high, very viscid in the inflorescence. Culms terete, smooth, 2.5–3.5 mm diam. Leaves thick-textured, canaliculate, septate-nodulose, as long as culms, to 5 mm wide, pale blue-green; marginal prickles very sparse to sparse, short, aculeate. Inflorescence compound or rarely decompound, 9–18 primary branches to 12 cm long; bracts 2–3 exceeding the inflorescence. Spikelets few in loose clusters 1.5–3.5 cm diam.; spikelets compressed, oblong, 10–35 mm long, 3.0–5.0 mm wide in side view, 6–28-flowered, very viscid; glume spacing 2.2–2.7 mm; spikelet usually persistent and glumes falling individually. Glumes elliptic to ovate, acute with straight mucro 0.2–0.5 mm long, stramineous to golden-brown, 2.5–3.0 mm long, 0.6–0.8 mm wide, sides 2–3-nerved. Anthers 0.7–1.3 mm long excluding appendage 0.05–0.1 mm long. Nut trigonous, broad-obovate to elliptic with broad-acute apex, faces concave, dark brown to black, tuberculate, shining, usually viscid, 1.6–2.1 mm long.

Key features for identification: Viscid spikelets; nut obovate to elliptic, dark brown to black, 0.8–1.3 mm diam.

Similar species: *C. cunninghamii* has nuts narrow-elliptic to narrow-ovate, red-brown to black, 0.3–0.7 mm diam.

Distribution: Scattered in the Pilbara region. Also occurs in N.T. and Qld.

Habitat: Red loam, sand and gravel soils in creeks.

Flowering period: February to July.

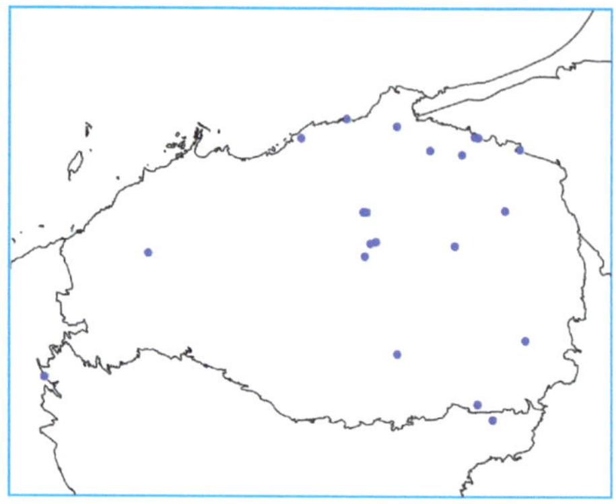

Cyperus ixiocarpus

Fruit/Seed collection

Fruit/Seed maturity: May to June.

Appearance of mature fruit/seed: The seeds, nutlet-like units with an indehiscent resinous glume, are arranged along many splaying spikes spreading from the apex of the flowering stems. When immature, each spike is light green in colour. During maturation the seeds turn black and the glumes light brown. Seeds detach easily from the main stem when collected by hand.

Cleaning requirements

Cleaning techniques: Collections should be relatively pure with occasional flower/spike stems present. Stem debris can be removed via sieving. Pure seeds can be obtained by passing stem material through a sieve.

Purity and viability: Viability high (>95%).

Seeds per gram: 806.

Germination requirements

Dormancy type: Non-dormant.

Embryo type: Capitate. **Embryo:seed ratio:** 0.34.

Pre-treatments: No pre-treatment requirements are known, but a distinct pattern in the germination response to temperature exists and warrants further investigation.

Germination: Maximum seed germination achieved with no pre-treatment is 85% (at 35°C). Germination decreases considerable at lower temperatures (55% at 30°C, and 30% at 25°C). Treatment with GA_3 or smoke increases germination by 10–15% at 25–30°C.

Notes: The narrow germination temperature window (germination only at 35°C) of fresh seeds (in water) is quite unique. Pre-treatments (e.g. dry after-ripening) may broaden the temperature range at which seeds will germinate. Notably, after a period of storage at -18°C, seeds failed to germinate and may have entered secondary dormancy.

Hibbertia glaberrima

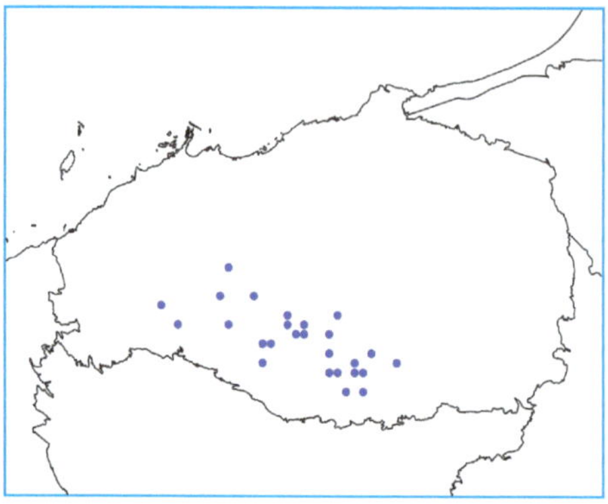

Common names: Pilbara guinea-flower.

Indigenous names: None recorded.

Description: Erect shrub, 40–100 cm high. Leaves narrowly ovate, acuminate to apiculate, 22–78 mm long, 6.5–18 mm wide, entire, tapering below the middle but lacking a distinct petiole, stem-clasping, densely hairy, becoming sparesely hairy with age. Peduncles axillary or terminal, 11–36 mm long. Innermost sepals 15–23 mm long, broadly ovate, the others gradually narrowing to the outermost, which is lanceolate, 12–14 mm long. Petals not much longer than the calyx, 13–16 mm long. Stamens numerous (*c.* 50–70), usually lacking staminodes outside, sometimes a few anthers not fully developing. Carpels 3 (or 4), glabrous, with 2–4 ovules in each. Seeds dark red.

Key features for identification: Large yellow flowers. Finely hairy leaves with pointed tips.

Similar species: Only a single species is currently recognised in the Pilbara. There are two forms, a broad, hairy-leaved form illustrated and described here, and a more slender, sparsely hairy- or glabrous-leaved form that is more similar to true *H. glaberrima* in the N.T. and S.A. Further study of this variation is required.

Distribution: Restricted to the southern Pilbara in the Hamersley Range.

Habitat: Stony soils on ridges and plateaux.

Flowering period: July to October.

Hibbertia glaberrima

Fruit/Seed collection

Fruit/Seed maturity: October to December.

Appearance of mature fruit/seed: Three (to four) thin-walled capsules turn from light green to light brown when mature. Each fruit segment contains two to four dark red seeds. A narrow collection window exists where seeds remain in the upright capsules. Nylon bags have been used successfully to trap seeds during capsule maturation.

Cleaning requirements

Cleaning techniques: Collections should largely contain pure seeds. If capsules are also collected, seeds can be separated easily by passing through a suitably sized sieve.

Purity and viability: Viability is variable with seeds prone to insect predation and abortion during maturation. Viability is 60–70%.

Seeds per gram: 381.

Germination requirements

Dormancy type: Morphophysiological.

Embryo type: Rudimentary. **Embryo:seed ratio:** 0.11.

Pre-treatments: Seeds require warm stratification for 8 weeks at 35°C and transfer to cooler incubation temperatures (25°C) for germination to proceed.

Germination: Generally 50–60% when seeds are treated with smoke, warm stratified for 8 weeks at 35°C, and then transferred to 25°C. Without smoke treatment, germination is limited to 30–40% following warm stratification.

Notes: Warm stratification and smoke exposure appears to be the best pre-treatment to alleviate physiological dormancy. Wet/dry cycling may also be effective. Germination takes 8–10 weeks as there is a requirement for embryo growth prior to germination (i.e. morphological dormancy[5]).

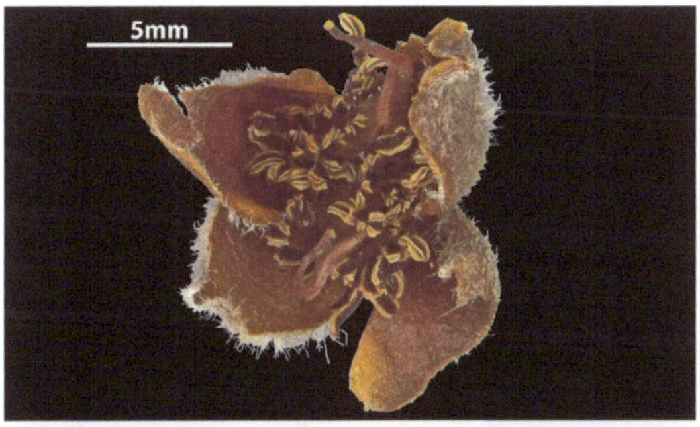

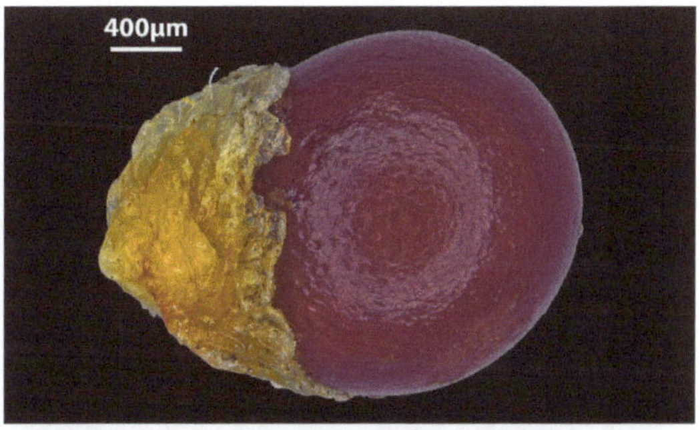

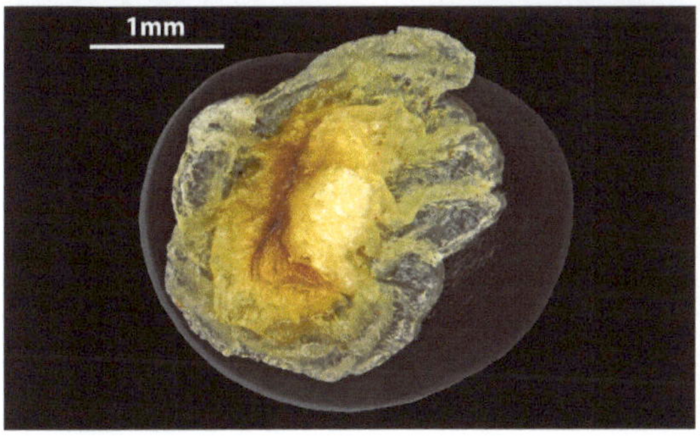

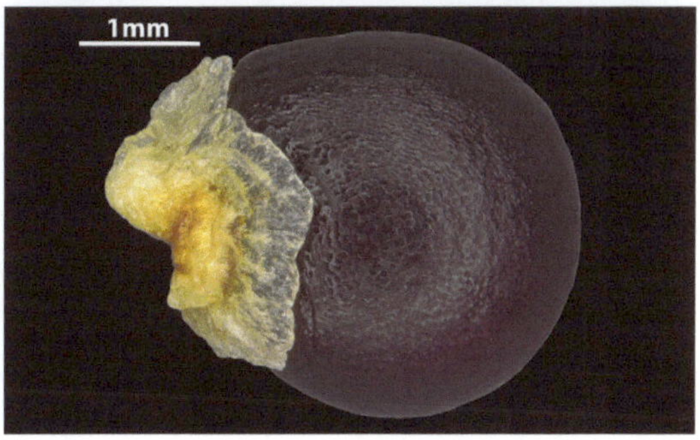

Acacia acradenia

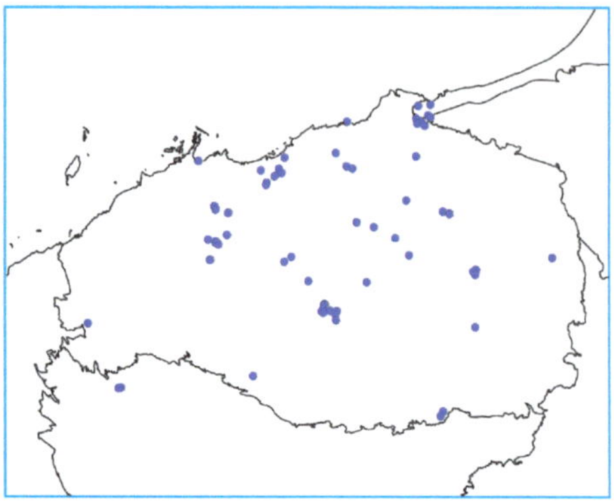

Common names: Silky wattle.

Indigenous names: Yirringan (Ngarluma).

Description: Spindly shrub to 3(–7.5) m high. Bark smooth or fibrous, grey, reddish grey or brown. Branchlets slightly flattened or angular towards apices, dark reddish, pubescent; ridges resinous-crenulated. Phyllodes elliptic to obovate, oblique, 55–150 mm long, 10–30 mm wide, rigid, hoary between resinous veins, with minor veins numerous, 7–9 per mm, not anastomosing, with 3 or more prominent veins confluent near base, with a basal gland to 2 mm above pulvinus and a minute gland at the acute to obtuse apex. Spikes 2–6.5 cm long, orange or golden. Flowers 5-merous, resinous; calyx 0.9–1.2 mm long, dissected for 1/5–1/2 their length, puberulous; corolla 1.8–2.2 mm long, glabrous; ovary puberulous. Pods linear, almost straight-sided, often curved, compressed-terete, 4–13 cm long, crustaceous, pubescent to glabrous, resinous, longitudinally furrowed. Seeds longitudinal, narrowly oblong- to oblong-elliptic, 3–6 mm long, dark brown; pleurogram with yellowish halo; areole open, depressed, olive-brown.

Key features for identification: Broad, dark green phyllodes with numerous fine veins. Terete pods slightly constricted between seeds.

Similar species: Related to both *A. gonoclada* and *A. laccata*, but is usually more hairy than either of these and has more terete pods.

Distribution: Common and widespread in the Pilbara to the southern Kimberley. Also in N.T. and Qld.

Habitat: Grows in skeletal sandy soils and red earths, on laterite, sandstone or calcrete, on rocky hillsides or in gorges, often along creeks with eucalypts and spinifex.

Flowering period: Recorded all year round. Fruiting most prolific about a month after rain events.

Acacia acradenia

Fruit/Seed collection

Fruit/Seed maturity: September to October.

Appearance of mature fruit/seed: When mature, pods turn light brown and split with gentle pressure.

Cleaning requirements

Cleaning techniques: Mature pods can be separated from seeds by passing through a sieve. Typically, the seeds separate readily from the pods. If not, the use of a rubber bung may be required. A vacuum separator can then be used to remove the pod debris.

Purity and viability: Generally high (>80%), but check for insect predation.

Seeds per gram: 110.

Germination requirements

Dormancy type: Physical.

Embryo type: Investing. **Embryo:seed ratio:** 1.00.

Pre-treatments: Wet heat for 2 mins at 80–95°C.

Germination: >95%. Unimpeded across a wide temperature range (10–35°C) once heat treated. Lower temperatures (10°C) impede germination by 30–50%.

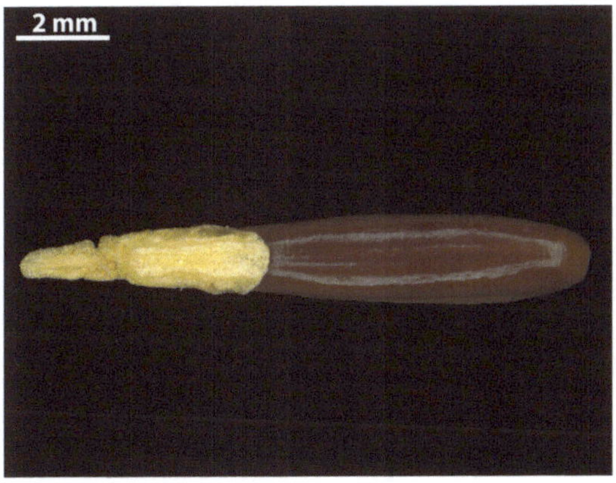

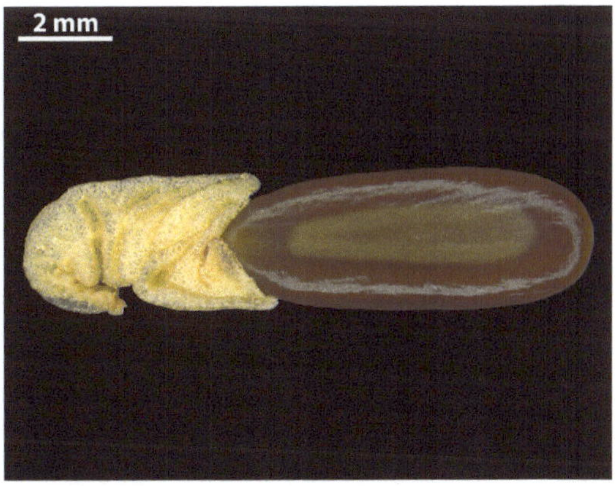

Acacia adoxa var. *adoxa*

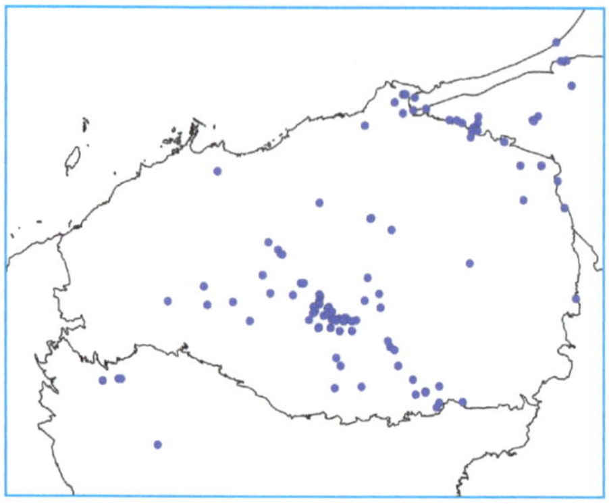

Common names: Grey whorled wattle.

Indigenous names: Pirarr-pirarr (Nyangumarta).

Description: Procumbent to spreading shrub to 0.7 m high. Stems tomentose or pilose with retrorse or erect hairs, or glabrous. Stipules linear, 0.8–1.2 mm long. Phyllodes 6–10 per whorl, ascending to erect, flattened, straight or slightly recurved, commonly 2–5 mm long, with or without oblique mucro to 0.2 mm long, pubescent or pilose to glabrous, with an impressed adaxial nerve and an obscure abaxial nerve. Peduncles 6–15 mm long, pubescent to glabrous. Heads commonly 25–35-flowered. Flowers 5- or 6-merous; calyx 0.5–0.8 mm long, hispid, pilose or glabrous, the lobes linear, slightly incurved, 0.3–0.4 mm long and thick; corolla lobes striate, hispid, pilose or glabrous. Pods sessile, 25–60 mm long, 6–7 mm wide, glabrous, viscid; margins thickened. Seeds transverse or slightly oblique, oblong, 4 mm long; pleurogram open.

Key features for identification: Whorled, grey-hairy phyllodes; seeds transverse or slightly oblique in pod.

Similar species: Stems tomentose; phyllodes pubescent; calyx and corolla lobes hispid: var. *adoxa*; stems, phyllodes and flowers glabrous to pilose: var. *subglabra*.

Distribution: Widespread in the Kimberley and Pilbara. Also in N.T.

Habitat: Grows in red sandy soils and ironstone gravel on stony plains and ridges.

Flowering period: April to October.

Acacia adoxa var. *adoxa*

Fruit/Seed collection

Fruit/Seed maturity: October.

Appearance of mature fruit/seed: Immature pods are highly resinous, light green, and positioned patchily along the plant branches. Mature pods turn light brown, with timing of maturation often variable.

Cleaning requirements

Cleaning techniques: Washing pods in lukewarm water with detergent or drying the pods in shaded outdoor conditions can assist in alleviating the stickiness of the resin. Multiple passes through a sieve may be required to separate seeds from the pods due to residual resin.

Purity and viability: Viability high (>95%).

Seeds per gram: 65 (range 63–75).

Germination requirements

Dormancy type: Physical.

Embryo type: Investing. **Embryo:seed ratio:** 1.00.

Pre-treatments: Wet heat for 1–2 mins at 90–95°C.

Germination: >90% once heat treated. Germination has only been tested at 25°C.

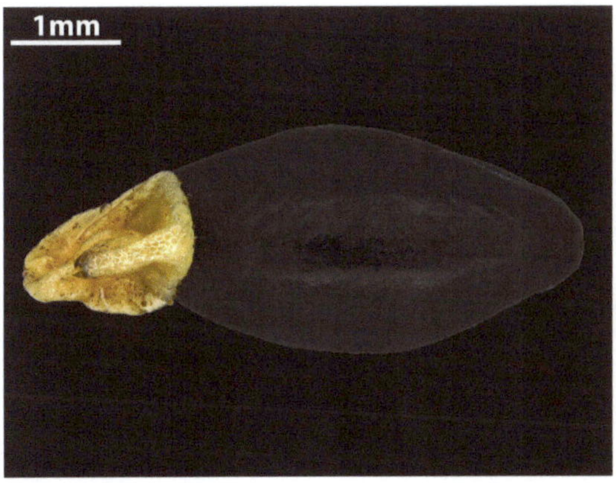

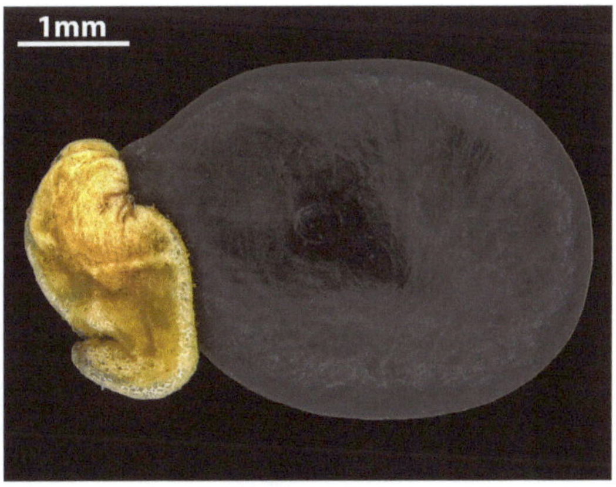

Acacia adsurgens

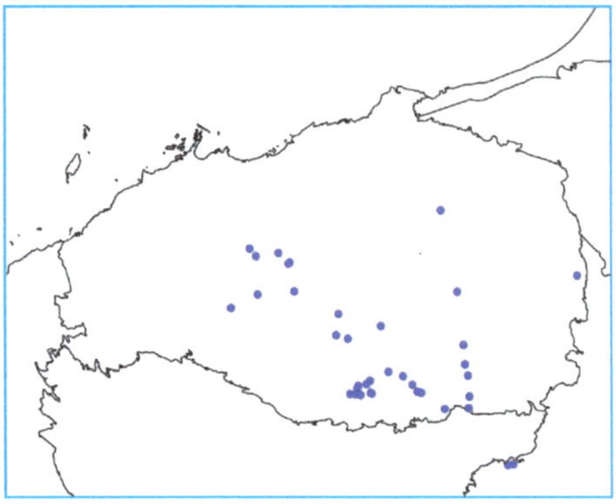

Common names: Whipstick wattle.

Indigenous names: Pilarri (Kurrama).

Description: Shrub or small tree to 4 m high. Bark smooth then fibrous, grey-brown, reddish beneath. Branchlets angular to flattened towards apices, light brown or reddish, glabrous, viscid. Phyllodes linear, straight or upwardly curved, flat, 60–180 mm long, 2–4.5 mm wide, coriaceous, glabrous, resinous when young, the midvein subprominent, with *c.* 8 obscure minor veins per mm; 1 basal gland, swollen, to 3 mm above pulvinus. Peduncles 5–15 mm long. Spikes 8–25 mm long, densely flowered, yellow. Flowers 5-merous; calyx 0.8–1.25 mm long, hardly dissected, with densely scurfy, viscid lobes, the midrib evident; corolla 1.2–1.8 mm long, glabrous; ovary glabrous or apex very finely tomentose. Pods linear, slightly raised over and constricted between seeds, straight or slightly curved, 35–120 mm long, 2–3.5 mm wide, coriaceous, light brown. Seeds longitudinal, narrowly oblong, compressed, 3.5–4.5 mm long, dark brown; pleurogram fine, surrounded by pale halo; areole open, pale.

Key features for identification: Smooth to fibrous bark, slender, dull-green phyllodes. Pods moderately constricted between seeds.

Similar species: Similar to *A. eriopoda* which has dark green leaves and pods very constricted between the seeds.

Distribution: Widespread in the Pilbara. Also in N.T., Qld and S.A.

Habitat: Grows in reddish sandy and gravelly soils, on flat plains and hillsides, commonly in spinifex grasslands.

Flowering period: Mostly May to September.

Acacia adsurgens

Fruit/Seed collection

Fruit/Seed maturity: October to November.

Appearance of mature fruit/seed: When mature, pods turn dusty brown and split with gentle pressure, or separate naturally. A short collection window persists following pod splitting as seeds remain attached to the pods via the aril.

Cleaning requirements

Cleaning techniques: Mature pods can be separated from seeds by passing through a sieve. Typically, the seeds separate readily from the pods. If not, the use of a rubber bung may be required. A vacuum separator can then be used to remove the pod debris.

Purity and viability: Generally high (>80%), but check for insect predation.

Seeds per gram: 122.

Germination requirements

Dormancy type: Physical.

Embryo type: Investing. **Embryo:seed ratio:** 1.00.

Pre-treatments: Wet heat for 2 mins at 90–95°C.

Germination: 100%. Unimpeded across a wide temperature range (10–35°C) once heat treated.

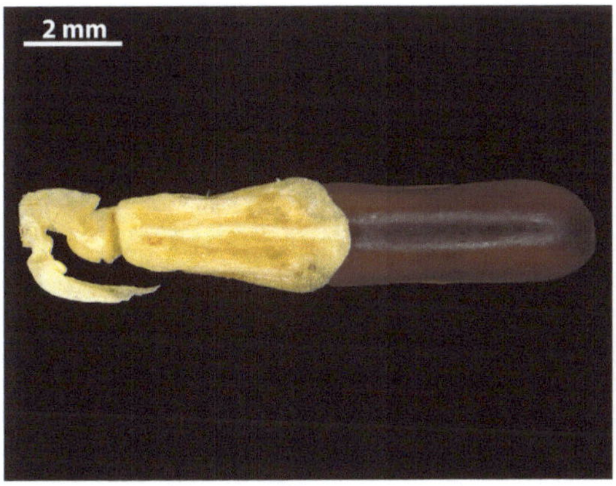

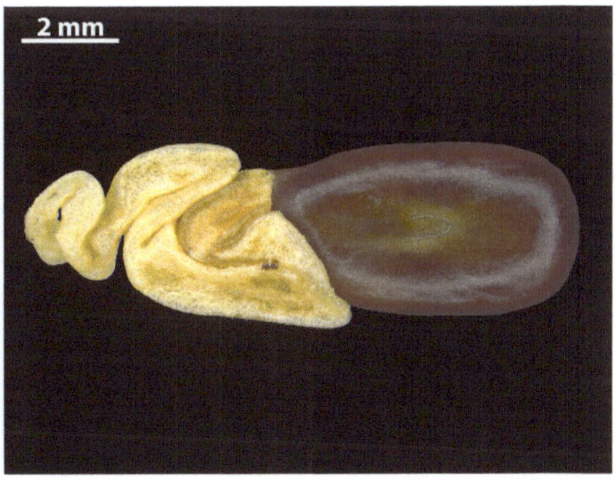

Acacia ancistrocarpa

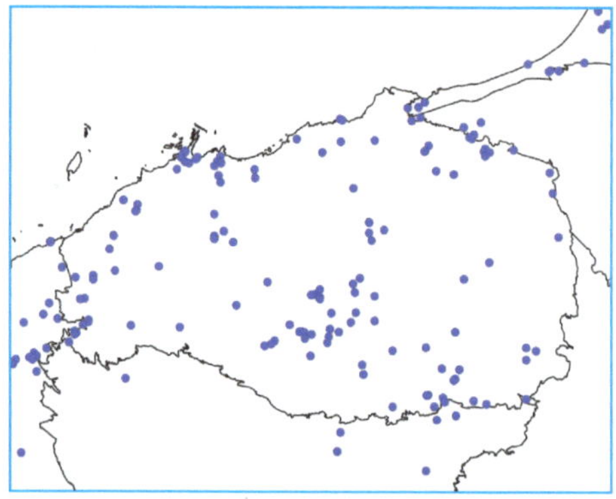

Common names: Fitzroy wattle.

Indigenous names: Wilharri (Thalanyji); Murnungu (Yindjibarndi & Ngarluma); Murnungu or Pulburing (Ngarluma); Balberin or Palpurin (Kurrama); Parlpirr (Putijarra); Parlpi (Nyangumarta).

Description: Multistemmed, resinous shrub or small tree 1–4 m high. Bark grey, smooth on upper trunk, dark and longitudinally fissured towards the base. Phyllodes linear to very narrowly elliptic, straight or slightly falcate, 45–180 mm long, 1.5–10 mm wide, coriaceous, bright olive-green, with 1–3 subprominent yellowish main veins. Flower spikes 25–45 mm long, golden. Pods narrowly oblong, tapered at both ends, 60–115 mm long, glabrous, resinous, dehiscing elastically. Seeds oblique, broadly oblong, 6–7 mm long.

Key features for identification: Long, narrow phyllodes, usually 1.5–10 mm wide. Pods sub-woody, seeds oblique.

Similar species: *Acacia eriopoda* is superficially similar, but has distinctive almost terete pods that are very constricted between the seeds.

Distribution: Widespread in the Pilbara and across tropical Australia. In N.T. and Qld.

Habitat: Grows in mallee-spinifex open woodlands and stony spinifex grasslands, also along watercourses, in deep red sands and skeletal soils on lateritic plains, along creeks, on flats or stony plains.

Flowering period: March to July.

Note: In the Pilbara region, this species is known to form hybrids with *A. trachycarpa* and possibly with *A. orthocarpa* and *A. stellaticeps*.

Acacia ancistrocarpa

Fruit/Seed collection

Fruit/Seed maturity: September to November.

Appearance of mature fruit/seed: Pods tend to form clusters of 1–4 pods and turn from lime green to brown when mature. Upon drying the pods coil heavily and drop seeds over a short period of time.

Cleaning requirements

Cleaning techniques: Seeds are generally released from the pods quite easily, even before cleaning. Hence, passing the seeds through a sieve without crushing the pods can make cleaning easier.

Purity and viability: Prone to insect predation, but viability is generally high (>80%).

Seeds per gram: 26 (range 22–31).

Germination requirements

Dormancy type: Physical.

Embryo type: Investing. **Embryo:seed ratio:** 1.00.

Pre-treatments: Wet heat for 2 mins at 90–95°C.

Germination: >95%. Unimpeded across a wide temperature range (10–35°C) once heat treated.

Note: Predated seeds can be removed with a vacuum separator or via floatation.

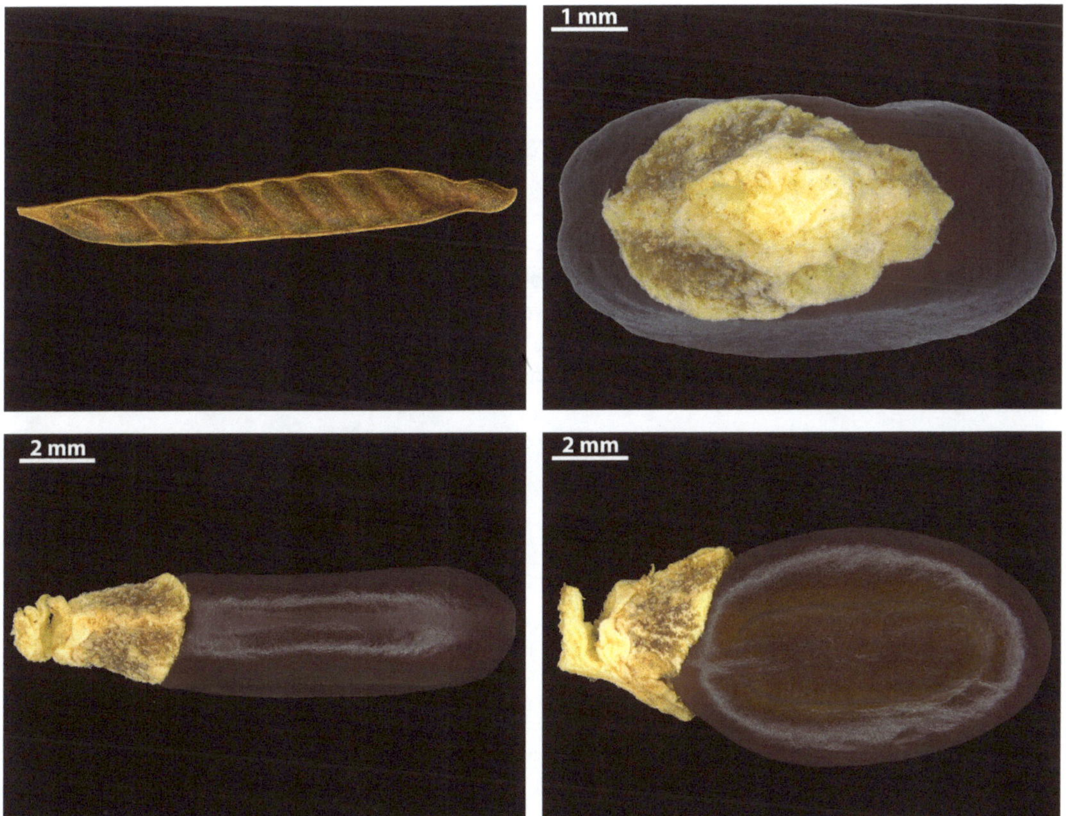

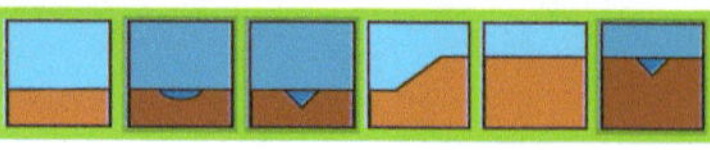

Acacia bivenosa

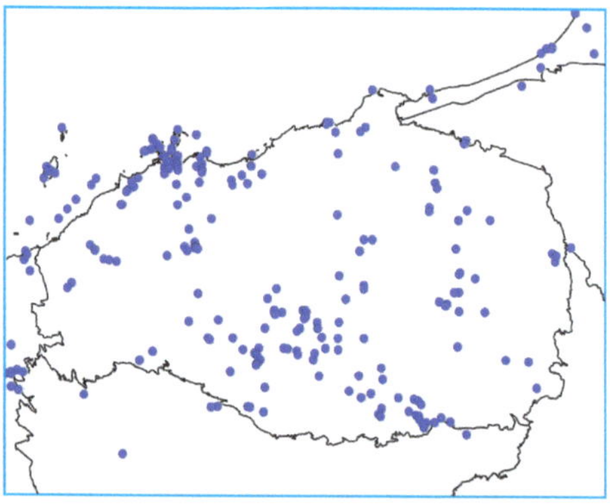

Common names: Two veined wattle, Cable Beach wattle.

Indigenous names: Garrbirri (Thalanyji); Murrurpa, Murruba, Murrurbaor or Morama (Kurrama); Mururru (Banyjima); Mururr (Nyangumarta).

Description: Single or multi-stemmed shrub 0.5–3 m high, glabrous. Branchlets often slightly pruinose. Phyllodes oblong-elliptic to obovate, 20–50 mm long, 6–25 mm wide, thin to subfleshy, green or glaucous, usually 2-nerved per face. Inflorescence a raceme with 6–7 globular flower heads. Pods erect, submoniliform, constrictions not pronounced, to 80 cm long, 5–9 mm wide. Seeds longitudinal, oblong-elliptic, 4–6 mm long, glossy, dark brown; funicle red or orange.

Key features for identification: Phyllodes usually with two prominent nerves. Pods submoniliform (slight constrictions between each seed).

Similar species: *A. ampliceps* and *A. sclerosperma* are both related to *A. bivenosa*, but both have only one prominent nerve per phyllode face.

Distribution: Widespread in the Pilbara, southern Kimberley and adjacent N.T.

Habitat: Grows on a variety of soils, most frequently on sandy soils.

Flowering period: April to November.

Notes: *A. bivenosa* is very variable. The phyllodes are usually distinctly 2-nerved, but the second nerve can be obscure or absent. A number of variants are known. Plants from the Hamersley Range and on islands of the Dampier Archipelago have phyllodes that are narrower than normal (5–12 mm wide); plants from Airlie Island have atypically long phyllodes, 5–9 cm long, that are 1- or 2-nerved; at Dampier and in the Tom Price–Newman area, there is a weeping variant.

Acacia bivenosa

Fruit/Seed collection

Fruit/Seed maturity: September to November.

Appearance of mature fruit/seed: Pods change colour from light green to light brown upon maturity. The pods tend to split wide open and the seeds are only held loosely, resulting in a fairly narrow collection window.

Cleaning requirements

Cleaning techniques: Mature pods can be separated from seeds by passing through a sieve. Typically, the seeds separate readily from the pods. If not, the use of a rubber bung may be required. A vacuum separator can then be used to remove the pod debris.

Purity and viability: Generally high (>80%).

Seeds per gram: 58 (range 31–58).

Germination requirements

Dormancy type: Physical.

Embryo type: Investing. **Embryo:seed ratio:** 1.00.

Pre-treatments: Wet heat for 2 mins at 90–95°C.

Germination: 100%. Unimpeded across a wide temperature range (10–35°C) once heat treated.

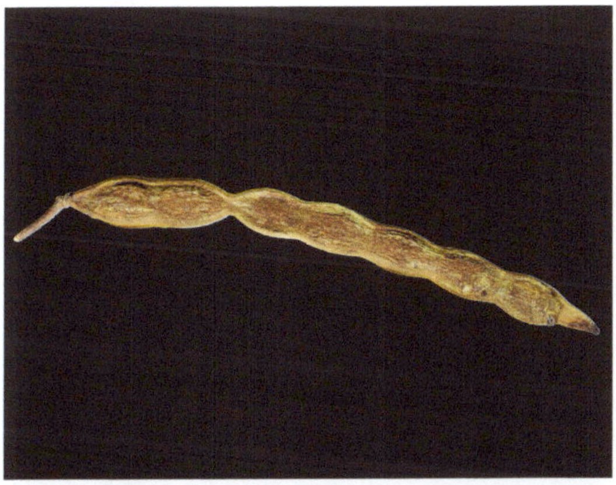

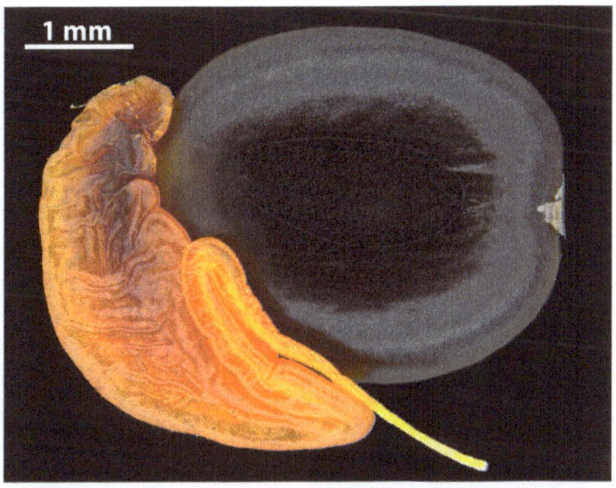

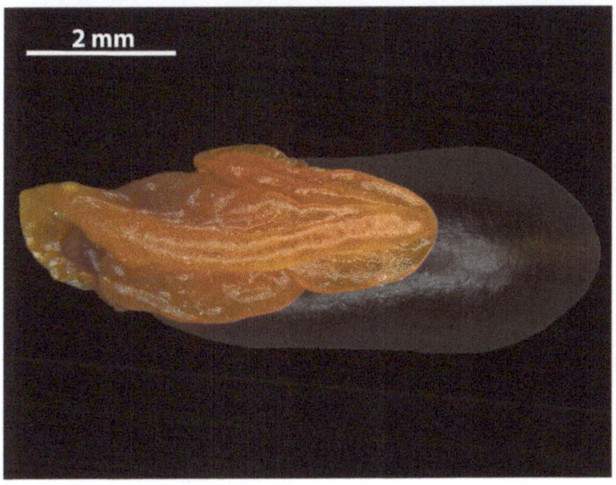

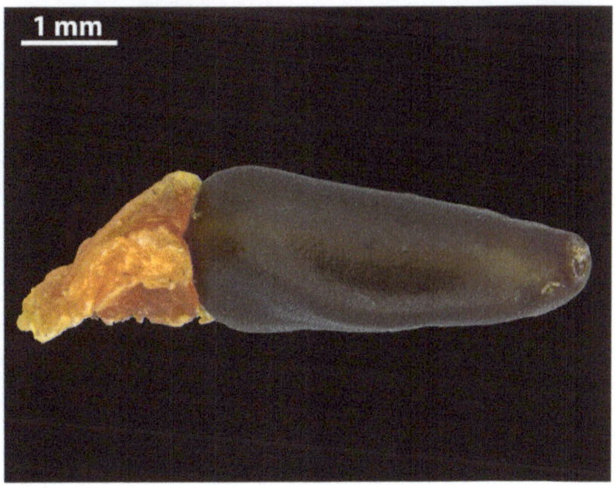

Acacia citrinoviridis

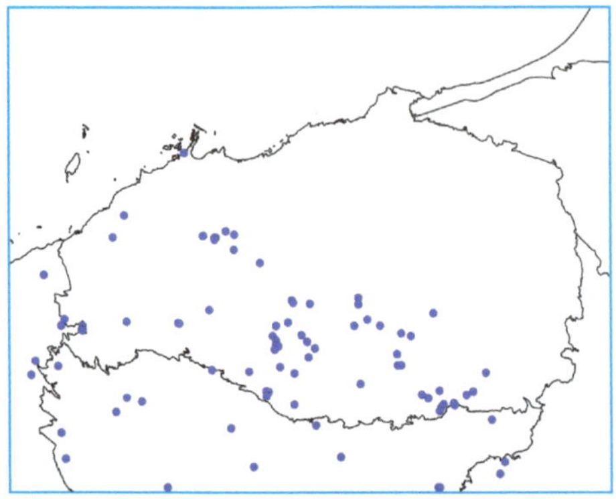

Common names: River jam, Black mulga, Pilbara jam.

Indigenous names: Warlun (Thalanyji); Jarrawayi (Yindjibarndi, Ngarluma & Kurrama); Yilyurru (Kariyarra).

Description: Small tree 3–5(–15) m high. Bark fissured on stem and branches, grey. Branchlets with pale longitudinal ridges, red-brown, shiny. Young shoots densely clothed with iridescent citron-green hairs (aging silvery white). Phyllodes linear or narrowly elliptic, flat, falcate, 80–160 mm long, 4–13 mm wide, coriaceous, densely sericeous to glabrescent, with prominent midvein and sometimes 2 lateral subprominent veins, the minor veins 7–10 per mm; gland basal, small. Spikes 8–32 mm long, dense, bright yellow. Flowers 5-merous; calyx *c.* 0.9 mm long, dissected to base, the sepals linear-spathulate, villous; corolla *c.* 1.2 mm long, dissected to *c.* 1/2, densely villous; ovary densely villous. Pods linear, straight, 25–120 mm long, 7–15 mm wide, hard, brittle, rugose, velvety with citron-green hairs (especially when young). Seeds longitudinal, broadly elliptic to circular, 5–6.5 mm diam., black or dark brown; pleurogram surrounded by pale halo; areole elliptic, closed, cream to dark green; funicle filiform, cream.

Key features for identification: New shoots densely silky, citron yellow hairs; phyllodes densely silvery sericeous.

Similar species: *A. hamersleyensis* has straighter, more coriaceous, less hairy phyllodes; broader pods with spreading hairs; receptacles not densely golden hairy.

Distribution: From the central Pilbara to Wiluna.

Habitat: Usually found along creeks and rivers in sandy rocky beds, also grows in stony soils in open forests.

Flowering period: April to June.

Acacia citrinoviridis

Fruit/Seed collection

Fruit/Seed maturity: October to November

Appearance of mature fruit/seed: When mature, pods are light brown in colour and covered in fine, velvety hairs. Seeds split and release from the pod after applying low-moderate pressure.

Cleaning requirements

Cleaning techniques: Mature pods can be separated from seeds by passing through a sieve. Typically, pods are thicker in this species and may require greater force to separate the pods when sieving. The use of a rubber bung may be required. A vacuum separator can then be used to remove the pod debris.

Purity and viability: Prone to insect predation, but viability is generally high (>80%).

Seeds per gram: 29 (range 29–44).

Germination requirements

Dormancy type: Physical.

Embryo type: Investing. **Embryo:seed ratio:** 1.00.

Pre-treatments: Wet heat for 2 mins at 90–95°C.

Germination: >95%. Unimpeded across a wide temperature range (10–35°C) once heat treated.

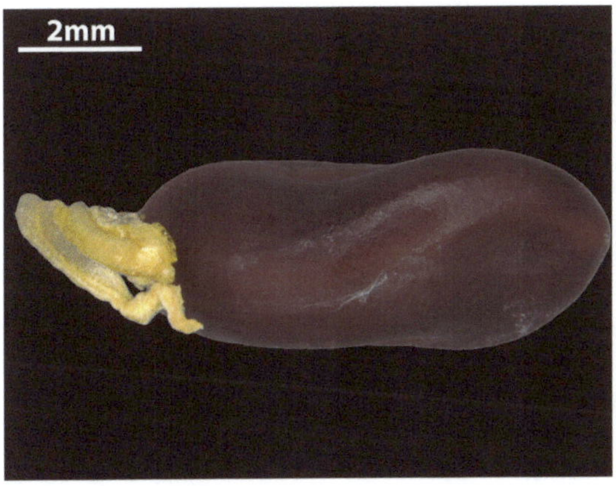

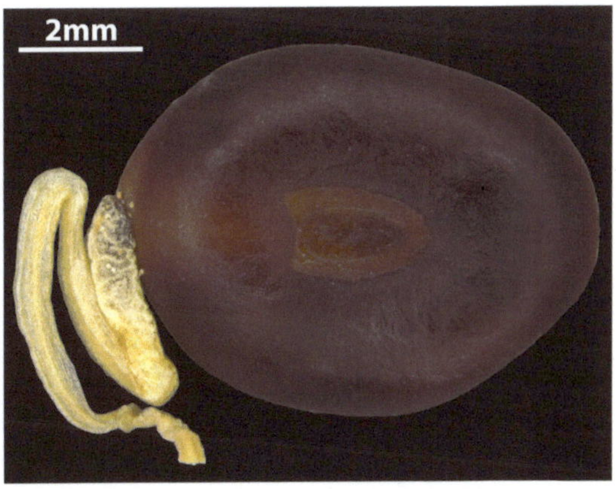

Acacia cowleana

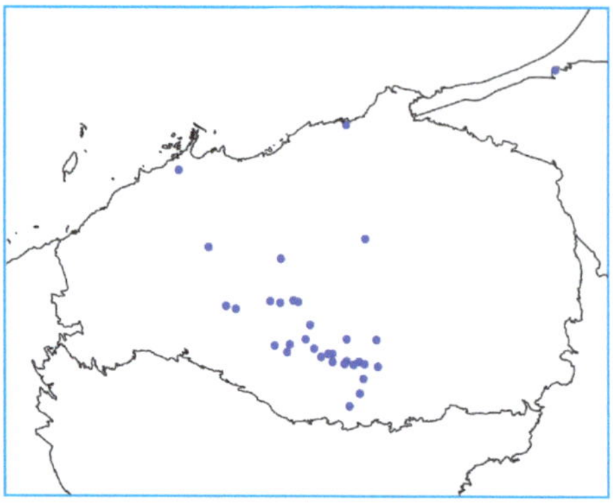

Common names: Halls Creek wattle.

Indigenous names: Wathamalu or Waddamullu (Kurrama).

Description: Shrub or tree to 4(–8) m high; crown with dense foliage. Branchlets prominently angled at extremities, becoming terete with age, minutely sericeous, rarely glabrous. New shoots at initiation encrusted with dark brown resin that obscures the underlying appressed indumentum. Phyllodes shallowly to moderately falcate, rarely dimidiate, 80–200 mm long, 14–36 mm wide, coriaceous, minutely sericeous, rarely glabrous; longitudinal nerves numerous, 2 or 3 per mm, with 3 more evident than the rest and the lowermost 2 running together at base of phyllode, minor nerves occasionally anastomosing. Inflorescences 1 or 2 per axil, simple or vestigial binate racemes with axes to 2 mm long; peduncles 3–10 mm long, sericeous, rarely glabrous; spikes 3–5 cm long, golden; bracteoles spathulate, 0.7 mm long. Flowers 5-merous; sepals united; ovary hairy, rarely glabrous. Pods linear, straight to shallowly curved, 3.5–5 mm wide, chartaceous to coriaceous, very sparsely minutely hairy or glabrous. Seeds longitudinal, oblong, 3.5–5 mm long, glossy, dark brown to black; funicle yellow to cream.

Key features for identification: New shoots resinous; phyllodes large, mostly 120–200 mm long, 14–36 mm wide.

Similar species: Close to *A. elachantha*, but more robust; new shoots encrusted by reddish brown resin; larger and coarser phyllodes; longer spikes; larger flowers; wider pods.

Distribution: In the southern Pilbara. Also in N.T., Qld and N.S.W.

Habitat: Grows mainly along seasonal creek banks.

Flowering period: June to August.

Acacia cowleana

Fruit/Seed collection

Fruit/Seed maturity: September to October.

Appearance of mature fruit/seed: Pods change colour from lime green to light brown upon maturity. The curled nature and clustering of the drying pods allows for easy collection in bulk as seeds tend to be retained in the cluster for a longer period of time.

Cleaning requirements

Cleaning techniques: Once pods are collected the seeds are loosely retained. Seeds can therefore be separated from pods via a sieve with minimal use of a rubber bung.

Purity and viability: Generally high (>90%).

Seeds per gram: 105 (range 76–105).

Germination requirements

Dormancy type: Physical.

Embryo type: Investing. **Embryo:seed ratio:** 1.00.

Pre-treatments: Wet heat for 2 mins at 80–95°C.

Germination: >95%. Unimpeded across a wide temperature range (10–35°C) once heat treated.

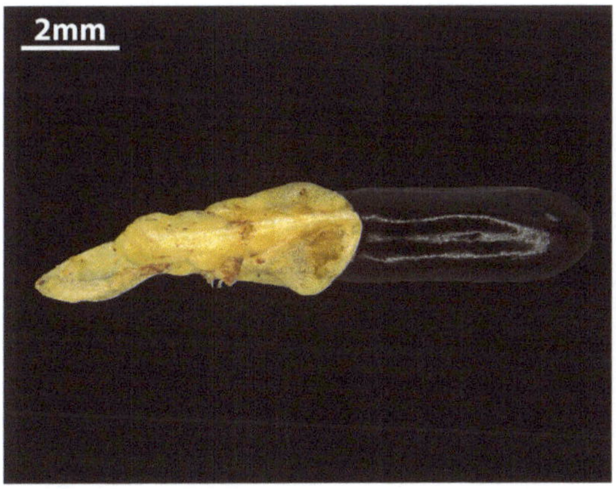

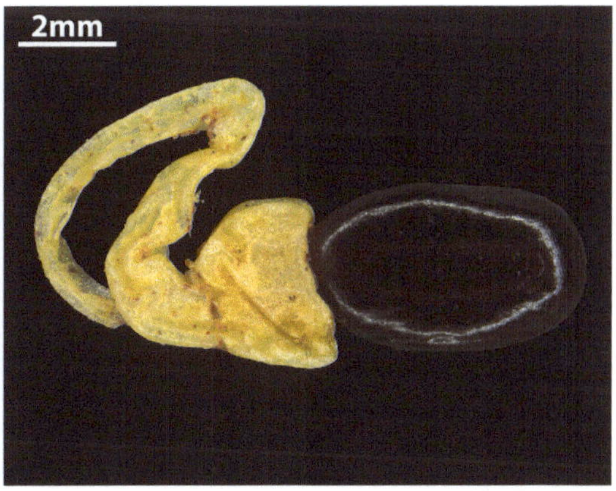

Acacia dictyophleba

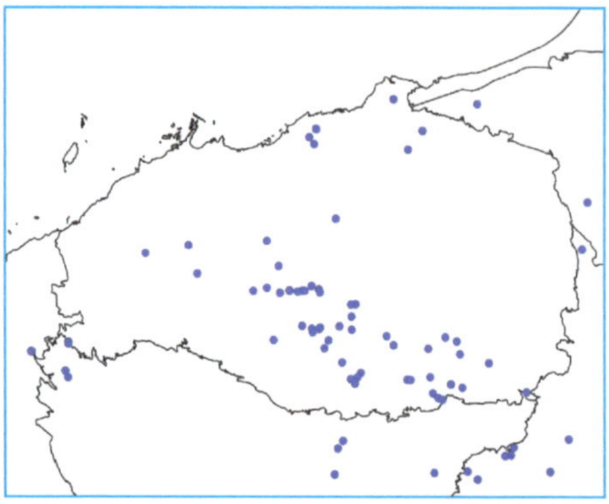

Common names: Sandhill wattle.

Indigenous names: Jabarndi (Thalanyji); Langkur or Lungkun (Nyangumarta).

Description: Glabrous, resinous shrub 1–4 m high. Branchlets often sparsely tuberculate. Phyllodes oblanceolate, 50–70 mm long, 10–20 mm wide, obtuse, mucronulate, thickly coriaceous, smooth or asperulate, green, grey-green or glaucous, with 2 or 3 prominent longitudinal nerves, the secondary nerves forming a coarse, open reticulum; basal gland prominent, elongate, 1–3 mm long. Inflorescences simple, 1–3 per axil; peduncles 10–20 mm long, the base ebracteate; heads globular or obloid, 9–13 mm diam. when dry, densely 40–60-flowered, golden. Flowers 5-merous; sepals united almost to apices. Pods stipitate, narrowly oblong, flat, raised over seeds alternately on each side, to 90 mm long, 10–16 mm wide, firmly chartaceous, often vernicose. Seeds transverse, *c.* 4 mm long, funicle small, pale.

Key features for identification: Small, finely veined, smooth phyllodes, small, few-flowered heads, small seeds.

Similar species: *A. jensenii* and *A. sabulosa* differ in more elongate and predominantly 1-nerved phyllodes. *A. melleodora* is recognised by smaller flowers and flower heads and generally smaller, less coarsely nerved phyllodes.

Distribution: Widespread from the Pilbara region in W.A. Also in S.A., N.T. and Qld.

Habitat: Grows mainly in deep red or red-brown siliceous sand, on dunes, interdunal areas or shallow stony soils.

Flowering period: March to September.

Acacia dictyophleba

Fruit/Seed collection

Fruit/Seed maturity: September to November.

Appearance of mature fruit/seed: The flat, elongated pods turn from green to brown upon maturity. Pods split and release seed easily with gentle pressure.

Cleaning requirements

Cleaning techniques: Mature pods can be separated from seeds by passing through a sieve. Typically, the seeds separate readily from the pods. If not, the use of a rubber bung may be required. A vacuum separator can then be used to remove the pod debris.

Purity and viability: Generally high (>90%).

Seeds per gram: 65 (range 64–95).

Germination requirements

Dormancy type: Physical.

Embryo type: Investing. **Embryo:seed ratio:** 1.00.

Pre-treatments: Wet heat for 1 min at 90–95°C.

Germination: >95%. Unimpeded across a wide temperature range (10–35°C) once heat treated.

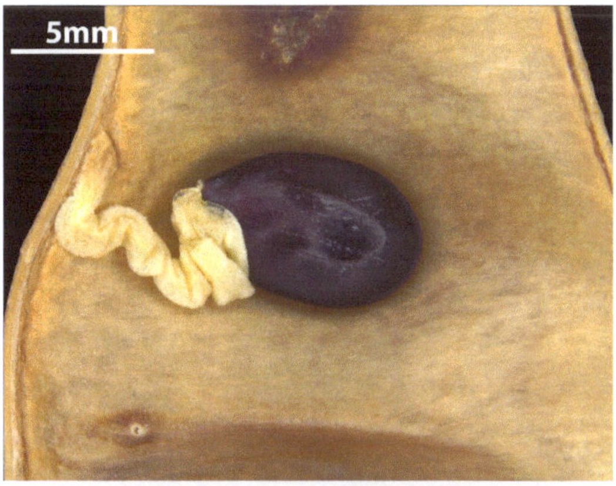

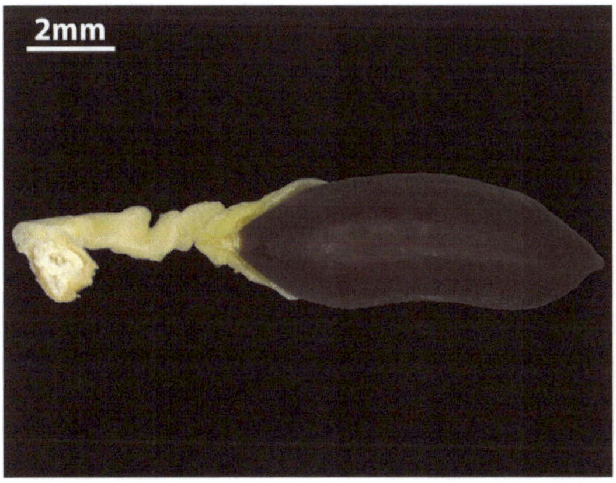

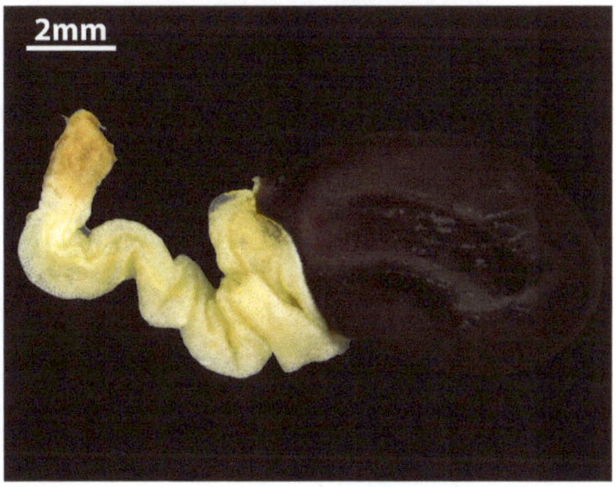

Acacia eriopoda

Common names: Broome pindan wattle.

Indigenous names: Yirrakulu (Nyangumarta).

Description: Slender, erect shrub or small tree, 1.5–6 m high. Bark smooth to fissured and fibrous, grey-brown. Branchlets light to reddish brown, often resinous. Phyllodes linear to linear-elliptic, flat, straight to slightly curved, 70–230 mm long, 1.5–5 mm wide, coriaceous, glabrous, scurfy, with a prominent midnerve and often a subprominent longitudinal nerve either side of it; gland inconspicuous, basal. Spikes 15–48 mm long, densely flowered, yellow. Flowers 5-merous; calyx 0.5–0.75 mm long; corolla 1.25–1.5 mm long. Pods submoniliform, with segments fusiform, straight to curved, 60–150 mm long, 3–4 mm wide, thinly woody, longitudinally furrowed, glabrous, slightly resinous when young. Seeds longitudinal, elliptic to broadly elliptic, dorsiventrally flattened, 4–6.5 mm long, dark brown; areole slightly raised, open, pale.

Key features for identification: Linear, erect phyllodes; hairy peduncles; villous and partly dissected calyces; rounded sepals; a gradual increase in hairiness of the ovary from base to apex; longitudinally wrinkled pods.

Similar species: *A. adsurgens* has dull-green phyllodes and pods only moderately constricted between the seeds. It is more closely related to the broad-phylloded *A. tumida* species group.

Distribution: Common from Broome and the northern Great Sandy Desert south through the Pilbara.

Habitat: Usually found in red sand, also in stony sand or sandy loams in savannah grasslands associated with *Triodia*, also along creeks.

Flowering period: April to August.

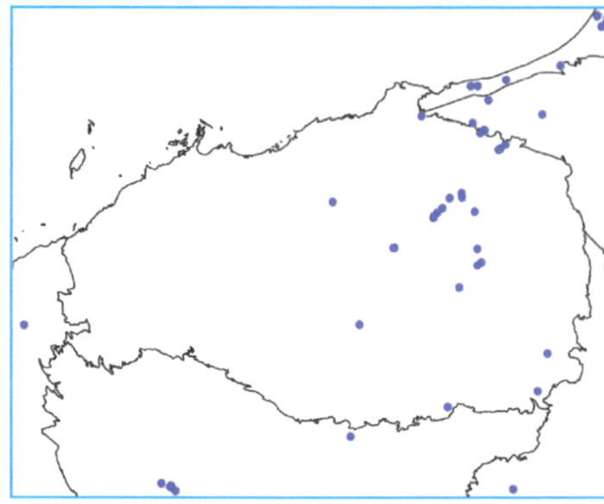

Acacia eriopoda

Fruit/Seed collection

Fruit/Seed maturity: October (limited observations).

Appearance of mature fruit/seed: Pods turn light brown and spilt easily when mature. Seeds are held weakly in the pod via the aril.

Cleaning requirements

Cleaning techniques: Once pods are collected the seeds are loosely retained. Seeds can therefore be separated from pods via a sieve with minimal use of a rubber bung.

Purity and viability: Generally high (>90%).

Seeds per gram: 62.

Germination requirements

Dormancy type: Physical.

Embryo type: Investing. **Embryo:seed ratio:** 1.00.

Pre-treatments: Wet heat for 1 min at 90–95°C.

Germination: >95%. Unimpeded across a wide temperature range (15–35°C) once heat treated. Lower temperatures (10°C) impede germination by 30–50%.

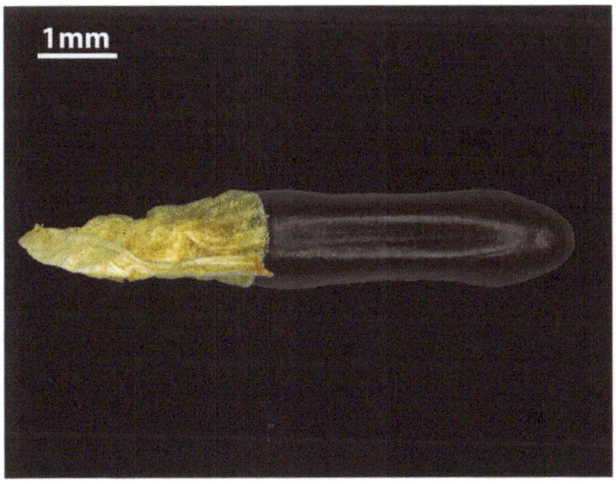

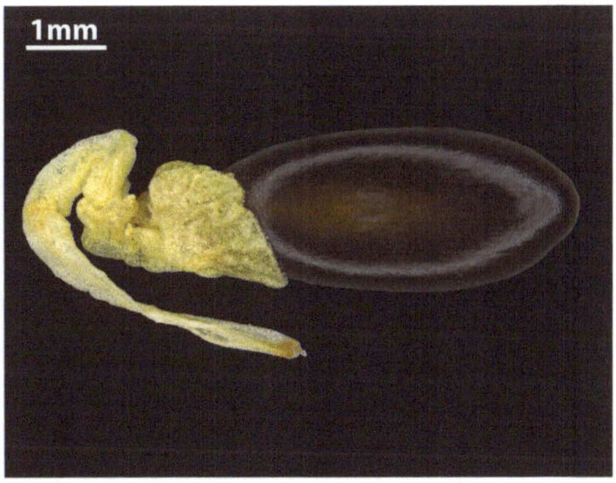

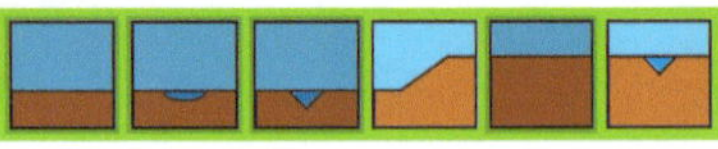

Acacia hamersleyensis

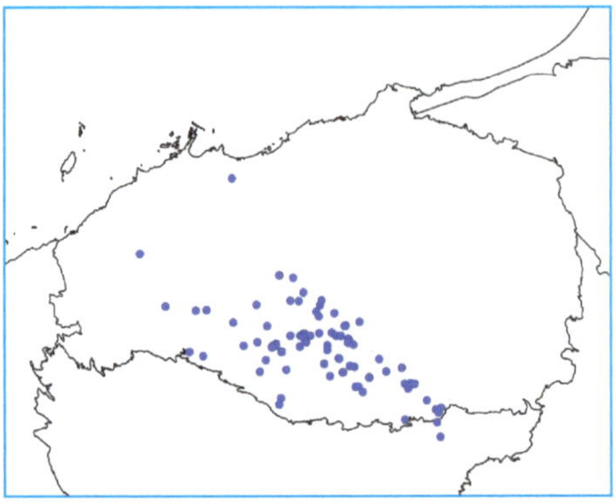

Common names: Karijini wattle, Hamersley Range wattle.

Indigenous names: None recorded.

Description: Spreading, shrubby tree to 4 m high. New shoots pale citron-sericeous, soon glabrous. Branchlets sometimes pruinose. Phyllodes narrowly elliptic, slightly asymmetric with upper margin more convex than lower, straight or shallowly falcate, 80–140 mm long, 7–15 mm wide, acuminate, coriaceous, glaucous to subglaucous, glabrous, multistriate by fine, close, non-anastomosing nerves with 1–3 slightly more prominent than the rest. Inflorescences simple, 1 or 2 per axil, often in false terminal racemes due to phyllode reduction; peduncles 3–10 mm long, thick, sparsely puberulous; spikes 30–60 mm long, 6–8 mm diam., dense, golden. Flowers 5-merous; the calyx tube white-villous. Pods narrowly oblong, to 80 mm long, 5–8 mm wide, firmly chartaceous, slightly undulate, short-pilose with hairs pale golden ageing white. Seeds oblique, oblong to elliptic, 4–4.5 mm long, glossy, grey-brown; funicle pale yellow.

Key features for identification: Yellow to orange branchlets; phyllodes rarely exceeding 20 mm wide; nerves never anastomosing; peduncles 3–7 mm long.

Similar species: Similar to *A. bromilowiana* but multi-stemmed; simple inflorescences; hairy peduncles and pods; sericeous new shoots and oblique seeds.

Distribution: Restricted to the Hamersley Range area, from Newman to Paraburdoo and to near Mt Brockman.

Habitat: Grows in iron-rich soils on ridges and upper slopes of ranges, sometimes forming groves along watercourses descending from ranges.

Flowering period: July to August.

Acacia hamersleyensis

Fruit/Seed collection

Fruit/Seed maturity: September to October.

Appearance of mature fruit/seed: Pods are covered in dense hairs. When mature, the thin, flat, elongated pods split along the long edge. Between 10 to 15 pods can be clustered together, making bulk collections easier.

Cleaning requirements

Cleaning techniques: Mature pods can be separated from seeds by passing through a sieve. Typically, the seeds separate readily from the pods. If not, the use of a rubber bung may be required. A vacuum separator can then be used to remove the pod debris, particularly the dense hairs.

Purity and viability: Generally high (>90%).

Seeds per gram: 67.

Germination requirements

Dormancy type: Physical.

Embryo type: Investing. **Embryo:seed ratio:** 1.00.

Pre-treatments: Wet heat for 1 min at 90–95°C.

Germination: >95%. Unimpeded across a wide temperature range (10–35°C) once heat treated.

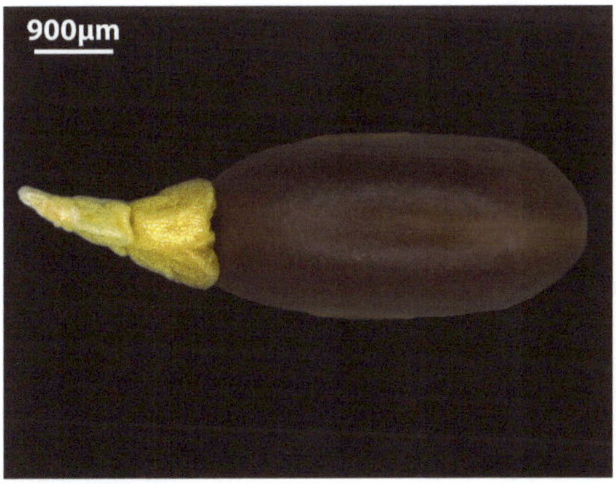

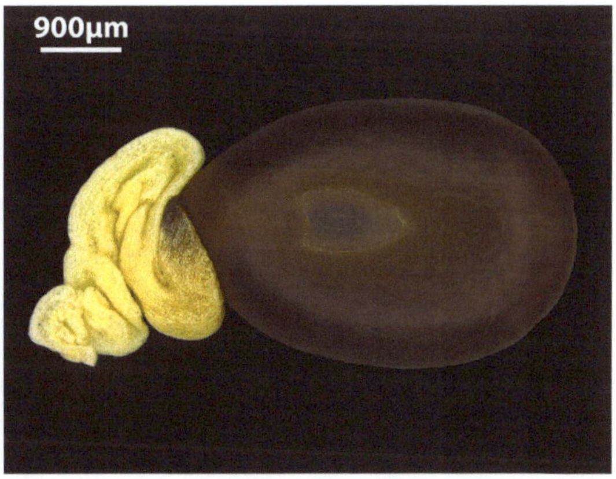

Acacia hilliana

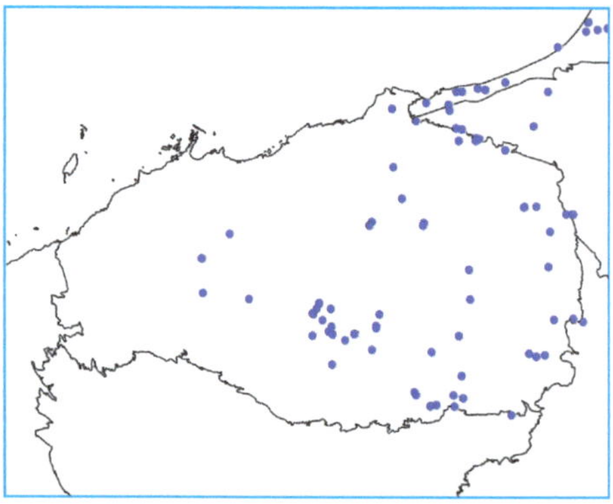

Common names: Hill's tabletop wattle, Sandhill wattle.

Indigenous names: Bundaljingu (Banyjima); Puntanungu (Nyangumarta).

Description: Semi-prostrate, flat-topped or domed, resinous shrub mostly to 1.5 m high, 1.2–4.5 m wide. Bark splitting into flaky strips, grey to dark brownish grey. Branchlets angular to flattened towards apices, red-brown, tuberculate; ridges inconspicuous. Phyllodes solitary or in clusters of 2 or 3, linear, straight or up-curved, terete to compressed, mostly 17–71 mm long, 0.6–1.4 mm wide, coriaceous, glaucous when mature, glabrous, usually tuberculate, obscurely multistriate; gland usually 1, basal, inconspicuous, 2.8–13 mm above pulvinus. Spikes 10–45 mm long, golden. Flowers 5-merous; calyx 0.5–1 mm long, dissected for 2/3–3/4, glabrous; corolla 1.1–2 mm long, dissected for 1/4–1/3, glabrous; ovary non-capitate or rarely capitate, glabrous. Pods erect, very narrowly oblanceolate, narrowed basally, flat, 20–77 mm long, 2.5–5 mm wide, woody, with oblique nerves, opening elastically from apex; margins and seed-partitions thick. Seeds oblique to almost longitudinal, laterally compressed, 2.8–5.5 mm long, chestnut to brown; areole light brown to brown, with pale halo; funicle narrowly turbinate.

Key features for identification: Low, flat-topped shrubs; young growth resinous; phyllodes terete; spikes long (mostly 20–45 mm) and erect; pods erect.

Similar species: The closely allied *A.* sp. Barklys (J.L. Egan 124) has flat phyllodes.

Distribution: Widespread and common in the Pilbara and southern Kimberley. Also in N.T. and Qld.

Habitat: Grows in skeletal, red loamy sand or sand on laterite or quartzite, on slopes or ridges, near creeks, inland sand dunes or rocky plains, in open woodland, shrubland or grassland with eucalypts, acacias and spinifex.

Flowering period: Flowers recorded all year round.

Acacia hilliana

Fruit/Seed collection

Fruit/Seed maturity: September to October.

Appearance of mature fruit/seed: Immature pods are highly resinous, light green, and positioned in upright clusters of 1–5. Mature pods turn light brown, with timing of maturation often variable.

Cleaning requirements

Cleaning techniques: Washing pods in lukewarm water with detergent or drying the pods in shaded outdoor conditions can assist in alleviating the stickiness of the resin. Multiple passes through a sieve may be required to separate seeds from the pods due to (residual) resin.

Purity and viability: Generally high (>90%), but occasional batches of low viability (60–70%).

Seeds per gram: 182 (range 90–295).

Germination requirements

Dormancy type: Physical.

Embryo type: Investing. **Embryo:seed ratio:** 1.00.

Pre-treatments: Wet heat for 2 mins at 90–95°C.

Germination: >95%. Unimpeded across a wide temperature range (10–35°C) once heat treated.

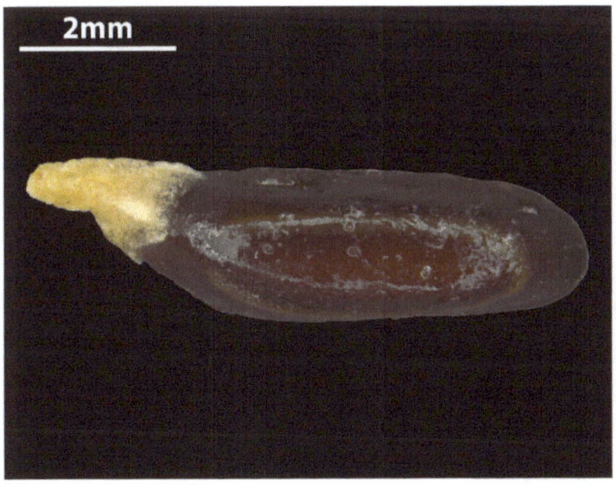

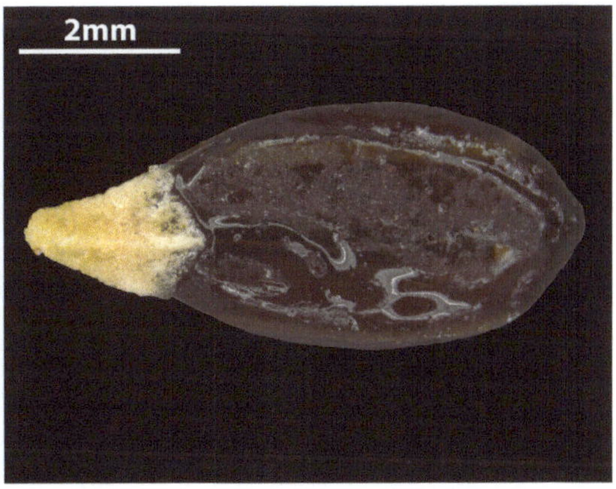

Acacia inaequilatera

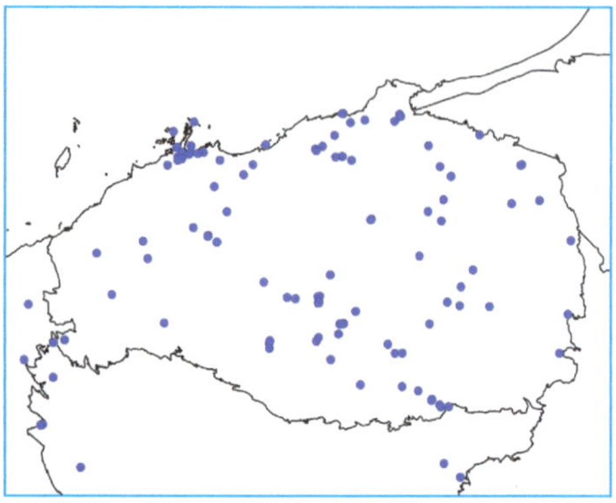

Common names: Camel bush, Corky kanji, Baderi.

Indigenous names: Bardirri, Budalee or Ganyji (Yindjibarndi); Garrany (Ngarluma); Ganyji (Banyjima); Partirri (Nyangumarta & Kurrama); Kanjipirn (Kariyarra); Parnaparna (Putijarra); Kanyji (Marthudunera).

Description: Gnarled, glabrous tree 1.5–5 m high with thick and corky, bark. Phyllodes inequilaterally ovate to elliptic or obovate, sometimes obliquely orbicular, 20–50 mm long, 15–35 mm wide, variably undulate, normally acute to acuminate, pungent, coriaceous, normally grey-green, glabrous; midrib prominent and excentric. Inflorescences 50–180 mm long, with globular heads in racemes, buds dark purple-red. Pods curved to openly coiled, to 110 mm long, 7–10 mm wide, pruinose. Seeds longitudinal, oblong to orbicular, 4.5–6 mm long, dull, brown, encircled by funicle.

Key features for identification: Bark corky; phyllodes inequilaterally ovate to elliptic or obovate; buds dark purple-red; seeds encircled by funicle.

Similar species: Most similar to *A. trudgeniana*, but distinguished by pruinose branchlets, pruinose, often more blue-grey phyllodes and more flowers per head (30–40). *A. marramamba*, *A. pyrifolia* and *A. strongylophylla* are related but lack corky bark and differ in the shape of their phyllodes.

Distribution: Widespread in the Pilbara region, extending to the southern Kimberley and central N.T.

Habitat: Sandy and loamy, or occasionally stony soils. On plains, stony hillsides and rocky plateaus.

Flowering period: May to October.

Acacia inaequilatera

Fruit/Seed collection

Fruit/Seed maturity: October to November.

Appearance of mature fruit/seed: Pod ripening occurs during the onset of consistently dry, hot conditions. The colouration of pods changes from green to light brown (and/or pruinose). The coiled nature of individual pods and the manner in which they cluster together makes it easy to collect seeds in bulk.

Cleaning requirements

Cleaning techniques: Mature pods can be passed through sieves with a larger mesh diameter (10–25 mm). Crushing the pods by hand or by a rubber bung assists in separating the seeds from the pods. Once sieved, the woody pod debris can be removed via vacuum separation.

Purity and viability: Seeds can be prone to insect predation on occasion, but viability is generally high (>90%).

Seeds per gram: 17 (range 17–24).

Germination requirements

Dormancy type: Physical.

Embryo type: Investing. **Embryo:seed ratio:** 1.00.

Pre-treatments: Wet heat for 2 mins at 70–95°C.

Germination: 100%. Unimpeded across a wide temperature range (10–35°C) once heat treated.

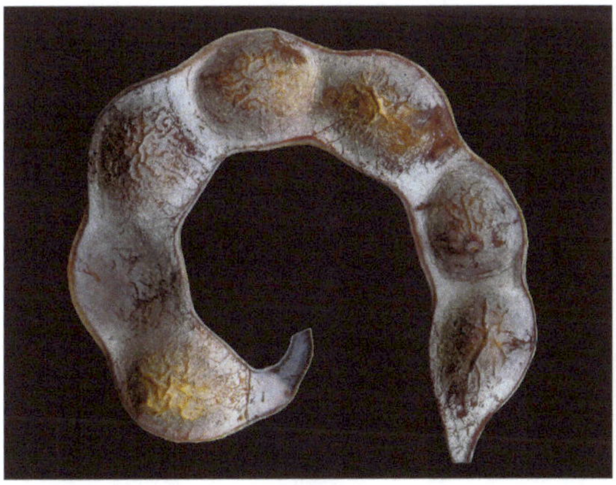

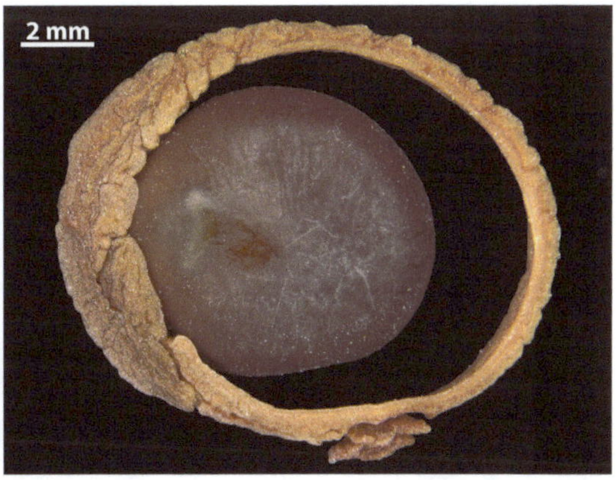

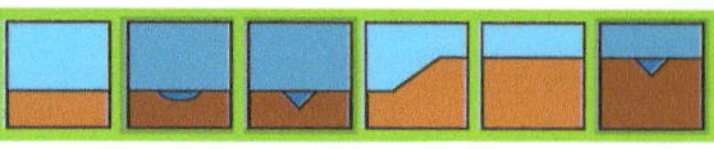

Acacia monticola

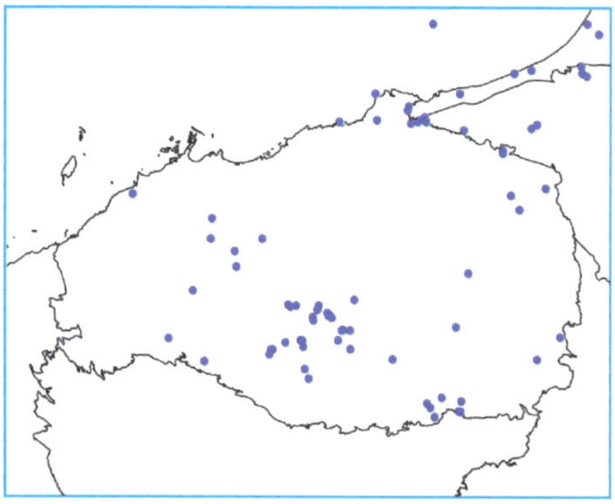

Common names: Red wattle, Curlybark tree.

Indigenous names: Mangkalangu (Kurrama); Burduwayi (Yindjibarndi & Ngarluma); Kawarr (Nyangumarta).

Description: Multistemmed resinous, shrub or tree to 3(–5) m high. Bark 'Minni Ritchi', red-brown or grey. Phyllodes rotund, broadly obovate, elliptic or narrowly elliptic, oblique, flat, 10–32 mm long, 5–15 mm wide, broadly rounded, retuse or acute, coriaceous, puberulous on nerves and margins, with 3–5 longitudinal nerves reaching apex and base; gland 1, prominent, to 3 mm above pulvinus. Heads globular to obloid, 6–10 mm diam., rarely spicate, to 22 mm long, yellow. Flowers 5-merous; calyx 0.6–1.4 mm long with fimbriolate margins; corolla 1.6–4 mm long, striate, glabrous; ovary hispid, densely papillose. Pods narrowly oblong, slightly indented, flat except slightly raised over seeds, 18–100 mm long, 7–14.5 mm, coriaceous, transverse to obliquely nerved, hispid, viscid; margins prominent, tuberculate. Seeds transverse or slightly oblique, broadly oblong-to oblong-elliptic, 4–6.2 mm long, dark brown.

Key features for identification: 'Minni Ritchi' bark; flower heads globular; corolla striate.

Similar species: *A. trachycarpa* and *A. effusa* have elongate flower heads.

Distribution: Widespread in the Pilbara, Great Sandy Desert and southern Kimberley. Also in N.T. and Qld.

Habitat: Grows mainly on ironstone, sandstone or quartzite in skeletal soils, often on ridges or steep slopes.

Flowering period: April to August.

Acacia monticola

Fruit/Seed collection

Fruit/Seed maturity: September to November.

Appearance of mature fruit/seed: When mature, the hairy, flat pods turn dark brown and spilt along the long, flat edge easily. The pods tend to be positioned individually or in small clusters, making bulk collection more difficult.

Cleaning requirements

Cleaning techniques: The flat nature of the mature pods at times makes it difficult to separate the seeds from the pods during sieving. Crushing pods in a canvas bag by hand can assist in breaking pods apart. Seeds are then easily separated using a sieve (and rubber bung, if required).

Purity and viability: Prone to insect predation, so viability can be variable (70–100%).

Seeds per gram: 33 (range 29–35).

Germination requirements

Dormancy type: Physical.

Embryo type: Investing. **Embryo:seed ratio:** 1.00.

Pre-treatments: Wet heat for 1 min at 90–95°C.

Germination: >90%. Unimpeded across a wide temperature range (15–35°C) once heat treated. Lower temperatures (10°C) impede germination by 30%.

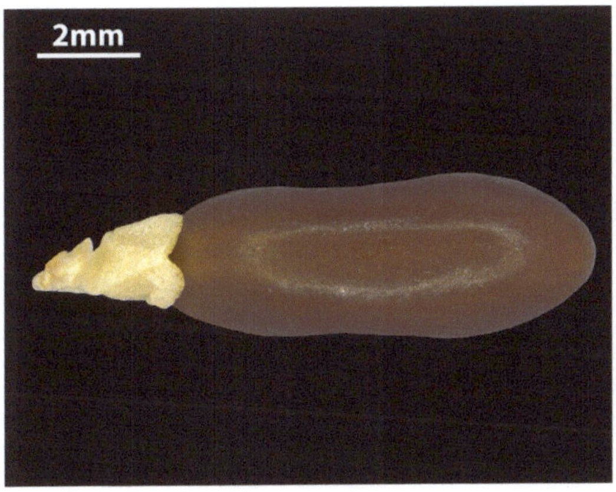

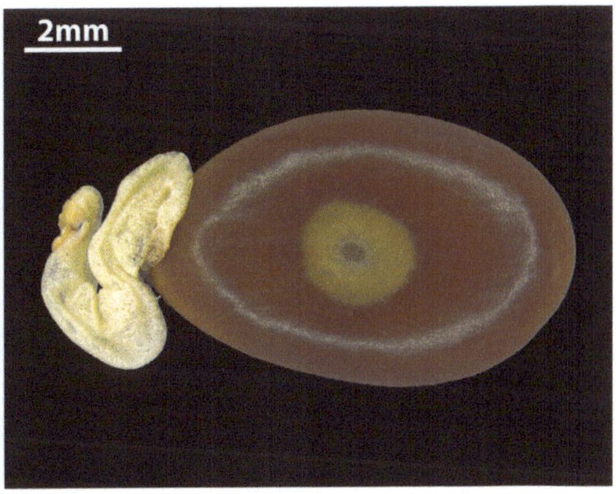

Acacia ptychophylla

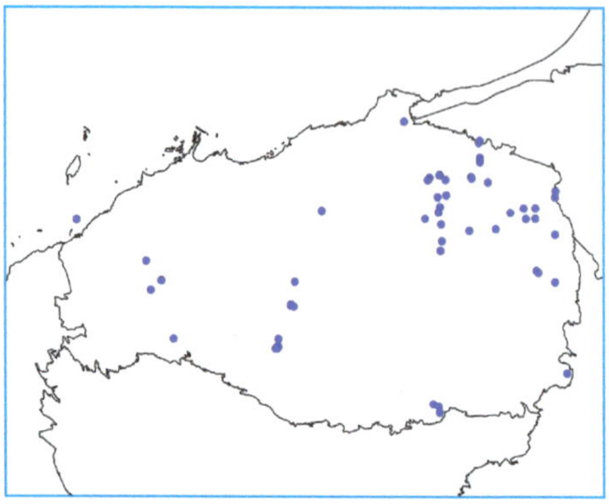

Common names: Curly-pod wattle.

Indigenous names: None recorded.

Description: Erect, resinous shrub to 1.5 m high. Bark splitting, grey-brown. Branchlets angular, becoming terete with crenulated resinous ridges. Phyllodes oblong- to narrowly oblong-elliptic, flat, straight, sometimes slightly incurved in lower half, 20–45 mm long, 3–10 mm wide, with a small callose tip at apex, coriaceous, glabrous, with 6–8 prominent, parallel, longitudinal nerves with resinous coatings; 1 small basal gland just above or near pulvinus and 1–3 additional, variably spaced glands on the upper margin. Spikes 20–45 mm long. Flowers with calyx 0.8–1 mm long, dissected to 2/3–3/4, glabrous; corolla 1.4–1.6 mm long, dissected to 1/2, glabrous; ovary densely villous. Pods narrowly elliptic to slightly narrowly oblanceolate, basally narrowed, straight or falcate, 5–7 cm long, woody, obliquely nerved; margins distinct, light brown. Seeds oblique, oblong-elliptic, 4–5 mm long, brown.

Key features for identification: Low, flat-topped shrubs; branchlets resin-ribbed; phyllodes narrowly oblong-elliptic; spikes long, bright golden; pods erect.

Similar species: Differs from *A. drepanocarpa* subsp. *latifolia* by its reddish brown penultimate branchlets.

Distribution: An uncommon species with a disjunct distribution in the eastern Pilbara and southeastern Kimberley regions of W.A., and extends just into N.T.

Habitat: Grows mostly on rocky slopes or sand plains, associated with spinifex communities.

Flowering period: April to July.

Acacia ptychophylla

Fruit/Seed collection

Fruit/Seed maturity: October to November.

Appearance of mature fruit/seed: Immature pods are light green and positioned in upright clusters of 1–5 pods. Mature pods turn light brown, split in half, and curl prominently. During maturation, the pods split rapidly and disperse seeds ballistically, resulting in a shorter collection window when compared to other *Acacia* species.

Cleaning requirements

Cleaning techniques: Seeds can be easily separated from the pods by passing the collected material through a sieve with a relatively large mesh size (15–30mm). Once sieved, smaller pod debris that passes through the sieve can be removed via vacuum separation.

Purity and viability: Generally high (>95%).

Seeds per gram: 81 (range 79–82).

Germination requirements

Dormancy type: Physical.

Embryo type: Investing. **Embryo:seed ratio:** 1.00.

Pre-treatments: Wet heat for 1–2 mins at 90–95°C.

Germination: >90% once heat treated. Germination has only been tested at 25°C.

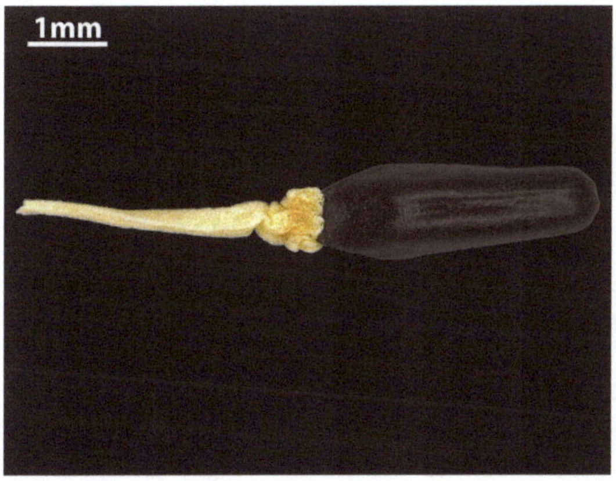

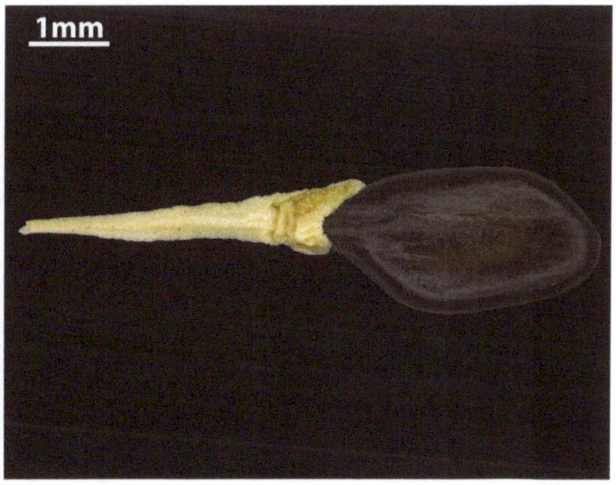

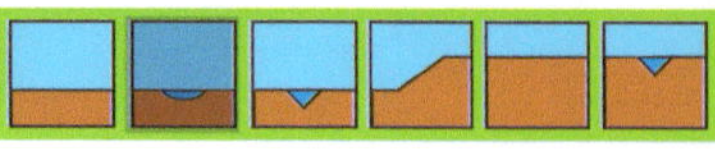

Acacia pyrifolia var. *pyrifolia*

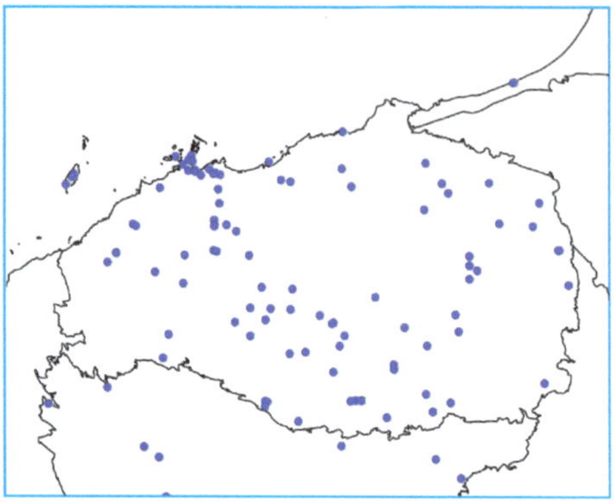

Common names: Kanji bush, Ranji bush.

Indigenous names: Gadarn (Thalanyji); Ganyji (Yindjibarndi, Banyjima & Ngarluma); Jirbarli (Kurrama); Kartarn (Jiwarli), Munturru (Putijarra & Nyangumarta).

Description: Straggling and open glabrous, often pruinose shrub 1–4 m high. Phyllodes variable, usually elliptic to orbicular or obovate, 25–60 mm long, 15–40 mm wide, slightly undulate, pungent, blue-grey to grey-green or glaucous, coriaceous, with prominent and central midrib. Inflorescences forming racemes 3–18 cm long; heads globular. Pods narrowly oblong, rounded over seeds, shallowly curved to openly once-coiled, to 8 cm long, 8–15 mm wide, firmly chartaceous. Seeds transverse to longitudinal, broadly elliptic to ovate, 4–6 mm long, dull, dark brown, funicle small.

Key features for identification: Non-corky bark; symmetrically elliptic to obovate phyllodes; pale peduncles, raceme axes and flower buds; short funicles.

Similar species: *A. pyrifolia* var. *morrisonii* has absent or slender stipules that are scarcely spinose; non-pruinose branchlets and green, elliptic to oblanceolate, thinly coriaceous phyllodes, 15–20 mm wide. *A. marramamba* is closely related, differing by non-pruinose branches. *A. inaequilatera* and *A. trudgeniana* differ in having corky bark.

Distribution: Widespread in the Pilbara, south to Carnarvon and north to the southern Kimberley.

Habitat: Grows in sand or loam, along watercourses, on slopes and stony plains, disturbed sites such as roadsides.

Flowering period: July to August.

Acacia pyrifolia var. *pyrifolia*

Fruit/Seed collection

Fruit/Seed maturity: September to November.

Appearance of mature fruit/seed: Pod ripening occurs during the onset of consistently dry, hot conditions. The colouration of pods changes from green to light brown. The coiled nature of individual pods and the manner in which they cluster together makes it easy to collect seeds in bulk.

Cleaning requirements

Cleaning techniques: Prior to crushing mature pods, a large proportion of the spherical seeds can be separated by passing seeds through a sieve with a larger mesh diameter (10–25 mm). Crushing the pods by hand or by a rubber bung assists in separating the remaining seeds from the pods. Once sieved, the woody pod debris can be removed via vacuum separation.

Purity and viability: Seeds can be prone to insect predation on occasions. But, viability is generally high (>90%).

Seeds per gram: 23 (range 17–27).

Germination requirements

Dormancy type: Physical.

Embryo type: Investing. **Embryo:seed ratio:** 1.00.

Pre-treatments: Wet heat for 2 mins at 80–95°C.

Germination: >95%. Unimpeded across a wide temperature range (10–35°C) once heat treated.

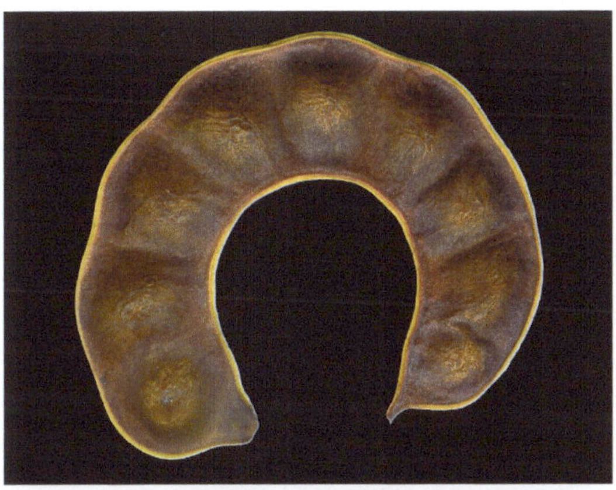

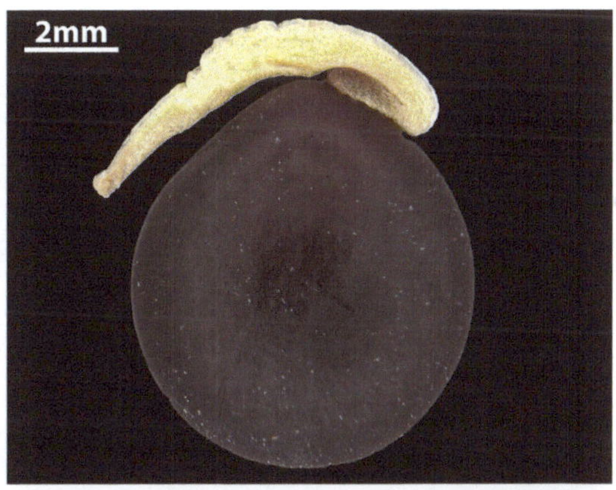

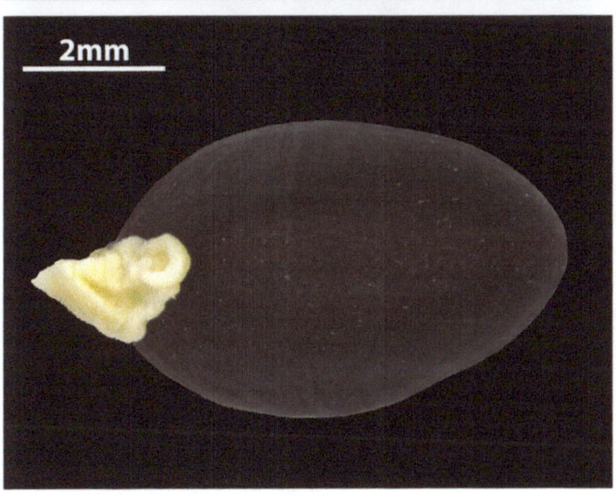

Acacia sclerosperma subsp. *sclerosperma*

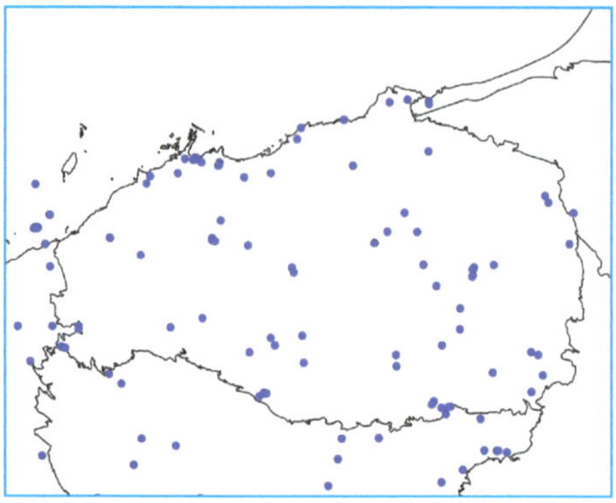

Common names: Limestone wattle, Rattle tree.

Indigenous names: Bagurda (Banyjima & Kurrama); Purtarrpa (Nyangumarta); Warntayin (Kariyarra).

Description: Dense, commonly rounded shrub or tree mostly to 4 m high. Bark smooth and light grey to branchlet extremities. Branchlets glabrous. Phyllodes narrowly linear, terete to flat, 45–140 mm long, 1–4 mm wide, thick, smooth, often wrinkled when dry, green to subglaucous, glabrous, 4-nerved in all, 1-nerved per face when flat; glands 2 or 3, not prominent. Inflorescences 2–5-headed racemes; raceme axes 3–30 mm long, glabrous; peduncles 4–15 mm long; heads globular, sub-dense, 15–20-flowered, golden; buds bright green. Flowers 5-merous; sepals united into a truncate calyx. Pods moniliform to submoniliform, 60–120 mm long, 10–20 mm wide, woody, golden brown, glabrous. Seeds longitudinal, spherical, 7–10 mm long, glossy, dark brown to black; aril red, hemispherical to clavate.

Key features for identification: Spreading shrub or small trees; bark light grey; phyllodes long and narrow; inflorescences short racemes (3–30 mm); pods large, moniliform seeds large (7–10 mm long) with a red funicle.

Similar species: Distinguished from most other members of the '*A. bivenosa* group' by its large, woody pods and large seeds.

Distribution: Occurs mainly in the Pilbara and Murchison districts.

Habitat: Grows on coastal dunes, along creeks and flood plains, in sand, limestone, loam and clay, in scrub, shrubland and riparian woodland.

Flowering period: April to October.

Acacia sclerosperma subsp. *sclerosperma*

Fruit/Seed collection

Fruit/Seed maturity: October to November.

Appearance of mature fruit/seed: The pods are highly constricted around the seeds and turn from lime green to light brown upon maturity. The pods tend to split wide open and the seeds are held only loosely within pod via the aril, resulting in a fairly short collection window.

Cleaning requirements

Cleaning techniques: Seeds can be easily separated from the pods by passing the collected material through a sieve with a relatively large mesh size (15–30 mm). Once sieved, smaller pod debris that passes through the sieve can be removed via vacuum separation.

Purity and viability: Generally high (>80%), but can be prone to insect predation.

Seeds per gram: 6 (range 5–6).

Germination requirements

Dormancy type: Physical.

Embryo type: Investing. **Embryo:seed ratio:** 1.00.

Pre-treatments: Wet heat for 2 mins at 90–95°C.

Germination: Variable (60–90%), due to inconsistent response to heat treatments. Germination appears unimpeded across a broad temperature range (15–35°C) once heat treatment is effective. Lower temperatures (10°C) reduce germination by 10–20%.

Notes: Seeds in this species are one of the largest observed in the *Acacia* genus, as a result of a thicker seed coat. The reasons for the inconsistent results of heat treatments used to alleviate physical dormancy requires further investigation. An alternative treatment to relieve physical dormancy may be immersion in concentrated (98%) sulphuric acid for 1–4 hours, but this treatment has yet to be tested.

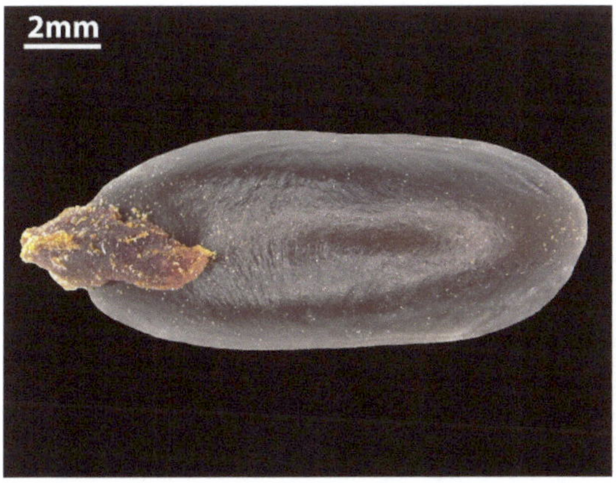

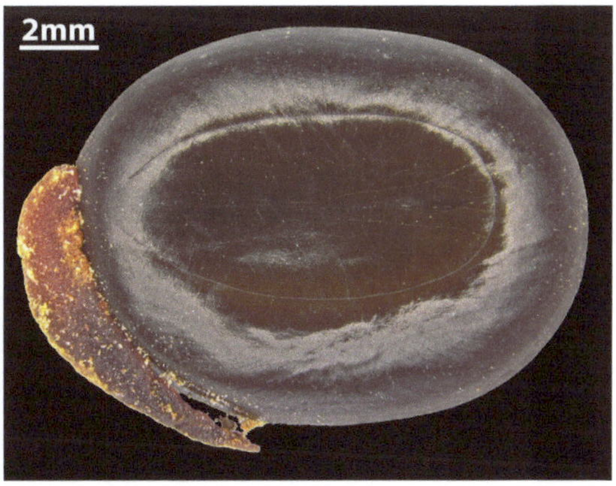

Acacia stellaticeps

Common names: Northern star wattle, Glistening wattle.

Indigenous names: Pirrnyur, Pirrinyurru, Pitinyurru or Panmangu (Nyangumarta); Panmangu (Ngarla).

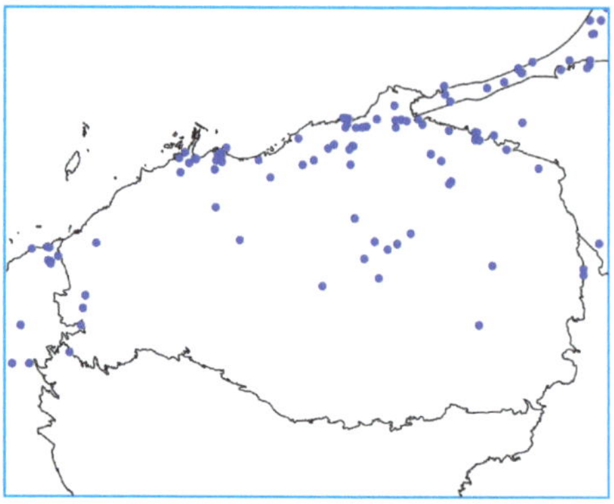

Description: Rounded or flat-topped, glabrous, resinous shrub to 2 m high and wide. Bark smooth, grey. Branchlets angular towards apices, pale green, brown, or yellowish; ridges resin-crenulated. Phyllodes elliptic-obovate to narrowly elliptic-obovate, oblique, slightly curved to straight, 8–25 mm long, 4–10 mm wide, rigid, the apex with a small knob-like mucro, with 0–3 faintly visible main nerves, the minor nerves obscure; basal gland minute, to 2 mm above pulvinus. Peduncles 5–30 mm long. Capitula 7.5–16.5 mm wide, 7–25-flowered, golden; buds spreading widely. Flowers 5-merous; calyx cupular, 0.7–1.1 mm long, barely dissected; corolla 2.4–3.5 mm long, dissected by 2/3, the petals striated; ovary villous. Pods erect, narrowly elliptic to linear or linear-oblanceolate to narrowly oblanceolate, basally tapered, straight-sided but sometimes indented between some seeds, flat, mostly 25–100 mm long, 5–10 mm wide, thinly woody, scurfy, obliquely striate, opening elastically from apex; margins slightly thickened. Seeds oblique, oblong- to narrowly oblong-elliptic, 4–6 mm long, light to dark brown; funicle narrowly conical.

Key features for identification: Low shrubs; new shoots resinous; phyllodes small, asymmetric; heads globular; peduncles long; flower buds large; pods held erect.

Similar species: Related to *A. newmanii* and *A. translucens*, but distinguished by its prominently spreading, large flower-buds, flowers with striated petals, and a erect knob-like to conical mucro on the phyllodes.

Distribution: Widespread in the Pilbara, southern Kimberley and south to Exmouth Gulf. Also in N.T.

Habitat: Grows in red sand over quartzite, limestone, laterite or ironstone, on hills or sandplains.

Flowering period: February to September.

Acacia stellaticeps

Fruit/Seed collection

Fruit/Seed maturity: September to October.

Appearance of mature fruit/seed: Immature pods are light green and positioned in upright clusters of 1–5 pods. Mature pods turn light brown, split in half, and curl prominently. During maturation, the pods split rapidly and disperse seeds ballistically, resulting in a short and sporadic collection window.

Cleaning requirements

Cleaning techniques: Once pods are collected the seeds are loosely retained. Seeds can therefore be separated from pods via a sieve with minimal use of a rubber bung.

Purity and viability: Generally high (>80%).

Seeds per gram: 75 (range 73–96).

Germination requirements

Dormancy type: Physical.

Embryo type: Investing. **Embryo:seed ratio:** 1.00.

Pre-treatments: Wet heat for 1 min at 90–95°C.

Germination: >95%. Unimpeded across a wide temperature range (10–35°C) once heat treated. A slight (10%) reduction in germination has been observed at low temperatures (10°C).

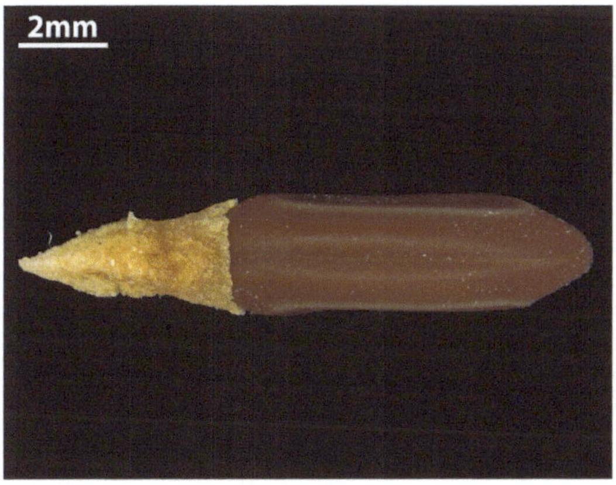

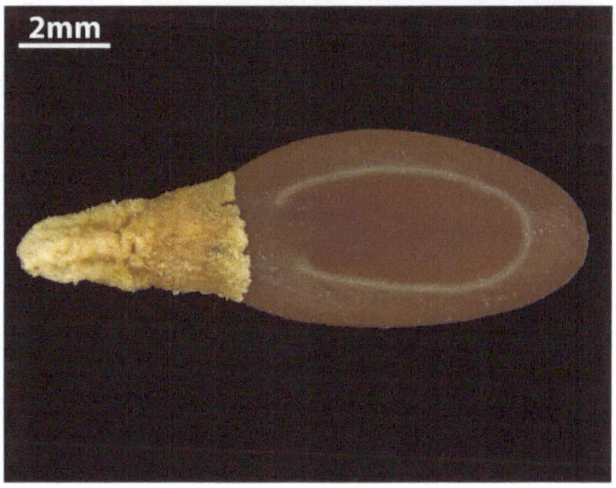

Acacia tenuissima

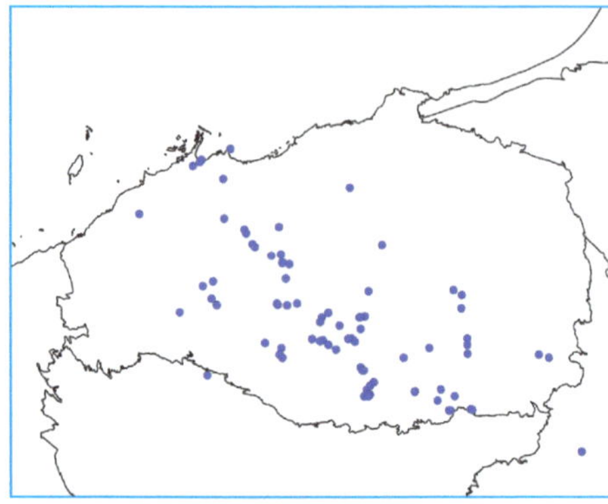

Common names: Broom wattle, Narrow-leaved wattle.

Indigenous names: Janangungu (Kurrama), Murruthurru (Banyjima).

Description: Multistemmed, resinous shrub to *c.* 4 m high. Bark smooth, grey purplish or brown. Branchlets terete, glabrous; ridges yellow resin-crenulated. Phyllodes filiform, terete to compressed, 60–150 mm long, 0.7–1.3 mm wide, glabrous, multistriate, the nerves usually inconspicuous, often some nerves and margins resin-crenulated when dry; gland 1, basal, just above pulvinus. Spikes 5–10 mm long, pale yellow to bright yellow. Flowers 5-merous; calyx 0.8–1 mm long, truncate, thin, dissected irregularly into lobes, glabrous; corolla 1.2–1.4 mm long, dissected to 1/2, glabrous; ovary mealy. Pods linear, raised over and constricted between seeds, straight, later curved or coiled, 25–50 mm long, 2–3.5 mm wide, chartaceous, scaly when mature. Seeds longitudinal, 2.5–4 mm long, black; areole small, open.

Key features for identification: Multi-stemmed, resinous shrubs; branchlet apices with resin-ribs; phyllodes long and slender; spikes pale yellow; pods linear to submoniliform.

Similar species: *A. adsurgens* is distinguished by having slender, terete phyllodes. Distinguished from *A. orthocarpa* most its readily by its striate (not puncticulate) phyllodes and chartaceous, submoniliform pods.

Distribution: Widespread from the Pilbara to the Tanami Desert. Also in N.T., Qld and S.A.

Habitat: Common in eucalypt woodland, often with spinifex, in sandy soils.

Flowering period: March to August.

Acacia tenuissima

Fruit/Seed collection

Fruit/Seed maturity: September to October.

Appearance of mature fruit/seed: When mature, the thin pods turn light brown and split with gentle pressure. A short collection window persists following pod splitting as seeds remain attached to the pods via the aril.

Cleaning requirements

Cleaning techniques: Once pods are collected the seeds are loosely retained. Seeds can therefore be separated from pods via a sieve with minimal use of a rubber bung.

Purity and viability: Generally high (>90%), with occasional batches of low viability due to insect predation (70%).

Seeds per gram: 153 (range 132–168).

Germination requirements

Dormancy type: Physical.

Embryo type: Investing. **Embryo:seed ratio:** 1.00.

Pre-treatments: Wet heat for 1 min at 90–95°C.

Germination: >95%. Unimpeded across a wide temperature range (10–35°C) once heat treated.

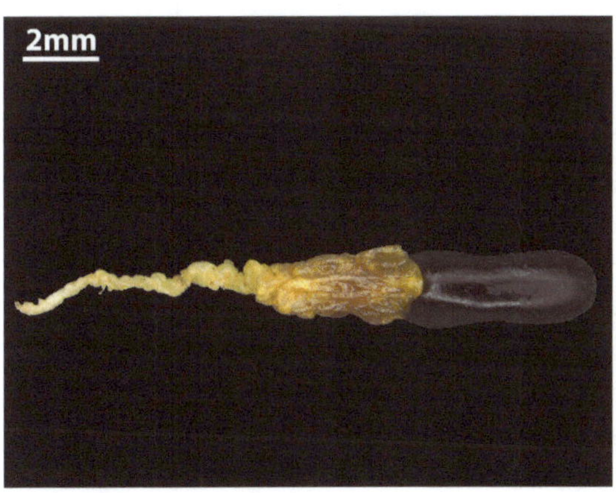

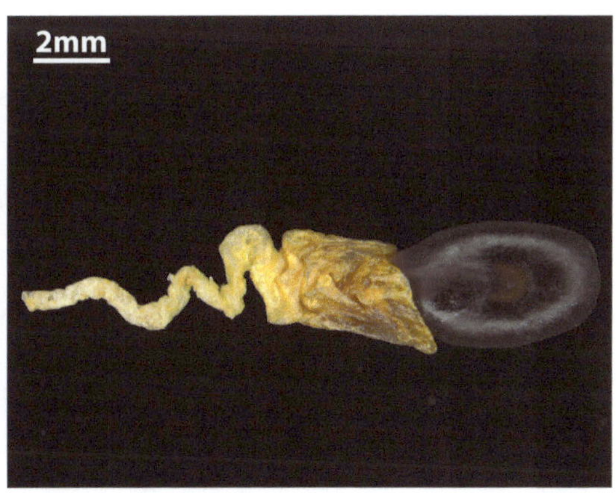

Acacia tumida var. *pilbarensis*

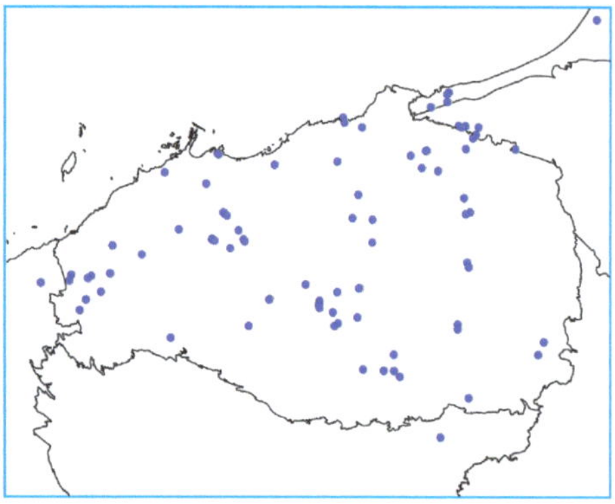

Common names: Pilbara pindan wattle.

Indigenous names: Muwarlingu (Yindjibarndi); Mugarli (Ngarluma); Mukarti, Mukarli or Pulpa (Nyangumarta); Pilamurka, Billamurga or Bilamurga (Kurrama).

Description: Shrub or tree 2–5(–15) m tall. Bark hard, grey. Branchlets glabrous or hairy, orange-yellow or pruinose. Phyllodes falcate to sub-falcate or dimidiate, 60–250 mm long, 7–60 mm wide, coriaceous, glabrous, pruinose or non-pruinose, green to glaucous; longitudinal nerves numerous, parallel. Inflorescences usually racemose (racemes 5–15 mm long) or axillary or terminal panicles (to 200 mm long), rarely some simple; peduncles 2–10 mm long, glabrous; spikes 20–60 mm long, yellow to light golden. Flowers 5-merous; calyx deeply dissected, pubescent; ovary pubescent. Pods narrowly oblong, sometimes constricted between and over the seeds, terete or slightly flattened, straight to curved and sometimes twisted, 30–120 mm long, 5–10 mm wide, woody, glabrous, pruinose or non-pruinose, longitudinally wrinkled or grooved. Seeds oblique, 4–9 mm long, 2–5 mm wide, glossy, black; funicle twice-folded, cream.

Key features for identification: Shrubs or trees; bark smooth and pruinose when young; branchlets glabrous, usually pruinose; phyllodes large; inflorescences axillary or terminal open racemes; pods narrowly oblong, terete to sub-terete or flattened, striate, usually moderately pruinose.

Similar species: *A. hamersleyensis* differs in having generally straighter phyllodes, simple (not racemose) inflorescences, broader spikes and hairy pods.

Distribution: Occurs in the Pilbara, southern Kimberley, Great Sandy Desert and Little Desert regions.

Habitat: Red sandplains and dunes.

Flowering period: April to October.

Acacia tumida var. *pilbarensis*

Fruit/Seed collection

Fruit/Seed maturity: September to November.

Appearance of mature fruit/seed: The slightly longitudinally wrinkled pods turn pale brown upon maturity. Clusters of pods in groups of 5–10 make bulk collections easier. Once split, the collection window is fairly short as the seeds remain attached to the pods via the aril.

Cleaning requirements

Cleaning techniques: The larger 'woody' pods do not break down readily, make separating seeds easier as most seeds will pass through the sieve without the need for crushing the pods using a rubber bung.

Purity and viability: High (>90%).

Seeds per gram: 16 (range 16–23).

Germination requirements

Dormancy type: Physical.

Embryo type: Investing. **Embryo:seed ratio:** 1.00.

Pre-treatments: Wet heat for 2 mins at 90–95°C.

Germination: >90%. Unimpeded across a wide temperature range (10–35°C) once heat treated. A slight (10%) reduction in germination has been observed at low temperatures (10°C).

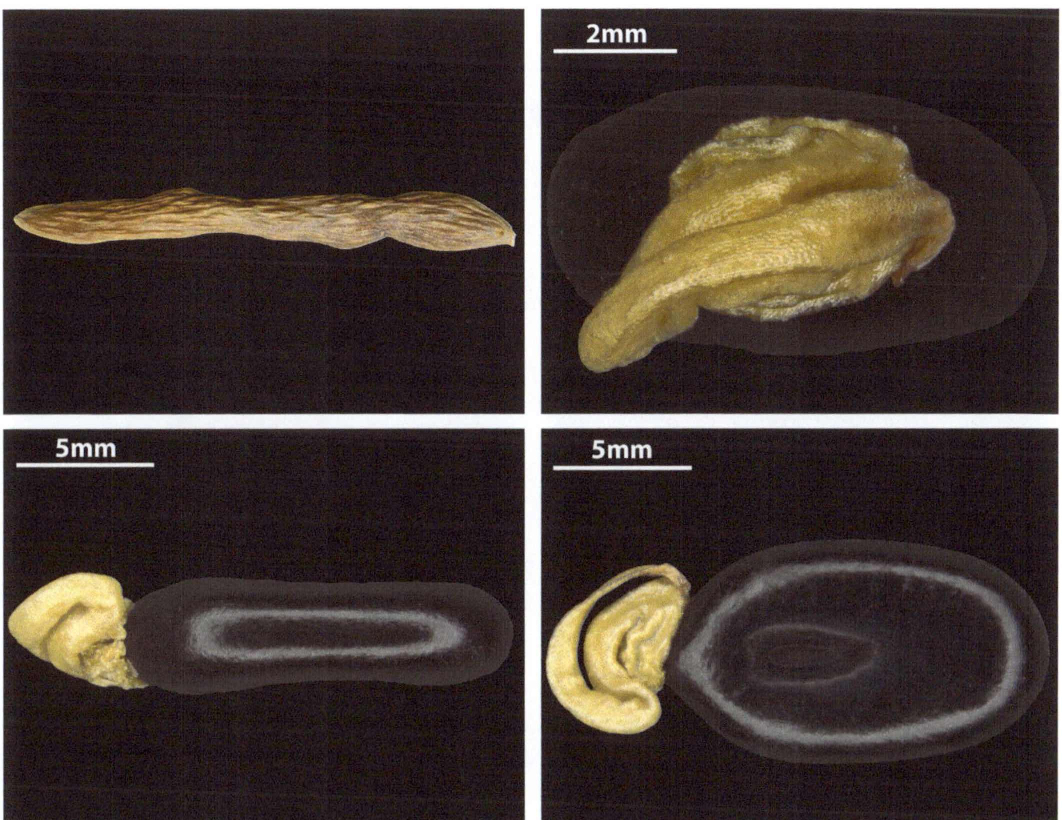

Gompholobium oreophilum

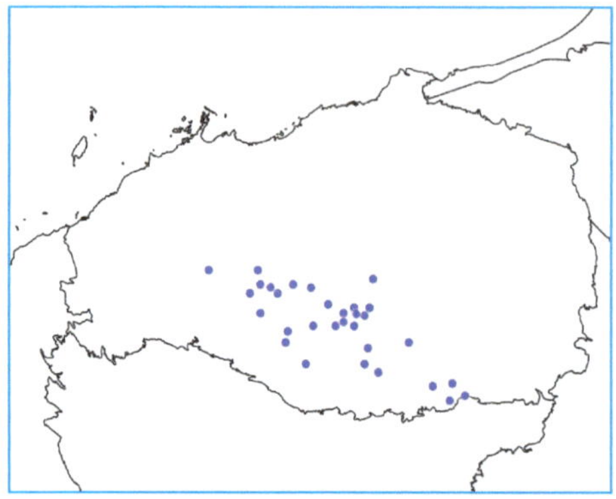

Common names: Pilbara wedge-pea.

Indigenous names: None recorded.

Description: Erect shrub 0.4–0.6 m high, 0.6–1 m wide with a dense indumentum of spreading hairs. Leaves imparipinnate with (4–)9–10 pairs of opposite leaflets, with a sparse to moderately dense indumentum; petiole 1.5–2.6 mm long. Leaflets elliptic, 4–8 mm long, 2–5 mm wide, discolorous, mid green to grey-green. Flowers 13–30 in racemes 3.8–15 cm long. Peduncles 2–6.6 mm long. Pedicels 5–9 mm long. Bracts subulate, 4–5 mm long. Bracteoles subulate, 3.5–6 mm long. Hypanthium 1.5–1.6 mm long. Calyx with adaxial lobes fused at base for 1.3–1.8 mm, asymmetrical, with one straight edge, 5.7–6 mm long, 2–2.6 mm wide, the abaxial lobes fused at base for 1.3–1.4 mm, symmetrical, 5.3–5.7 mm long, 1.8–2.5 mm wide. Standard with a claw 1.7–2.4 mm long and a broadly ovate lamina 10.2–12 mm long, 12–16.5 mm wide, yellow to yellow-orange with cream eye markings; wings straight, narrowly obovate, 10.3–12 mm long, 2.5–4.3 mm wide, yellow to yellow-orange; keel 10–12 mm long, 4.5–6 mm wide, creamish yellow. Ovary 4–4.5 mm long; style 6–6.7 mm long; ovules 2. Fruit ellipsoid, 6–6.2 mm long, 6–7 mm wide, densely hairy. Seeds ellipsoid, 2.6–2.8 mm long, tan-brown with black spots.

Key features for identification: Small, bipinnate leaves in 5–10 pairs; hairy calyx.

Similar species: *G. karijini* has almost glabrous leaves and *G. polyzygum* has leaflets in >10 pairs.

Distribution: Restricted to the Hamersley and Chichester Ranges in the southern Pilbara.

Habitat: Ironstone ridges, with *Triodia* and other shrubs.

Flowering period: July to September.

Gompholobium oreophilum

Fruit/Seed collection

Fruit/Seed maturity: October (to November).

Appearance of mature fruit/seed: Many individual pods are clustered at the apex of flowering stems. When mature, the ovoid to spherical pods turn from light green to light brown and spilt, releasing 1or 2 seeds. Mature seeds are black or brown in colour. Bulk collections can be carried out using buckets/trays if the timing is correct.

Cleaning requirements

Cleaning techniques: Seed collections will most likely include leaf and pod debris. Seeds can be separated from the plant debris by using a suitably sized sieve. If required, unopened pods can be split open by using a rubber bung. Remaining debris can be removed by vacuum separation.

Purity and viability: Seed viability/quality can be variable due to the contribution of immature seeds (i.e. typically seeds that are green or yellow in colour). However, viability of mature seeds is generally high (>90%).

Seeds per gram: 105 (range 105–116).

Germination requirements

Dormancy type: Physical (or potentially combinational [physical + physiological]).

Embryo type: Bent. **Embryo:seed ratio:** Not determined.

Pre-treatments: Wet heat for 1–2 mins at 90–95°C.

Germination: 40–60% once heat treated. Germination has only been tested at 25°C.

Notes: After heat treating to remove physical dormancy, occasionally it has been detected that some fraction of the seedlot does not germinate. This is potentially due to the current heat treatments not alleviating 100% of the impermeability or that there is some level of physiological dormancy in seeds of this species (i.e. combinational dormancy). If combinational dormancy is confirmed, treatments such as dry after-ripening is likely to assist in promoting germination once seeds are rendered permeable to water[9].

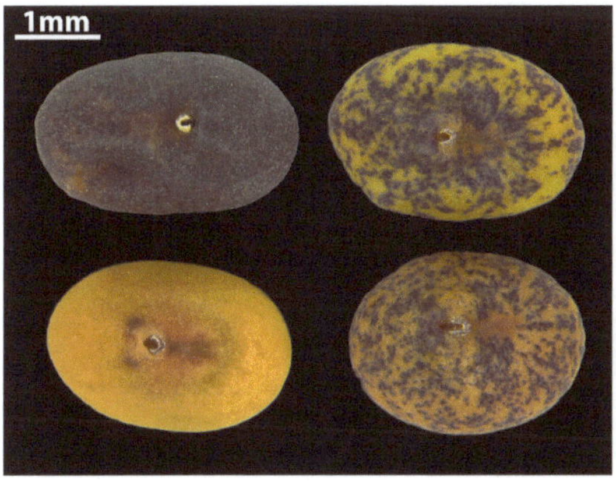

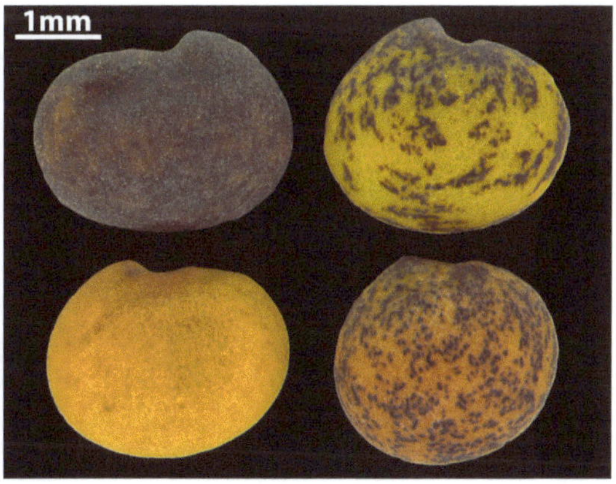

Indigofera monophylla

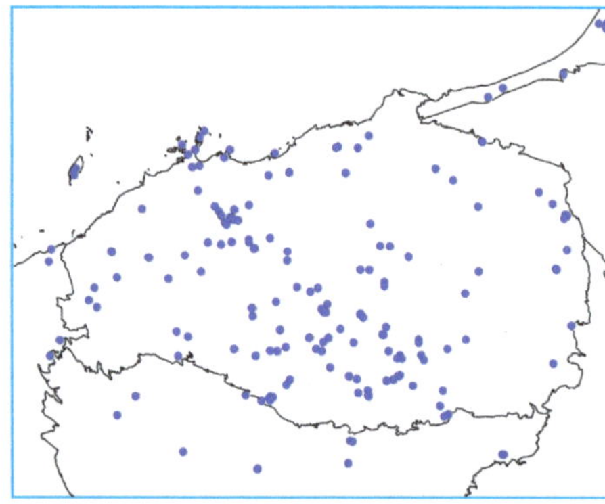

Common names: Single-leaf indigo.

Indigenous names: None recorded.

Description: Spreading shrub to 1.5 m high; stems very densely hairy, the hairs equally 2-branched. Stipules linear, 2–4 mm long. Leaves grey or grey-green, 1-foliolate; petiole 2.5–9 mm long, with a distinct articulation at the top and prominent stipels 0.5–2 mm long; petiolule 2.5–3.5 mm long; blade ovate to obovate, 15–38 mm long, 12–25 mm wide, sparsely to densely hairy, apex obtuse and apiculate or emarginate. Inflorescence 35–160 mm long, usually exceeding the leaves. Calyx 2–2.5 mm long, grey to dark red-brown tomentose, the lobes equal to or shorter than the tube. Corolla red to purple, 6–7.5 mm long, the standard with a white or yellow centre. Pod cylindric, 16–32 mm long, 3–4 mm wide, terete with a pointed apex, tomentose; endocarp spotted. Seeds 5–7.

Key features for identification: Broad, 1-foliolate leaves; dark red to purple corolla; densely hairy pods.

Similar species: There are many *Indigofera* species in the Pilbara but most are smaller herbs or subshrubs. *I. georgei* is a small shrub but has bipinnate leaves with broad leaflets.

Distribution: From the southern Kimberley to the Gascoyne River and Meekatharra in the west and to Warburton in the east. Also occurs in N.T.

Habitat: On rocky hills, sandplains or alluvial sands.

Flowering period: May to August.

Note: This species is quite variable in indumentum and degree of prominence of the veins on the under-surface of the blade. It is likely to be divided into multiple taxa in the future.

Indigofera monophylla

Fruit/Seed collection

Fruit/Seed maturity: September to November.

Appearance of mature fruit/seed: When immature, the pods appear light green (with a red-tinge) and are soft and flexible when pinched. During maturation, the pods will 'harden' off and turn brown. Collections can be made when the pods easily split with gentle pressure. Seed colour will change from green to brown-black when mature.

Cleaning requirements

Cleaning techniques: Mature pods can be separated from seeds by passing through a sieve. Typically, the seeds separate readily from the pods. If not, the use of a rubber bung may be required. A vacuum separator can then be used to remove the pod debris.

Purity and viability: Viability is generally high (>95%).

Seeds per gram: 313 (range 189–313).

Germination requirements

Dormancy type: Physical.

Embryo type: Bent. **Embryo:seed ratio:** 1.78.

Pre-treatments: Wet heat for 1–2 mins at 70–90°C. Higher treatment temperatures and longer exposure durations result in loss of seed viability.

Germination: Maximum seed germination achieved on water after alleviation of physical dormancy is 70–95% at 25°C.

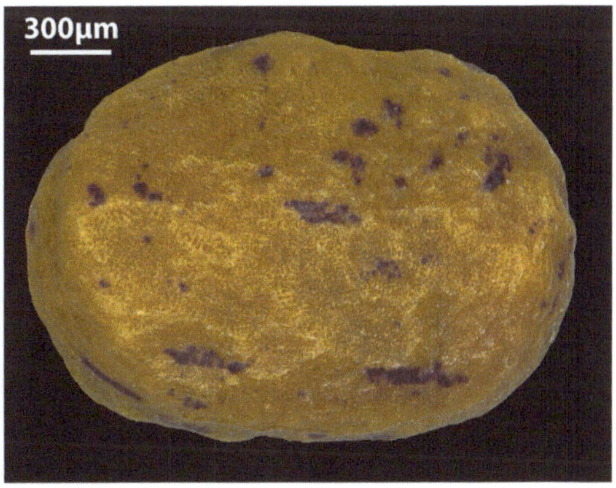

Mirbelia viminalis

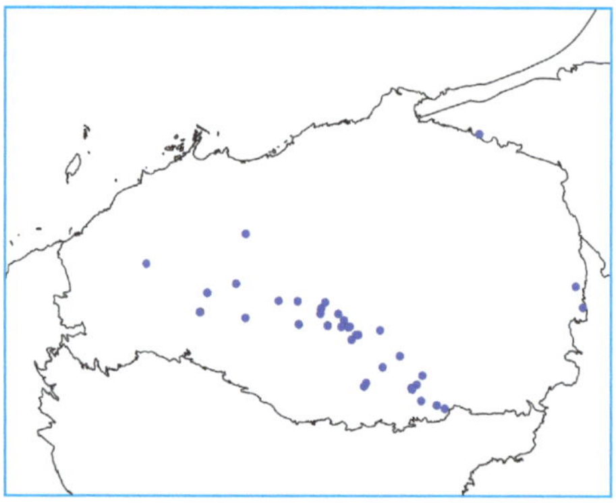

Common names: Leafless mirbelia.

Indigenous names: None recorded.

Description: Much-branched spiny shrub to 1.5 m high, sparsely sericeous or glabrous apart from the flowers; branches and branchlets ribbed, ultimate branchlets pungent. Leaves reduced to scales, exstipulate, appressed, narrowly ovate to subulate, 1–3 mm long, rigid and more or less pungent. Flowers solitary in the leaf scale axils, scattered along the upper branchlets, shortly pedicellate; bracteoles on the pedicel just below the calyx, subulate and concave, *c.* 1 mm long, persistent. Calyx 4–5.5 mm long, sericeous; lobes triangular, 2–3 mm long, margins with a little wool inside, acute. Corolla yellow; standard 4.5–6 mm long, the limb broadly ovate, emarginate; wings 3.5–5 mm; keel 3–4 mm long, more or less obtuse. Ovary stipitate, glabrous to villous; ovules 4. Style short, thick and uncinate towards apex. Pod stipitate, light brown, obliquely ovoid-ellipsoid, 7 mm long, 4 mm wide, adaxially grooved, the sutures thickened, glabrous or sparsely hairy.

Key features for identification: Leafless, spinescent branchlets that are finely hairy; yellow flowers; inflated pods.

Similar species: There are no other *Mirbelia* species in the Pilbara. *Jacksonia aculeata* has shorter, twisted branchlets with spinescent stipules.

Distribution: From the north Kimberley south to the Hamersley Range, Great Sandy Desert and the Rawlinson Range. Also occurs in N.T. and Qld.

Habitat: Often on sand or sandstone, usually associated with ridges and slopes.

Flowering period: March to August.

Mirbelia viminalis

Fruit/Seed collection

Fruit/Seed maturity: September to November.

Appearance of mature fruit/seed: The pods are very small in diameter (<10 mm) and ovoid to ellipsoid in shape. At maturity, the pod colour will change from light green to light brown and split in half. A small collection window exists due to the fragile nature of the small pods. Each pod contains 1–2 seeds.

Cleaning requirements

Cleaning techniques: Once pods are collected the seeds are loosely retained. Seeds can therefore be separated from pods via a sieve with minimal use of a rubber bung.

Purity and viability: Viability high (>90%).

Seeds per gram: 416 (range 264–416).

Germination requirements

Dormancy type: Physical.

Embryo type: Bent. **Embryo:seed ratio:** 1.96.

Pre-treatments: Wet heat for 1–2 mins at 70–90°C. Higher treatment temperatures and longer exposure durations result in loss of seed viability.

Germination: >85%. Unimpeded across a wide temperature range (15–35°C) once heat treated. Germination is reduced to 60% at 10°C.

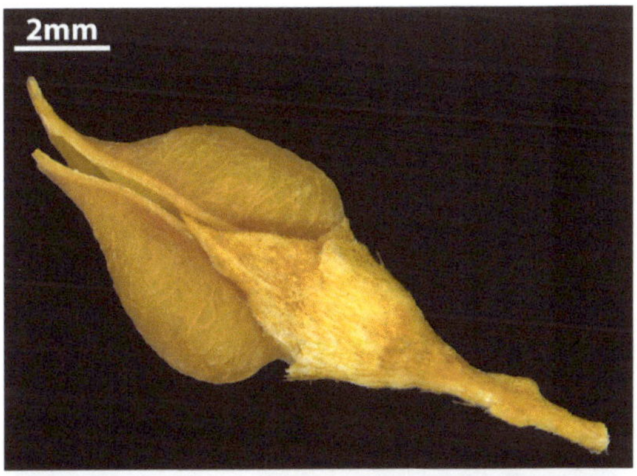

Petalostylis labicheoides

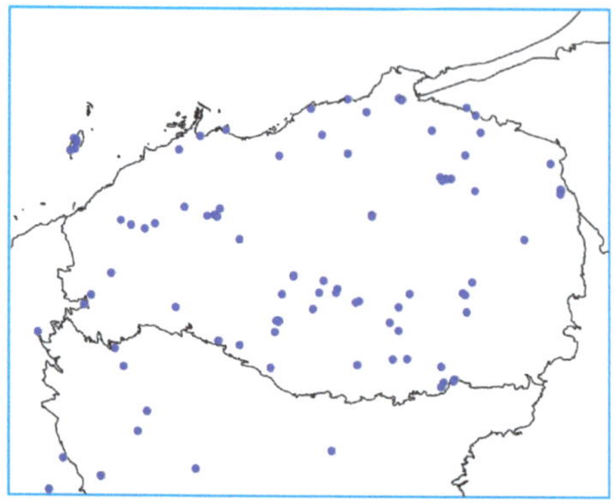

Common names: Butterfly bush, Slender petalostylis.

Indigenous names: Mirndunggaa (Yindjibarndi & Ngarluma).

Description: Shrub to 3 m tall; branchlets usually sparingly appressed-pubescent but sometimes glabrous, often pruinose. Leaves: stipules narrowly ovate, to 3.5 mm long, 1 mm wide; rachis 0.5–6 cm long, sparingly appressed-pubescent or glabrous; leaflets 3–19, lanceolate, elliptic or narrowly obovate-oblong, acuminate apically, 8–30 mm long, 3–9 mm wide, glabrous or with scattered, uncinate hairs above, subglabrous or with scattered, appressed hairs below. Sepals 10–16 mm long, with appressed or spreading hairs, sometimes erect, uncinate hairs present. Petals 15–20 mm long, 10–15 mm wide. Pod 20–34 mm long, 7–9 mm wide, glabrous or with scattered, appressed or uncinate hairs on margins.

Key features for identification: Flower regular, not pea-shaped, with a red mark at the base of the upper lobe; stamens curved; leaves bipinnate; pods flattened.

Similar species: Distinguished from *P. cassioides* by short leaves with few large leaflets in contrast to *P. cassioides* which tends to have longer leaves with more numerous, smaller leaflets.

Distribution: This species has a disjunct distribution, being found in northwestern W.A., the Flinders Ranges, S.A., central and eastern Qld and in N.S.W.

Habitat: Occurs on sandplains, stony ridges, in skeletal soils, along dry watercourses and in creek beds.

Flowering period: March to December.

Petalostylis labicheoides

Fruit/Seed collection

Fruit/Seed maturity: August to October.

Appearance of mature fruit/seed: Pods tend to be spread throughout the foliage, ripen sporadically, and have a semi-ballistic dispersal strategy, making bulk collections difficult. Pod colour changes from light green to dark brown when mature.

Cleaning requirements

Cleaning techniques: Once pods are collected the seeds are loosely retained. Seeds can therefore be separated from pods via a sieve with minimal use of a rubber bung.

Purity and viability: Viability high (>95%).

Seeds per gram: 45 (range 45–52).

Germination requirements

Dormancy type: Physical.

Embryo type: Investing. **Embryo:seed ratio:** Not determined.

Pre-treatments: Wet heat for 1 min at 80–90°C.

Germination: >90% once heat treated. Germination has only been tested at 25°C.

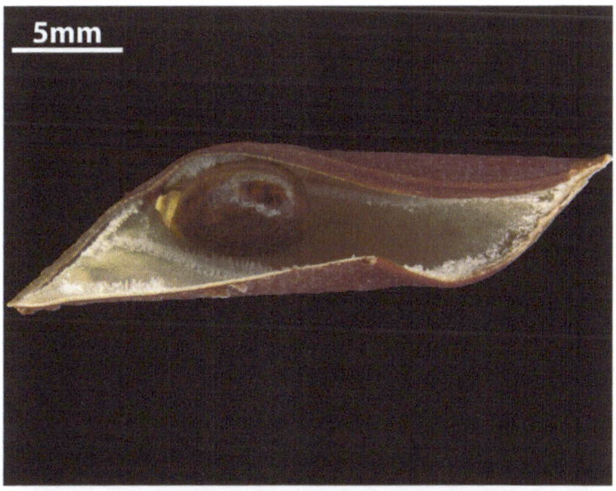

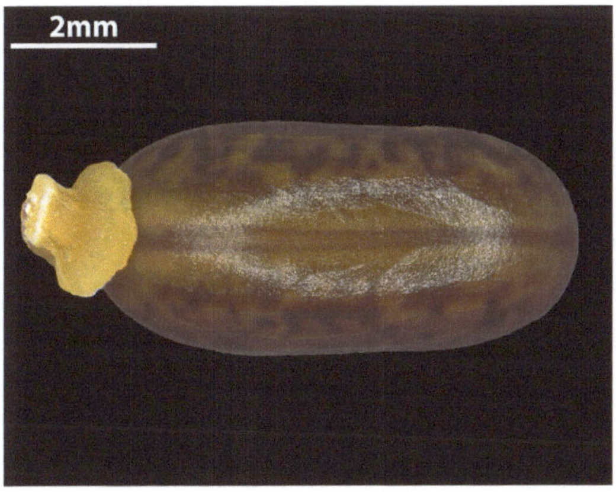

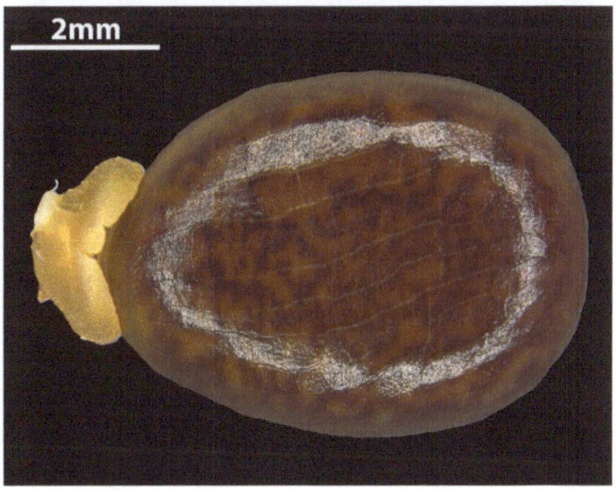

Senna artemisioides subsp. *oligophylla*

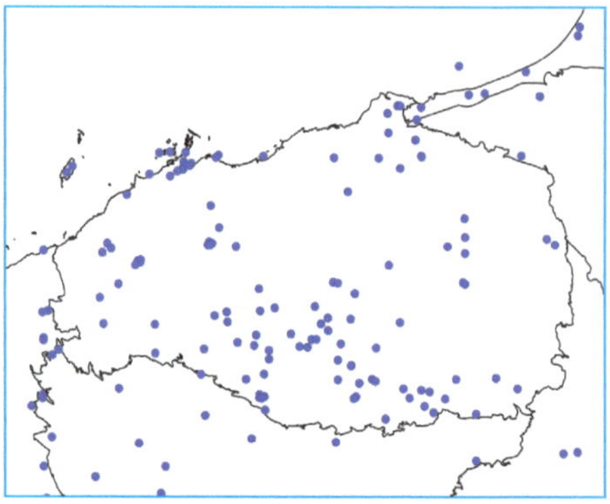

Common names: Few-leaved senna.

Indigenous names: Nyirri (Yindjibarndi & Ngarluma).

Description: Erect or spreading shrub, 0.6–2 m tall, sparsely pubescent with appressed hairs, often glaucous. Leaves 20–50 mm long; leaflets in 2–3 pairs spaced 15–25 mm apart, ovate to obovate, 10–40 mm long, 10–20 mm wide, flat or rarely recurved at the margin, obtuse to emarginate; glands 1–3, between all or the lowest leaflet pairs, sessile, flat, dark. Inflorescence of 4–12 flowers, in upper axils. Petals 8–10 mm long. Fertile stamens 10; filaments subequal, 1–2 mm long or with 1 abaxial filament 2–3 mm long; anthers subequal, 3–4 mm long. Pod flat, to 70 mm long, 15–25 mm wide, straight or slightly curved, entire.

Key features for identification: Plant more than 0.5 m tall, erect or spreading; leaflets in 2–3 pairs, presented horizontally; leaflet pairs 15–25 mm apart, dorsiventrally flattened and channelled above, <5 times longer than wide, 10–20 mm wide, sparsely pubescent; petiole terete, at least 6 mm long; petals 8–10 mm long; fertile anthers 10; pod without ridges on the inner faces of the valves, flat, not crenate; seeds dull or rarely lustrous.

Similar species: This subspecies freely intergrades and forms hybrids with related species and subspecies making accurate identification of the many resulting forms very difficult. See the key features listed above and details of known hybrids are provided in the *Flora of Australia Online*.

Distribution: Widespread in the Pilbara region and much of the interior of Australia.

Habitat: Red sandy or gravelly soils, stony areas and creeklines.

Flowering period: April to October.

Senna artemisioides subsp. *oligophylla*

Fruit/Seed collection

Fruit/Seed maturity: September to November.

Appearance of mature fruit/seed: The long flat pods change colour from light green-yellow to brown when mature. When ripe, the pods split with gentle pressure. Ripe seeds are shiny and olive-green in colour.

Cleaning requirements

Cleaning techniques: Mature pods can be separated from seeds by passing through a sieve. Typically, the seeds separate readily from the pods. If not, the use of a rubber bung may be required. A vacuum separator can then be used to remove the pod debris.

Purity and viability: Seeds are prone to insect predation. However, seed viability is generally high (>80%).

Seeds per gram: 41 (range 31–55).

Germination requirements

Dormancy type: Physical.

Embryo type: Investing. **Embryo:seed ratio:** 0.98.

Pre-treatments: Wet heat for 1 min at 90°C. Higher treatment temperatures and longer exposure durations result in loss of seed viability. Conversely, dormancy alleviation is variable at treatment temperatures <80°C.

Germination: 75–80%. Unimpeded across a wide temperature range (15–35°C) once heat treated. No germination recorded at 10°C.

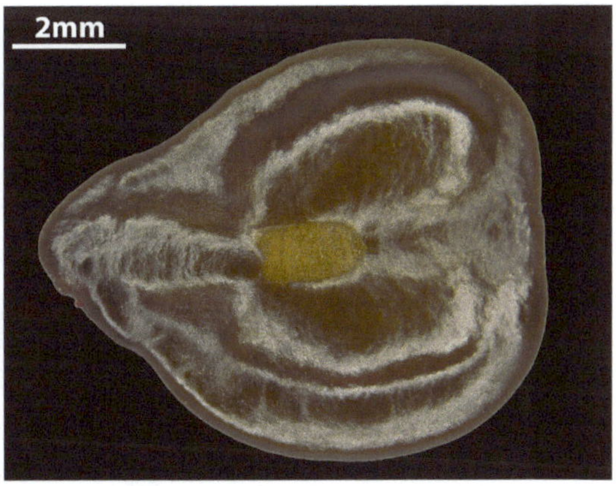

Senna glutinosa subsp. *pruinosa*

Common names: White senna.

Indigenous names: None recorded.

Description: Shrub, erect or straggling, to 2 m tall, almost glabrous, usually pruinose. Leaves 50–100 mm long including a terete petiole 10–15 mm long; stipules broadly falcate, persistent; leaflets in 3–5 pairs spaced 8–10 mm apart, elliptic, 10–20 mm long, 4–7 mm wide, broadly acute to shortly rounded; glands 1–3, between the lowest leaflet pairs, sessile, flat. Inflorescence sub-umbellate, of 4–6 flowers, in upper axils; peduncle 20–30 mm long; pedicels 10–20 mm long; bracts somewhat persistent. Petals 12 mm long, sometimes sparsely hairy outside. Fertile stamens 10; filaments subequal, subsessile, or 1–2 longer (to 2 mm); anthers subequal, 4–5 mm long. Pod flat, 30–60 mm long, 10 mm wide, straight, entire.

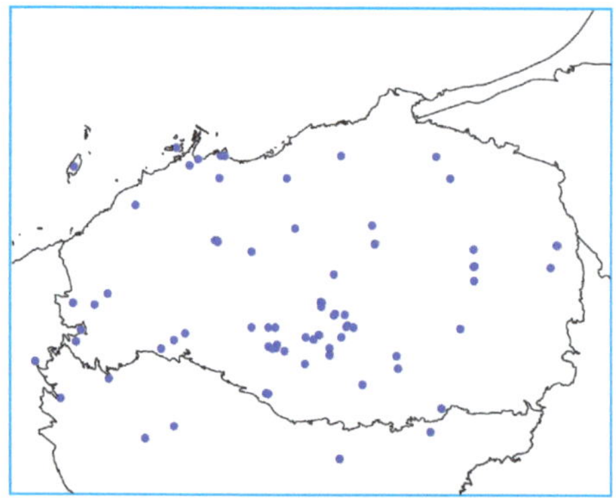

Key features for identification: Leaves flat, sclerophyllous, pruinose or green; petiole less than 45 mm long; leaflets in 3–7 pairs, 4–7 mm wide, elliptic to broadly elliptic; leaf rachis with glands between the lower leaflet pairs; petals 11–19 mm long; fertile anthers 10; pod flat, without ridges on the inner faces of the valves; seeds lustrous.

Similar species: Intergrades to other taxa are common. See note under *Senna* form taxon '*luerssenii*' in the *Flora of Australia Online.*

Distribution: Widespread in the Pilbara and arid zone of W.A. Also in N.T., S.A., Qld and N.S.W.

Habitat: Grows in arid shrublands.

Flowering period: April to September.

Senna glutinosa subsp. *pruinosa*

Fruit/Seed collection

Fruit/Seed maturity: October to November.

Appearance of mature fruit/seed: The long flat pods change colour from light green to brown (and pruinose) when mature. When ripe, the pods split with gentle pressure. Ripe seeds are shiny and yellow-green in colour.

Cleaning requirements

Cleaning techniques: Mature pods can be separated from seeds by passing through a sieve. Typically, the seeds separate readily from the pods. If not, the use of a rubber bung may be required. A vacuum separator can then be used to remove the pod debris.

Purity and viability: Seeds are prone to insect predation. However, seed viability is generally high (>90%).

Seeds per gram: 51 (range 51–58).

Germination requirements

Dormancy type: Physical.

Embryo type: Investing. **Embryo:seed ratio:** 0.94.

Pre-treatments: Wet heat for 1–2 mins at 70–90°C. Higher treatment temperatures and longer exposure durations result in loss of seed viability.

Germination: >85% across a wide temperature range (15–35°C) once heat treated. Germination decreases by 40–70% at 10°C and 30°C.

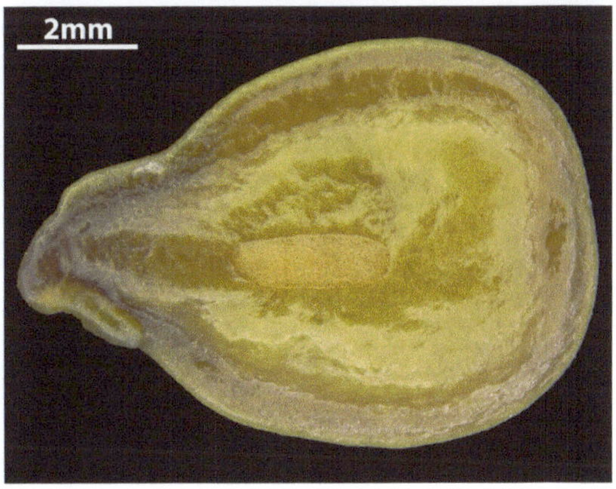

Senna notabilis

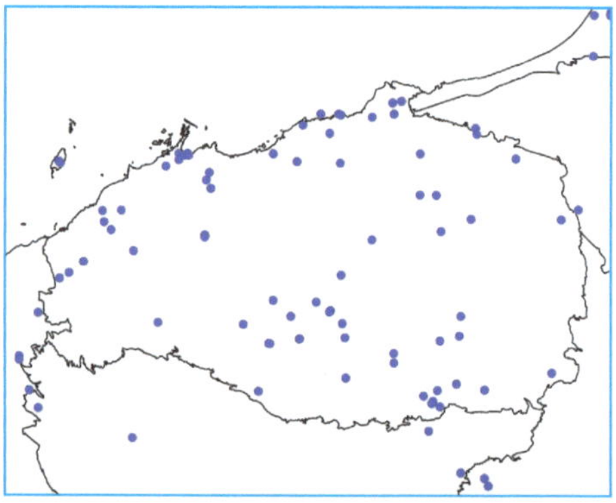

Common names: Cockroach bush, Cockroach senna.

Indigenous names: Yarringandayin (Yindjibarndi); Yarringarndalyi (Ngarluma).

Description: Spreading shrub, annual or perennial, 0.5–1.5 m high, pubescent with appressed hairs. Leaves compound, 100–200 mm long including a terete petiole 15–20 mm long; leaflets in 6–12 pairs spaced 10–15 mm apart, lanceolate, 15–30 mm long, 8–12 mm wide, acute to aristate; glands between most leaflet pairs, small. Inflorescence dense, of 20–30 flowers. Flowers yellow, on spikes. Petals about 7 mm long. Fertile stamens 7, with 3 adaxial staminodes; filaments subequal, 1 mm long; anthers subequal, the largest two 3 mm long, the others 2 mm long. Pod brown, flat, with cross-ridges in centre, 25–40 mm long, 10–15 mm wide, straight, entire.

Key features for identification: Leaves distinctly hairy; foliar glands present, small leaf rachis usually eglandular; floral bracts pubescent, acuminate, caducous, green; inflorescences dense, axillary and/or terminal; fertile anthers 7, the 2 abaxial anthers markedly larger; pod flat or compressed, 25–40 mm long, without lateral wings.

Similar species: The small foliar glands are distinctive, shared only with *S. venusta* which is a taller shrub with finer hairs on the leaflets; the leaflets and pods are much larger, the floral bracts glabrous, obtuse, persistent and yellow.

Distribution: Widespread in the interior and northern parts of Australia and throughout the Pilbara region.

Habitat: Red sandy, clay or stony soils and deep desert sands, often with *Triodia* species.

Flowering period: April to October.

Senna notabilis

Fruit/Seed collection

Fruit/Seed maturity: July to November.

Appearance of mature fruit/seed: The flat-broad pods change colour from light green to brown when mature. The central face of the pod often has distinct black blotches where the seeds are located within the pod. When ripe, the pods split with gentle pressure and often rattle when touched. Ripe seeds are shiny and light brown to grey-black in colour.

Cleaning requirements

Cleaning techniques: Mature pods can be separated from seeds by passing through a sieve. Typically, the seeds separate readily from the pods. If not, the use of a rubber bung may be required. A vacuum separator can then be used to remove the pod debris.

Purity and viability: Viability high (>90%).

Seeds per gram: 74 (range 60–77).

Germination requirements

Dormancy type: Physical.

Embryo type: Investing. **Embryo:seed ratio:** 0.96.

Pre-treatments: Wet heat for 1 min at 70–90°C. Higher treatment temperatures and longer exposure durations result in loss of seed viability.

Germination: >90%. Unimpeded across a wide temperature range (10–35°C) once heat treated.

Senna venusta

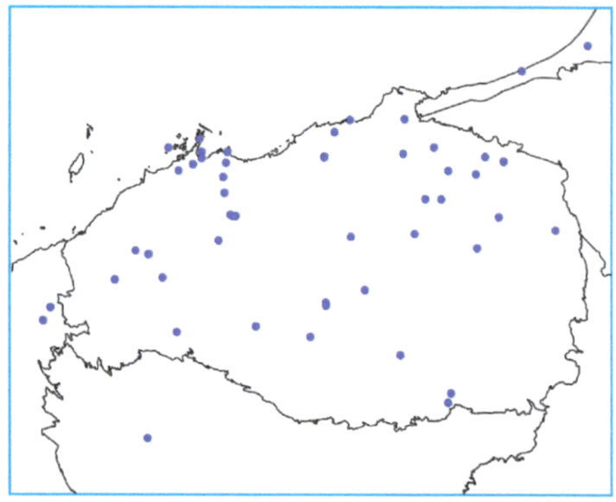

Common names: Candlestick senna.

Indigenous names: None recorded.

Description: Spindly shrub to 2 m high, glabrescent or puberulous. Stipules ovate-cordate, 12–20 mm long. Leaves 150–300 mm long; rachis with inconspicuous erect glands between pairs of leaflets; leaflets 18–28, shortly petiolulate, obliquely oblong to ovate-oblong, 30–65 mm long, 10–27 mm wide, base obtuse, apex aristate or mucronate. Racemes axillary, spike-like, appearing terminal, many-flowered; bracts yellow, scarious, ovate to broadly elliptic, 12 mm long, usually glabrous, obtuse, caducous; pedicels 10–12 mm long. Sepals elliptic, 8–10 mm long, glabrous, obtuse. Petals obovate, 7.5–15 mm long, clawed. Stamens 7 fertile and 3 reduced to staminodes, unequal, the larger 1 or 2 with curved anthers 8–10 mm long, the smaller 5 or 6 with straight anthers 3–5 mm long; filaments slender; anthers sagittate, beaked. Pod brown, narrowly oblong, 42–70 mm long, 10–15 mm wide, flat, straight, glabrous or minutely puberulous, reticulate. Seeds 6–12, brown to black, ovate-oblong, 6 mm long, 4.5 mm wide.

Key features for identification: Leaves finely hairy; foliar glands small; inflorescences dense, axillary; floral bracts yellow; fertile anthers 7; pod flat, 2.5–10 cm long, wingless.

Similar species: *S. notabilis* is a shorter plant with more hairy leaves and smaller, ribbed pods.

Distribution: Widespread in the Kimberley and Pilbara Regions, south to Carnarvon. Also in N.T. and Qld.

Habitat: Sandy plains and rocky plateaux.

Flowering period: March to August.

Senna venusta

Fruit/Seed collection

Fruit/Seed maturity: July to November.

Appearance of mature fruit/seed: The pods change colour from light green to brown when mature. When ripe, the pods split with gentle pressure and often rattle when touched. Ripe seeds are olive-green in colour (with an occasional dry-flaky covering).

Cleaning requirements

Cleaning techniques: Mature pods can be separated from seeds by passing through a sieve. Typically, the seeds separate readily from the pods. If not, the use of a rubber bung may be required. A vacuum separator can then be used to remove the pod debris.

Purity and viability: Viability high (>95%).

Seeds per gram: 27 (range 22–27).

Germination requirements

Dormancy type: Physical.

Embryo type: Investing. **Embryo:seed ratio:** 1.00.

Pre-treatments: Wet heat for 1 min at 70–90°C. Higher treatment temperatures and longer exposure durations result in loss of seed viability.

Germination: >80% across a wide temperature range (15–30°C) once heat treated. Germination decreases by >30% at 10°C.

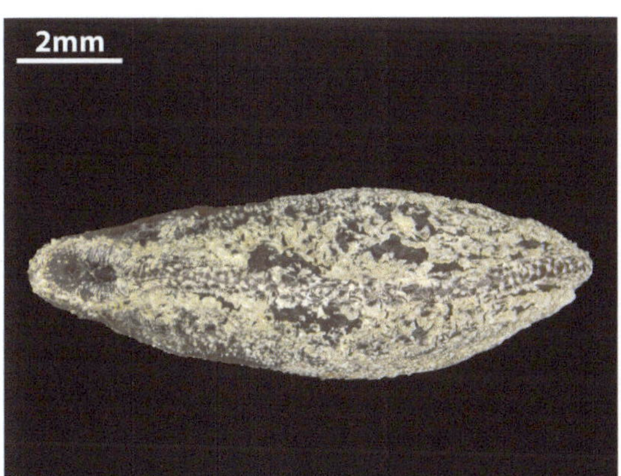

Tephrosia sp. Fortescue (A.A. Mitchell 606)

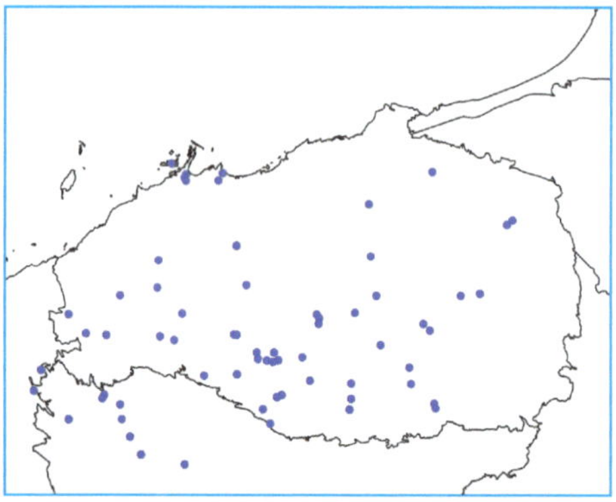

Common names: Pilbara poison pea.

Indigenous names: None recorded.

Description: Subshrub to 0.6 m high, with minute appressed hairs. Stipules linear to subulate, 7–10 mm long, persistent. Leaves petiolate, pinnately (1–)5–11-foliolate, the axis extended beyond the distal pair of leaflets; leaflets lanceolate to narrowly elliptic, 37–61 mm long, 6–8 mm wide, all equal, glabrous above, with minute appressed hairs below, veins conspicuous, apex acuminate. Inflorescence a raceme, minutely hairy, axis conspicuously scarred after the flowers have fallen; bracts subulate, *c.* 6 mm long, caducous; pedicels 4–6 mm long. Calyx 8–8 mm long, usually with dense appressed to spreading hairs; tube 1.5–2 mm long; lobes triangular; lobes of upper lip 2–2.5 mm long, connate for most of their length; lobes of lower lip narrow-triangular, unequal, with the central lobe 5–6 mm long, lateral lobes 3–5 mm long. Corolla orange, 10–12 mm long. Pod narrowly oblong to linear in outline, 22–42 mm long, 4–5 mm wide, densely appressed hairy, margins thickened. Seeds 3–6.

Key features for identification: Small, erect shrub. Closely spaced pinnate leaflets. Orange flowers. Hairy pod.

Similar species: *T. virens* is a taller shrub with glabrous pods and fewer, broader leaflets. *T.* sp. D Kimberley Flora (R.D. Royce 1848) is a decumbent shrub with fewer, larger leaflets that are often densely hairy.

Distribution: Widespread in the Pilbara, extending south of the region.

Habitat: Rocky hillsides.

Flowering period: November to February.

Tephrosia sp. Fortescue (A.A. Mitchell 606)

Fruit/Seed collection

Fruit/Seed maturity: November to March.

Appearance of mature fruit/seed: When immature, the pods appear light green and are soft and flexible when pinched. During maturation, the pods will 'harden' off and turn light brown to yellow. Collections can be made when the pods easily split with gentle pressure. Seed colour will change to mottled orange-black when mature.

Cleaning requirements

Cleaning techniques: Mature pods can be separated from seeds by passing through a sieve. Typically, the seeds separate readily from the pods. If not, the use of a rubber bung may be required. A vacuum separator can then be used to remove the pod debris.

Purity and viability: Viability high (>95%).

Seeds per gram: 102 (range 74–102).

Germination requirements

Dormancy type: Combinational (physical + physiological).

Embryo type: Bent. **Embryo:seed ratio:** 2.10.

Pre-treatments: Wet heat for 1–2 mins at 80–90°C will break dormancy. Higher treatment temperatures and longer exposure durations result in loss of seed viability.

Germination: Once treated to remove physical dormancy, germination of fresh seeds is <5% across a wide temperature range (10–35°C) with the physiological dormancy component and a small fraction of impermeable seeds contributing to low germination. Maximum germination of stored seeds (ranging between 3–24 months) on water after treating to remove physical dormancy is *c.* 60% at 25°C. An increase in germination after exposure to GA_3 or smoke has not been detected.

Notes: Treatments to overcome the physiological dormancy component have not been investigated. However, dry after-ripening is likely to assist in promoting germination once seeds are rendered permeable to water[9].

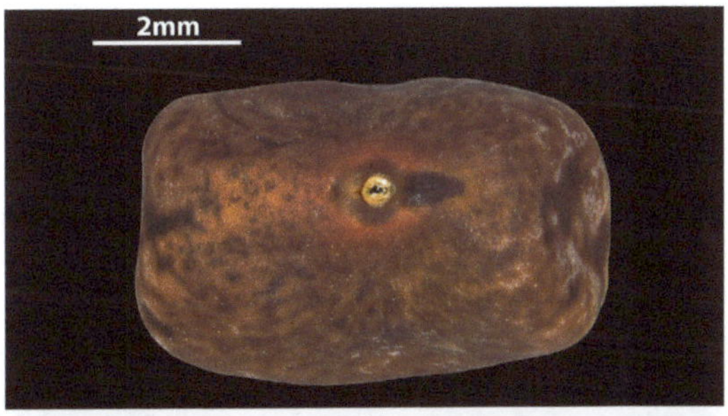

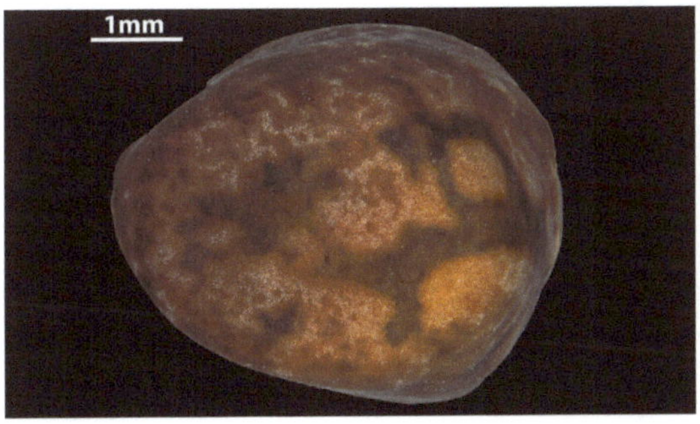

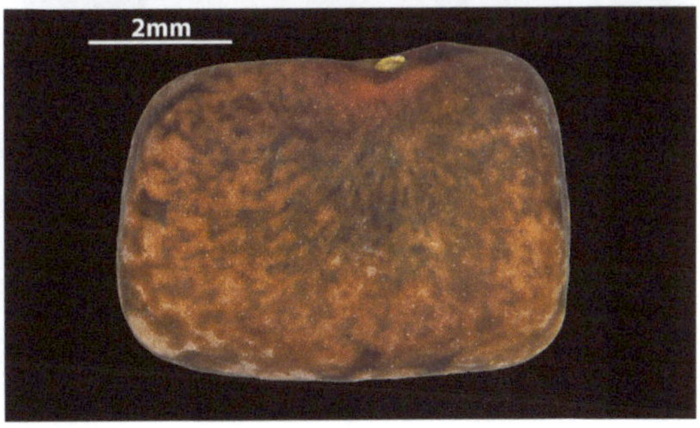

Goodenia armitiana

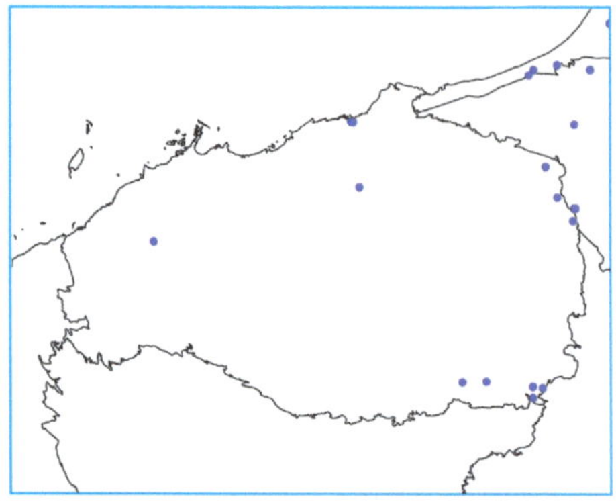

Common names: Armit's goodenia.

Indigenous names: None recorded.

Description: Erect multi-stemmed perennial to 0.5 m high, with long spreading eglandular hairs and minute glandular hairs. Basal leaves absent, cauline leaves sessile, linear, 8–65 mm long, 0.5–8 mm wide, margin entire, often recurved, usually with a tuft of eglandular hairs in the axil. Inflorescence of terminal leafy racemes; pedicels 5–12 mm long, ebracteolate, glandular-hairy. Floral tube 1.5–2.5 mm long; corolla spur prominent. Sepals elliptic, 2.5–6 mm long, densely glandular-hairy, acute. Corolla yellow, 9–15 mm long, with glandular and eglandular hairs; lobes narrowly winged, the 2 adaxial lobes unequal. Style hairy in upper half; indusium broader than long, hairy, orifice ciliate. Capsule globular to broadly ellipsoid, 5–8 mm long, 4–6 mm wide, compressed. Seeds circular to broadly elliptic, flat, 3–5 mm across including the wing; body elliptic, pitted, rimmed; wing 1–1.5 mm wide.

Key features for identification: Long, slender, cauline leaves; mixed long, spreading and minute glandular hairs.

Similar species: A variable species with several closely related species. *G. microptera* usually has more densely hairy basal and lower cauline leaves. *G. triodiophila* has longer leaves, branched stems. *G. virgata* has equally winged adaxial corolla lobes.

Distribution: From the southern Kimberley to the Gibson Desert, uncommon in the Pilbara. Also in N.T. and Qld.

Habitat: In grassland or shrubland.

Flowering period: March to November.

Goodenia armitiana

Fruit/Seed collection

Fruit/Seed maturity: October to November.

Appearance of mature fruit/seed: The flowers are located at the apex of this spindly shrub. During maturation, capsules will turn from green to light brown. Once mature, the capsules will split in half releasing flat disc-like seeds. A narrow collection window exists as the capsule opens quickly once mature. However, larger collections can be made with the use of a bucket/tray.

Cleaning requirements

Cleaning techniques: Some capsule material may be collected during bulk sampling. Passing seeds through a sieve or vacuum separation removes the capsule debris.

Purity and viability: Viability high (>85%).

Seeds per gram: 554.

Germination requirements

Dormancy type: Non-dormant.

Embryo type: Spatulate. **Embryo:seed ratio:** Not determined.

Pre-treatments: Exposure of fresh seeds to GA_3 and smoke increases germination by 30–60% at temperatures lower, or higher, than 25°C.

Germination: Maximum seed germination achieved with no pre-treatment is 85% at 25°C. Treatment with GA_3 or smoke increases germination to >75% at 15–30°C.

Notes: The broadening of the temperature range at which germination occurs in the presence of GA_3 and smoke suggests that a proportion of the seeds have some level of physiological dormancy. Therefore, additional dormancy break pre-treatments such as dry after-ripening or wet/dry cycling may assist in improving germination across a wider temperature range without the need for treatment with GA_3 or smoke[6].

Goodenia stobbsiana

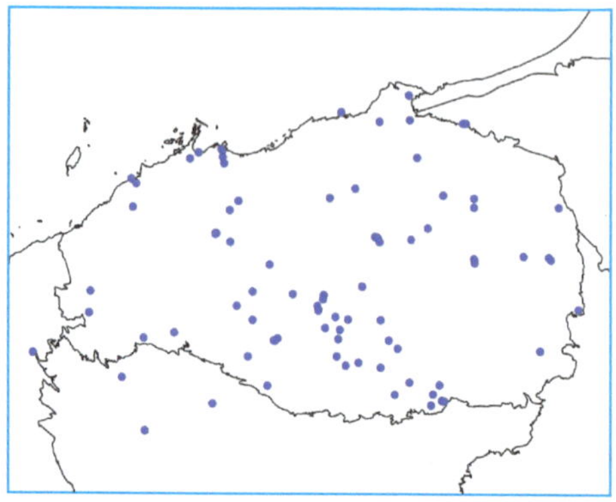

Common names: Stobbs' goodenia.

Indigenous names: None recorded.

Description: Usually several-stemmed, much-branched, perennial subshrub to 80 cm tall, glandular-viscid. Leaves basal and cauline, obovate to oblanceolate, sometimes dentate, glandular-pubescent to glabrescent; lamina 40–70 mm long, 20–30 mm wide. Flowers in loose thyrses to 400 mm long; bracts leaf-like, smaller apically; peduncle to 50 mm long; bracteoles usually linear, to 10 mm long; pedicel mostly to 4 mm long. Corolla blue–purple, hairy inside; enations in regular rows bearing conspicuous tuft of hairs; lobes unequal to 14 mm long. Indusium broadly obtriangular, 3 mm long, convex. Ovules 30–50. Fruit ovoid, 8–9 mm long, valves entire. Seed elliptic, 2 mm long, blackish brown.

Key features for identification: Corolla blue–purple; enations in regular rows.

Similar species: Similar to *G. scaevolina*, differing in the more unequal corolla lobes, the more numerous simple hairs on the outside of the corolla and the more conspicuous enations inside the corolla.

Distribution: Widespread in the Pilbara region.

Habitat: Usually on stony ground, appearing particularly after fire and on disturbed ground.

Flowering period: March to October.

Three photographs on right by Geoff & Ruth Byrne.

Goodenia stobbsiana

Fruit/Seed collection

Fruit/Seed maturity: September to November.

Appearance of mature fruit/seed: Flowers are located at the apex of this much-branched upright shrub. Post-anthesis, immature green capsules generally develop progressively from the lower branches. At maturity the capsules split and release numerous flat disc-like seeds, brown-black in colour. Bulk collections can be made easily as the capsules are held upright on stems, providing greater seed retention, and allowing whole stems to be upturned into buckets/bags.

Cleaning requirements

Cleaning techniques: Some capsule material may be collected during bulk sampling. Passing seeds through a sieve or vacuum separation should remove the capsule debris relatively easy.

Purity and viability: Viability variable (65–100%).

Seeds per gram: 1837 (range 958–1837).

Germination requirements

Dormancy type: Physiological.

Embryo type: Spatulate. **Embryo:seed ratio:** 0.66.

Pre-treatments: Exposure to GA_3 enhances germination of fresh seeds.

Germination: Maximum seed germination achieved on water and smoke is <5% across all incubation temperatures. Treatment with GA_3 increases germination to 70–85% at 20–25°C.

Notes: Additional dormancy break pre-treatments such as dry after-ripening or wet/dry cycling may assist in improving germination across a wider temperature range[6].

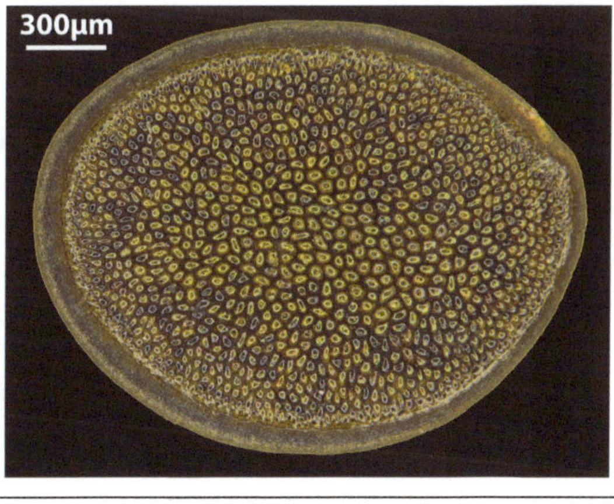

Goodenia triodiophila

Common names: Spinifex goodenia.

Indigenous names: None recorded.

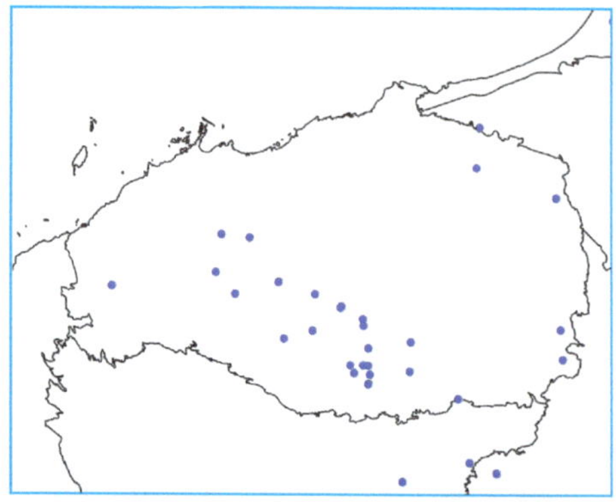

Description: Much-branched, stiff, ascending herb, to 40 cm tall, glabrous or with sparse, simple hairs; stems reddish or brownish, often hairy at base, green and glabrous above. Cauline leaves terete or linear, involute; lamina 40–100 mm long, 1 mm wide. Flowers in racemes to 150 mm long, or sub-umbels; bracts leaf-like; pedicel 8–30 mm long, articulate; bracteoles absent. Sepals lanceolate to narrowly elliptic, 1.5–3 mm long. Corolla 10–12 mm long, hairy outside and inside, without enations, auriculate; abaxial lobes 4–5 mm long; wings 0.5–1.8 mm wide, narrower above auricle. Indusium depressed-obovate, 1.5–1.8 mm long. Ovules 20–25. Fruit globular to ovoid, compressed, 6 mm diam. Seeds orbicular, 4–6 mm diam., transverse, colliculate, yellow-brown; wing 1–1.5 mm wide, yellowish.

Key features for identification: Corolla yellow with a brownish throat, with glandular and simple hairs outside, sparsely hairy inside; sepals adnate to ovary only in lower half. The corolla has a prominent pocket, and the wings of the adaxial lobes are narrower above the auricle.

Similar species: *G. armitiana* and *G. microptera* have short stocks not elongated basal stems, hairy leaves and lack a prominent pocket on the corolla.

Distribution: Occurs in arid regions of inland Australia, in W.A., N.T. and western Qld.

Habitat: Occurs particularly in *Triodia* communities.

Flowering period: April to October.

Goodenia triodiophila

Fruit/Seed collection

Fruit/Seed maturity: June (observations limited).

Appearance of mature fruit/seed: The flowers are located at the apex of this spindly shrub. During maturation, capsules will turn from green to light brown. Once mature, the capsules will split in half releasing flat disc-like seeds. A narrow collection window exists as the capsule opens quickly once mature.

Cleaning requirements

Cleaning techniques: Some capsule material may be collected during bulk sampling. Passing seeds through a sieve or vacuum separation should remove the capsule debris relatively easy.

Purity and viability: Viability high (100%).

Seeds per gram: 950.

Germination requirements

Dormancy type: Non-dormant.

Embryo type: Spatulate. **Embryo:seed ratio:** 0.88.

Pre-treatments: Exposure of fresh seeds to GA_3 and smoke increases germination by 10–20% at temperatures lower, or higher, than 25–30°C.

Germination: Maximum seed germination achieved with no pre-treatment is >95% (at 25–30°C). Treatment with GA_3 or smoke increases germination to >90% at 20–35°C.

Notes: Additional dormancy break pre-treatments such as dry after-ripening or wet/dry cycling may assist in improving germination across a wider temperature range without the need for treatment with GA_3 or smoke[6].

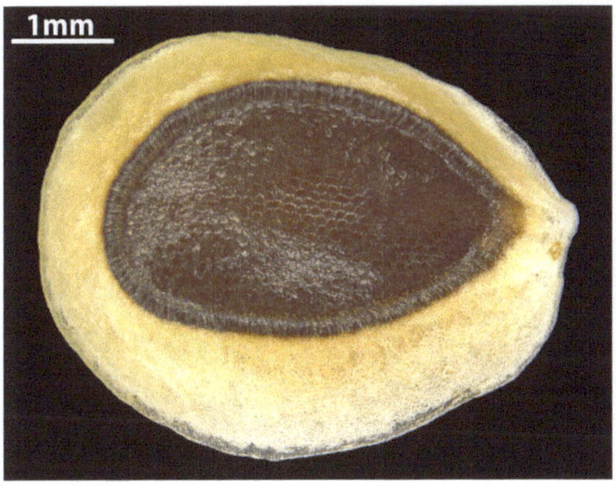

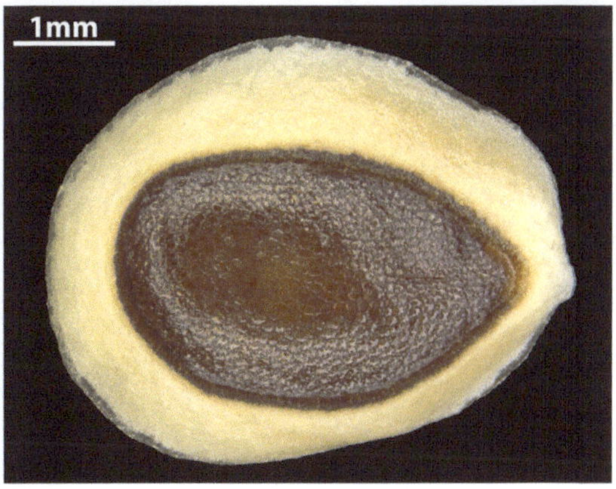

Goodenia vilmoriniae

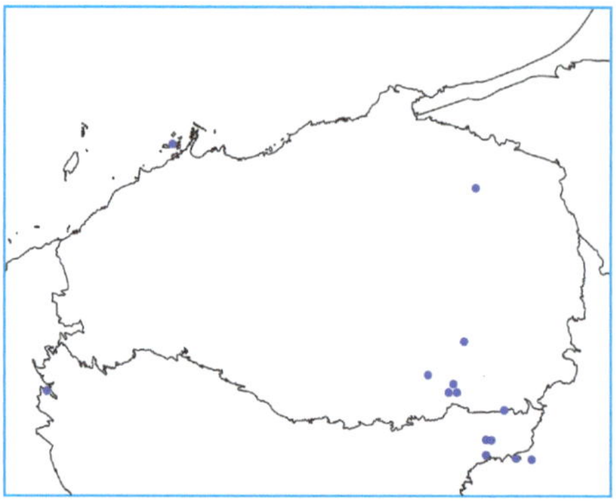

Common names: Purple goodenia.

Indigenous names: None recorded.

Description: Ascending to erect, annual herb to 40 cm tall, villous and glandular-hairy. Leaves mostly basal, linear to lanceolate, entire to dentate, sometimes pinnate-lobed; lamina 40–100 mm long, 3–8 mm wide. Flowers in racemes to 10 cm long, or subumbels; bracts leaf-like; pedicel 10–35 mm long, articulate; bracteoles absent. Sepals lanceolate to elliptic, 4–5 mm long. Corolla 15–22 mm long, pubescent in throat, without enations, auriculate; abaxial lobes 6.5–8.5 mm long; wings 2 mm wide. Indusium depressed-obovate, 2–2.5 mm long. Ovules 30. Fruit subglobular, usually compressed, 7–9 mm diam. Seeds orbicular, 4–4.5 mm diam., colliculate, dark brown to nearly black; wing 1 mm wide, hyaline, thin.

Key features for identification: Corolla pale blue, purple to lilac, with yellow in throat.

Similar species: The only other Pilbara *Goodenia* lacking bracteoles and with a bluish-purple corolla is *G. pallida*, which is not villous.

Distribution: From the eastern Pilbara to the Tanami Desert. Also in N.T., S.A. and Qld.

Habitat: Grows in sandy soil.

Flowering period: April to September.

Goodenia vilmoriniae

Fruit/Seed collection

Fruit/Seed maturity: June (observations limited).

Appearance of mature fruit/seed: The flowers are located at the apex of this multi-branched shrub. During maturation, capsules will turn from green to light brown. Once mature, the capsules will split in half releasing flat disc-like seeds. A narrow collection window exists as the capsule opens quickly once mature.

Cleaning requirements

Cleaning techniques: Some capsule material may be collected during bulk sampling. Passing seeds through a sieve or vacuum separation should remove the capsule debris relatively easy.

Purity and viability: Viability *c.* 75%.

Seeds per gram: 756.

Germination requirements

Dormancy type: Physiological.

Embryo type: Spatulate. **Embryo:seed ratio:** 0.79.

Pre-treatments: Exposure of seeds to GA_3 increases germination by 20–30%.

Germination: Maximum seed germination achieved with no pre-treatment is <5% at 10–35°C. Treatment with GA_3 increases germination to 20–35% at 15–35°C.

Notes: Additional dormancy break pre-treatments such as dry after-ripening or wet/dry cycling may assist in overcoming physiological dormancy[6].

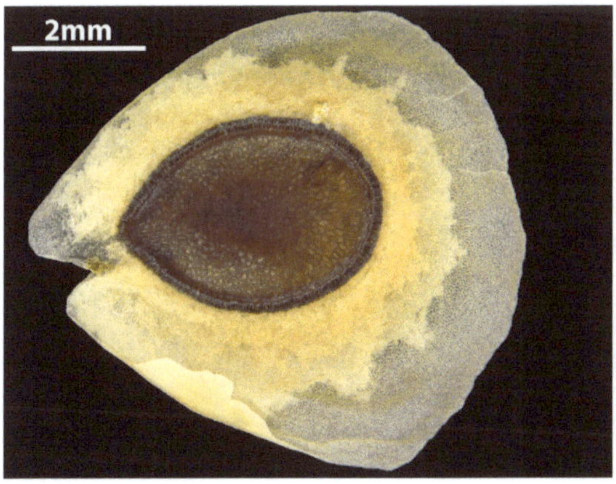

Velleia panduriformis

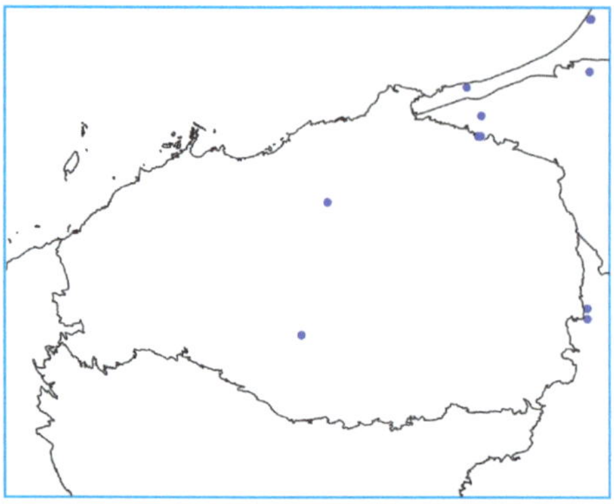

Common names: Pindan poison, Cabbage poison.

Indigenous names: None recorded.

Description: Perennial herb to 1.5 m high, glabrous. Leaves in an ascending rosette, obovate and dentate. Inflorescence: peduncles up to 1.5 m high, glabrous; bracteoles leaf-like, connate at least on one side, broadly ovate to semicircular, 20–100 mm long, 15–110 mm wide, decreasing in size, glaucous, glabrous or rarely minutely puberulous, dentate; flowers shortly pedicellate. Sepals 5, free, elliptic to ovate-elliptic, glabrous or puberulous; adaxial sepal 15–20 mm long, broader and more prominently dentate; other sepals 10–15 mm long. Corolla yellow, 18–25 mm long, glabrous to puberulous outside, somewhat villous inside; lobes 7–10 mm long, shortly and narrowly winged. Style villous; indusium broader than long, sparsely villous. Capsule ovoid to broadly ovoid, somewhat compressed, 11–13 mm long, 6–12 mm wide, glabrous. Seeds 12, circular to broadly elliptic, 7–9 mm long, 7–10 mm wide including a broad wing; body pitted to almost smooth; wing up to 3 mm wide.

Key features for identification: Semicircular, connate bracteoles which are leaf-like.

Similar species: *V. connata* differs in having much smaller bracts and a spreading habit.

Distribution: From the Kimberley south to Balfour Downs Station and the Gibson Desert. Also occurs in N.T.

Habitat: Often on sand or clay in woodland.

Flowering period: April to November (January).

Note: Reportedly toxic to stock.

Velleia panduriformis

Fruit/Seed collection

Fruit/Seed maturity: September to October.

Appearance of mature fruit/seed: When immature, seeds will be encased in an ovoid-shaped capsule surrounded by persistent sepals. Upon maturation, the capsules will split at the apex and progressively open releasing 10–15 flat disc-like seeds with a small wing.

Cleaning requirements

Cleaning techniques: If collected carefully by upturning old flower stems, near pure seeds can be collected as seeds easily fall from the open capsules. Unwanted plant debris and capsules can then be removed by vacuum separation.

Purity and viability: Viability high (100%).

Seeds per gram: 127.

Germination requirements

Dormancy type: Physiological.

Embryo type: Spatulate. **Embryo:seed ratio:** 0.69.

Pre-treatments: Dormancy alleviation treatments unknown.

Germination: Generally 0–10%. Very low germination of seeds occurs across all incubation temperatures (10–35°C) and after treatment with GA_3 and smoke.

Notes: Slightly more germination is evident after exposure to GA_3 and smoke, suggesting dormancy break pre-treatments such as dry after-ripening or wet/dry cycling may assist in overcoming physiological dormancy[6].

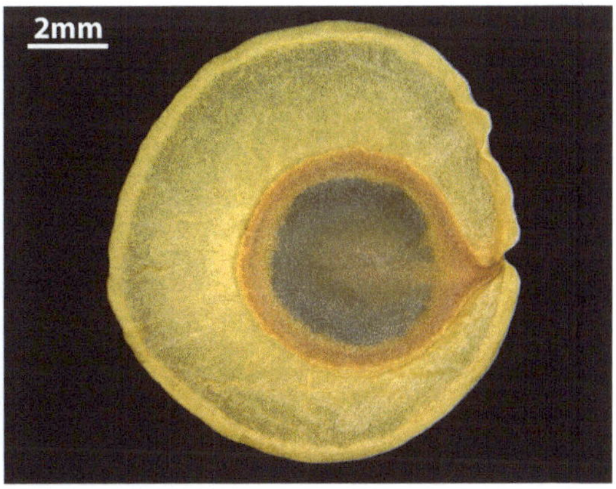

Newcastelia cladotricha

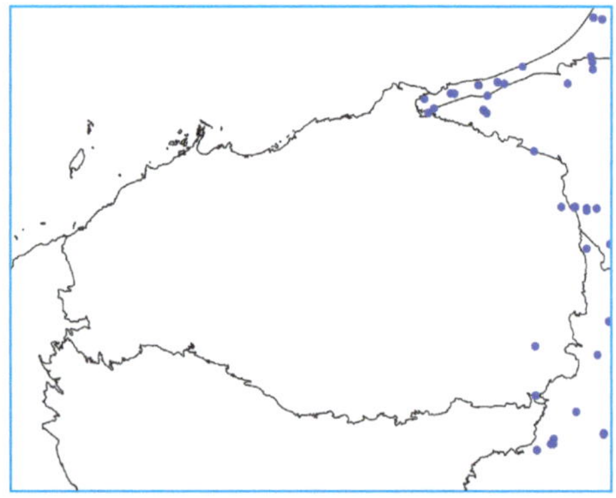

Common names: Lamb's tails.

Indigenous names: None recorded.

Description: Shrub, 0.3–1 m high; stems with grey-ferruginous hairs. Leaves opposite, sessile or subsessile, narrowly oblong-elliptic to elliptic, 8–43 mm long, 4–15 mm wide, with a revolute margin, obtuse. Inflorescence a simple terminal spike of opposite-decussate flowers, becoming elongate and interrupted, 35–120 mm long; axis with purplish hairs. Flowers sessile, mauve or purplish grey. Calyx 3–4.5 mm long, densely hairy; lobes 5, equalling or somewhat shorter than tube. Corolla purple, 4.5–6.5 mm long; lobes 5, 1.5–2 mm long, shortly acuminate. Stamens inserted below summit of corolla tube, reaching to the base of the corolla lobes, included or scarcely exserted. Ovary densely glandular. Style included, 1.5–2.5 mm long. Fruit almost globular, 2–3 mm in diameter.

Key features for identification: Flowers in elongated spikes; stamens and style included; corolla-tube densely villous inside and lobes ovate.

Similar species: Related to *N. velutina* and *N. interrupta* but distinguished by oblong to ovate-oblong leaves; calyx tomentum much longer and fluffy, purplish-grey; ovary glabrous but glandular.

Distribution: From the southern Kimberley south to the far-northern and eastern Pilbara. Also occurs in N.T.

Habitat: Occurs on red sand dunes and sand flats.

Flowering period: May–September.

Newcastelia cladotricha

Fruit/Seed collection

Fruit/Seed maturity: September to October.

Appearance of mature fruit/seed: Small woody fruits are covered in dense white hairs that form a spherically shaped fruiting unit. These fruits are located along an elongated flowering stem and form clusters of 3–4. When mature, the fruiting units will fall easily from the plant.

Cleaning requirements

Cleaning techniques: The dense hairs can be removed from the fruiting unit by rubbing the material through a sieve using a rubber bung. The lighter hairy material can be removed using vacuum separation. The small woody indehiscent fruits contain up to 4 seeds within the multi-loculed fruits.

Purity and viability: The hairs covering the fruits contribute up to 80% of collected material, leading to low fruit purity. Further, once the hairs are removed the proportion of fruits that contain a viable seed is very low.

Seeds per gram: 257 (indehiscent fruits).

Germination requirements

Dormancy type: Physiological (inferred from other multi-loculed fruits and studies conducted on the Lamiaceae family[1]).

Embryo type: Not determined. **Embryo:seed ratio:** Not determined.

Pre-treatments: Not tested.

Germination: Not tested.

Notes: Due to insufficient fruit material containing seeds, germination experiments have not yet been conducted. However, seeds enclosed in indehiscent fruits are commonly difficult to germinate[7,8]. Potential dormancy alleviation treatments include dry after-ripening, warm stratification, or wet/dry cycling.

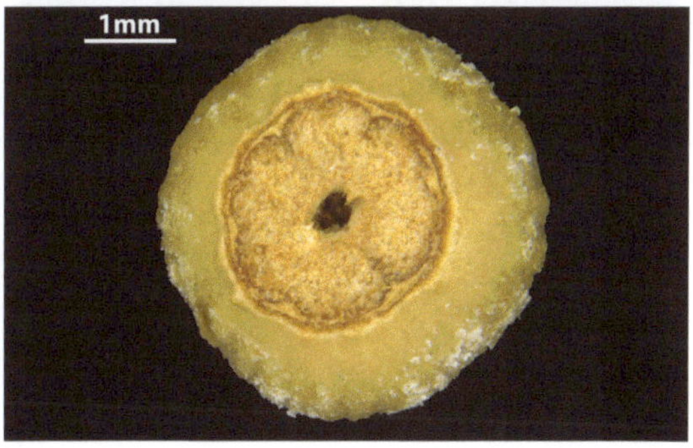

Newcastelia sp. Hamersley Range (S. van Leeuwen 4264)

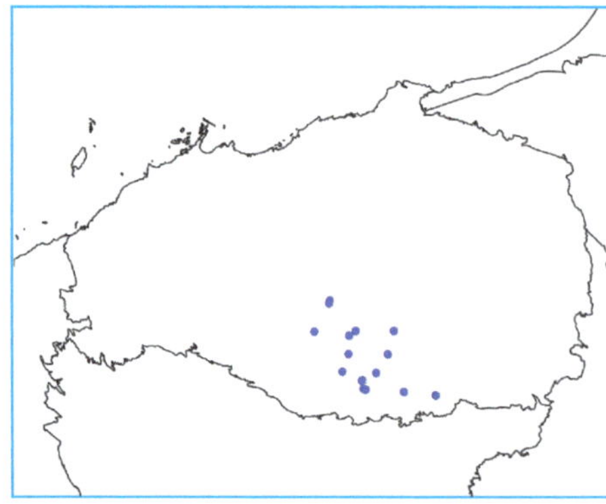

Common names: Hamersley Range lamb's tails.

Indigenous names: None recorded.

Description: Small shrub, 0.5–1 m tall, to 1 m wide. Stems woody, lower stems dull greyish green, middle stems pale green, upper stems dull reddish. Leaves opposite, decussate, densely covered in white hairs on both surfaces giving them a grey-white colour, prominently veined, 15–20 mm long, 8–9 mm wide. Flowers white, mauve or lilac with white throat and tube, pleasantly perfumed; arranged in dense terminal spikes 25 mm long, 15 mm wide. Anther filaments mauve. Style and stigma white, style exerted. Fruit densely hairy, 5–6 mm long, 3 mm wide.

Key features for identification: Flowering and fruiting heads compact. Leaves very finely hairy.

Similar species: Easily distinguished from *N. cladotricha* by the compact flower heads and white, mauve or lilac flowers with a white throat and tube.

Distribution: Restricted to the Hamersley Range.

Habitat: Grows on rock fissures on small cliffs high in the landscape.

Flowering period: August to September.

Newcastelia sp. Hamersley Range (S. van Leeuwen 4264)

Fruit/Seed collection

Fruit/Seed maturity: October to November.

Appearance of mature fruit/seed: Small woody fruits are covered in dense white hairs that form a spherically shaped fruiting unit. Collectively, numerous fruits are clustered at the tip of woody stems and form a larger fruiting unit. When mature, the fruiting units will fall easily from the plant.

Cleaning requirements

Cleaning techniques: The dense hairs can be removed from the fruiting unit by rubbing the material through a sieve using a rubber bung. The lighter hairy material can be removed using vacuum separation. Once the hairs are removed, small woody indehiscent fruits will remain and contain up to 4 seeds within the multi-loculed fruits.

Purity and viability: The hairs covering the fruits contribute up to 70% of collected material, leading to low fruit purity. Further, once the hairs are removed the proportion of fruits that contain a viable seed is very low (*c.* 5%).

Seeds per gram: 343 (indehiscent fruits).

Germination requirements

Dormancy type: Physiological (inferred from other multi-loculed fruits and studies conducted on the Lamiaceae family[1]).

Embryo type: Not determined. **Embryo:seed ratio:** Not determined.

Pre-treatments: Not tested.

Germination: Not tested.

Notes: Due to insufficient fruit material containing seeds, germination experiments have not yet been conducted. However, seeds enclosed in indehiscent fruits are commonly difficult to germinate[7,8]. Potential dormancy alleviation treatments include dry after-ripening, warm stratification, or wet/dry cycling.

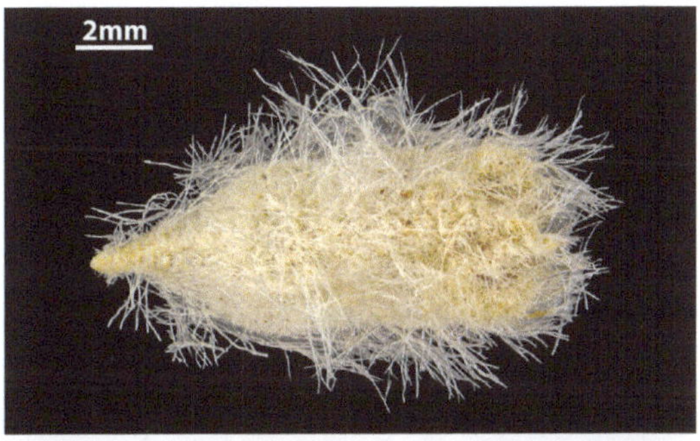

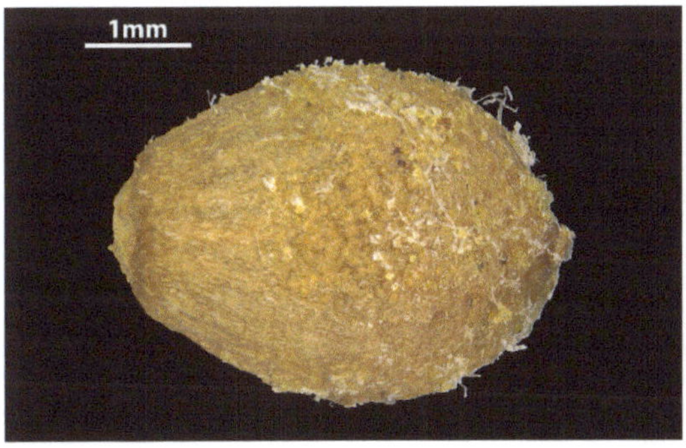

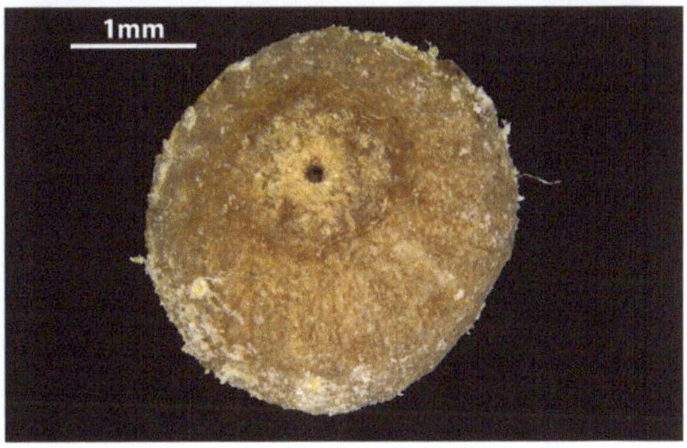

Abutilon otocarpum

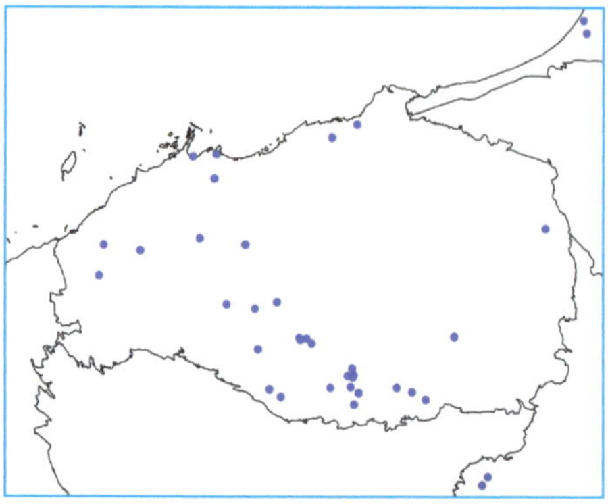

Common names: Desert Chinese lantern.

Indigenous names: None recorded.

Description: Subshrub to 1 m high; stems and petioles softly and densely hairy with long stellate hairs. Stipules linear, 5–12 mm long. Leaves ovate to very broadly ovate or almost circular, 10–70 mm long, 11–60 mm, densely stellate-hairy above and below, base cordate, margin crenate, apex obtuse to acuminate or acute. Calyx 9–15 mm long, very densely and softly stellate-hairy, conspicuously keeled and saccate at the base; lobes 6–13 mm long, longer than the tube, gradually tapered to a long acute apex. Petals yellow, 8–12 mm long, shorter than to scarcely exceeding the calyx. Schizocarp ovoid, 6–9 mm long, 7–11 mm diam., shorter than the fruiting calyx; mericarps 10–20, readily separating at maturity but enclosed by the calyx, stellate-hairy, outer edge obtuse. Seeds 3 per mericarp, brown, deeply rugose, glabrous.

Key features for identification: Short petals, not much longer than the sepals. Schizocarp densely stellate hairy.

Similar species: *Abutilon lepidum* has larger flowers and the schizocarp is glandular hairy.

Distribution: Widespread from the Kimberley, across the Pilbara, south to Glenorn Station, north of Kalgoorlie. Also occurs in all mainland states.

Habitat: Red sand, stony loam or alluvium on plains.

Flowering period: January to August.

Abutilon otocarpum

Fruit/Seed collection

Fruit/Seed maturity: May to July.

Appearance of mature fruit/seed: The whole fruiting body (made up of multiple thin-walled segments) turns from green to yellow/light brown. At maturity, the individual fruit segments begin to separate and open.

Cleaning requirements

Cleaning techniques: The whole fruit or individual fruit segments can be broken down by rubbing through a sieve using a rubber bung. The lighter fruit segments and debris can then be separated from the seed using a vacuum separator.

Purity and viability: Generally high (>80%).

Seeds per gram: 458.

Germination requirements

Dormancy type: Combinational (physical + physiological).

Embryo type: Folded. **Embryo:seed ratio:** 1.59.

Pre-treatments: Dry heat for 10 mins at 95–100°C. Alternatives include wet heat for 5 mins at 90–95°C or immersion in 98% sulphuric acid for 4 hours.

Germination: 50–70% across a wide temperature range (10–35°C).

Notes: Identification of combinational dormancy in this species means additional dormancy break pre-treatments such as dry after-ripening or wet/dry cycling may be required to overcome the physiological component of dormancy[9].

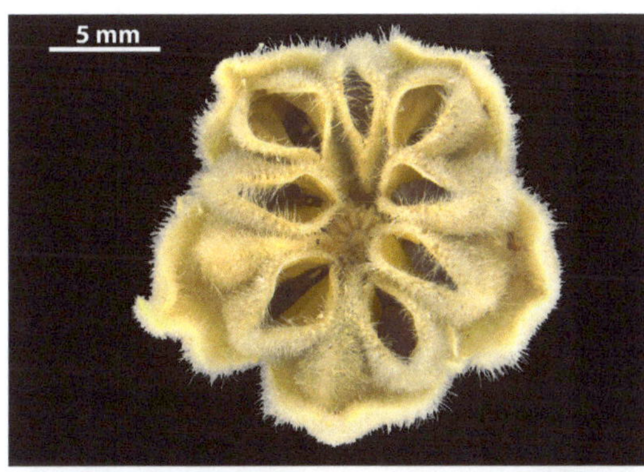

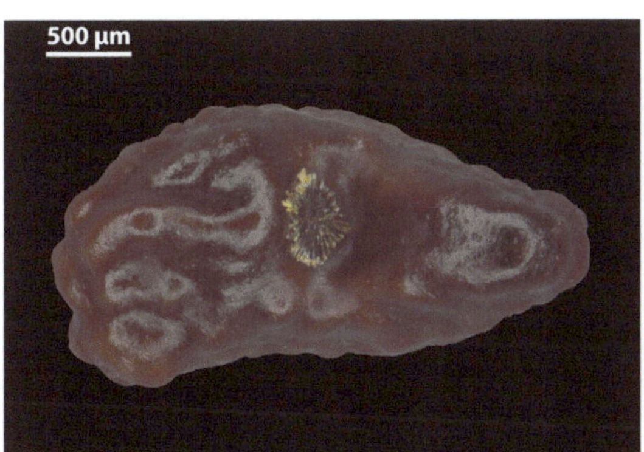

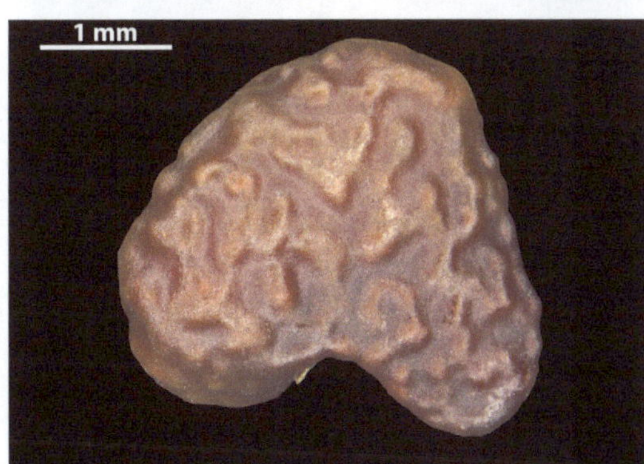

Androcalva luteiflora

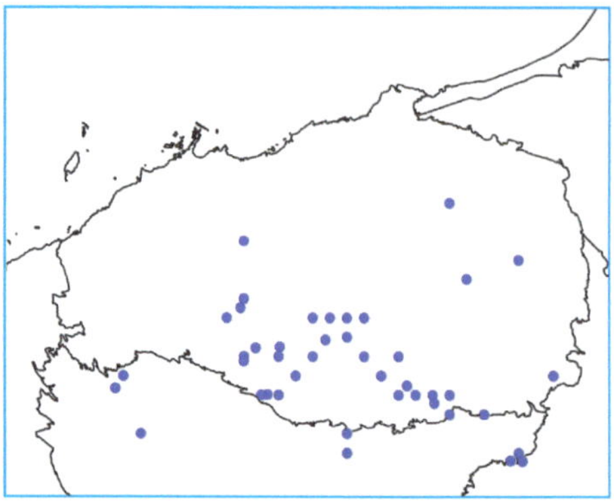

Common names: Yellow-flowered Rulingia.

Indigenous names: None recorded.

Description: Erect and spreading shrub, 1–4 m high, suckering from rhizomes. Leaves with petioles 0.5–6.1 mm long; blade undulate or flat, margin scarcely to moderately recurved, base cordate, obtuse or attenuate, margin irregularly serrulate or entire, apex obtuse; blade ovate, narrowly ovate, cuneate or oblong, 1.2–17 mm long, 1.5–11.5 mm wide. Inflorescence 7–23 mm long, 3–18-flowered. Calyx yellow with base green; 3.3–4.3 mm long. Petal base and ligule yellow, with 2 basal red ribs; 3.4–4.5 mm long, 1.5–3 mm wide. Anthers dark red. Fruit globose; with inconspicuous wings present, 1–3 mm wide; fruit 7–21 mm wide; outer surface of fruit with dense, white, stellate hairs beneath dense setae. Seed obovoid, exotesta dark brown, or blackish-brown with moderately raised tubercles, seeds 1.4–3.3 mm long, 0.9–1.9 mm wide with a white aril.

Key features for identification: Erect shrub; stellate hairy leaves; yellow flowers with red marks at base of petals; densely hairy fruit.

Similar species: *A. loxophylla* is a low-growing, spreading shrub to 0.5 m.

Distribution: Common in the southern Pilbara, south to the Bremer Range and east to the Northern Territory.

Habitat: Grows in a wide variety of soils including sandstone, ironstone, granite, sand, sandy loam or laterite, frequently along creek margins or on rock outcrops.

Flowering period: July to October.

Androcalva luteiflora

Fruit/Seed collection

Fruit/Seed maturity: October to November.

Appearance of mature fruit/seed: The semi-spherical to ovoid capsules are covered in dense hairs that turn from green to light brown when mature. Whole capsules containing 10–15 seeds can be collected as soon as the capsule begins to split.

Cleaning requirements

Cleaning techniques: Seeds can be separated from the capsules by shaking through a suitably sized sieve without the need to break the capsule apart. Use a rubber bung if the capsules do not release the seed. Capsule debris can be removed using a vacuum separator.

Purity and viability: Viability high (>95%).

Seeds per gram: 212 (range 182–212).

Germination requirements

Dormancy type: Physical.

Embryo type: Spatulate. **Embryo:seed ratio:** 0.96.

Pre-treatments: Wet heat for 2 mins at 90–95°C.

Germination: Germination is variable (55–85%), due to inconsistent response to heat treatments. Wet heat treatments appear to result in more consistent germination. Germination appears unimpeded across a broad temperature range (15–35°C) once effectively heat treated. Lower temperatures (10°C) reduce germination by 10–20%.

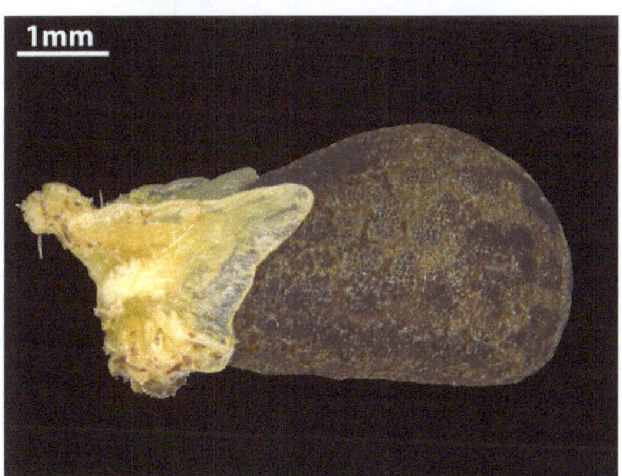

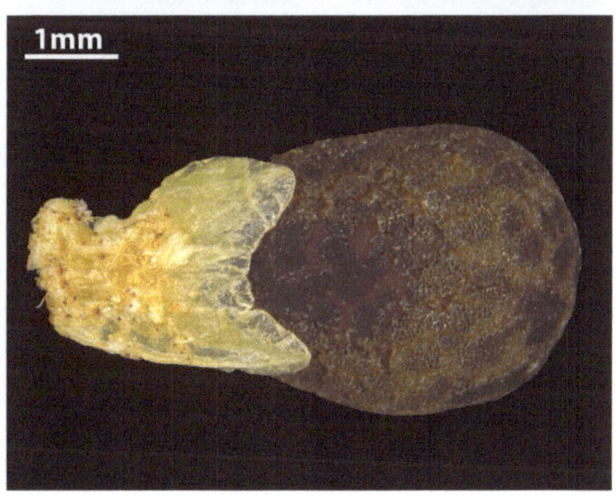

Corchorus crozophorifolius

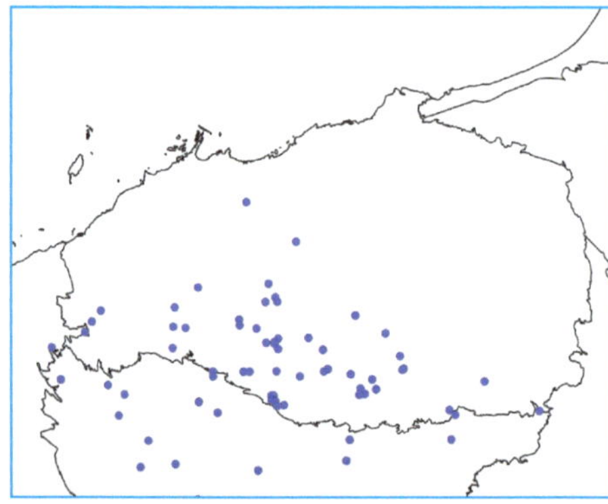

Common names: Felty Corchorus.

Indigenous names: None recorded.

Description: Shrub to 1 m high, spreading to erect. Indumentum ferruginous or rarely grey-white, very dense, floccose, comprised of mostly dendritic-stellate and stellate hairs but simple hairs also present. Stellate hairs sessile or stipitate, up to 1 mm across. Dendritic-stellate hairs up to 4 mm long. Simple hairs glandular, white, flexuous, up to 0.3 mm long. Leaves with petioles 10–25 mm long; stipules subulate-linear, 4–20 mm long; lamina ovate, 40–90 mm long, 20–60 mm wide, concolorous; margin dentate-serrate; apex acute to rounded. Inflorescence umbellate, 4–15-flowered; peduncles 5–20 mm long; pedicels 2–12 mm long. Sepals 5, persistent, narrowly obovate-elliptic, 10–18 mm long, 2–4 mm wide; abaxial surface with a dense indumentum of dendritic-stellate and stellate hairs. Petals 5; lamina obovate, 8–11 mm long, 3–6 mm wide, glabrous; claw 1–1.5 mm long, stellate-villose on abaxial surface and margins. Stamens 50–100; filaments 3–8 mm long; anthers 0.8 mm long. Ovary ovoid to cylindrical; style 5 mm long. Fruits ovoid, 12–17 mm long, 10–12 mm diam., circular in transverse section; indumentum of dendritic-stellate and stellate hairs, the largest hairs 4 mm long. Seeds compressed obovoid, 2 mm long.

Key features for identification: Floccose indumentum on the stems, petioles and inflorescences.

Similar species: Similar to *C. incanus* but easily distinguished by having a densely hairy annulus.

Distribution: From the central Pilbara south to Carnarvon and Wiluna.

Habitat: In *Acacia* shrubland and woodland communities, on sandy, limestone or granite soils on alluvial flats, along watercourses, on rocky rises and hills.

Flowering period: May to November.

Corchorus crozophorifolius

Fruit/Seed collection

Fruit/Seed maturity: November to December.

Appearance of mature fruit/seed: Densely covered in hairs, the elongated, ovoid capsules split into 4–5 segments at maturity. Optimum time to collect capsules is when the apex of the capsules first split. Each capsule contains many seeds (*c.* 10–30).

Cleaning requirements

Cleaning techniques: Capsules need to be manually opened or crushed over a sieve using a rubber bung to separate the seeds from the capsule. Vacuum separation assists in removing the lighter hairs and heavier capsule debris.

Purity and viability: The contribution of the capsule and dense hairs results in low (<15%) purity. Capsules and seeds are prone to insect predation. Once cleaned to pure seeds, viability is generally high (>90%). However, occasionally collections of this species have many seeds with no embryo/endosperm present.

Seeds per gram: 920.

Germination requirements

Dormancy type: Physical.

Embryo type: Spatulate (inferred from other *Corchorus* species). **Embryo:seed ratio:** Not determined.

Pre-treatments: Wet heat for 10 secs to 1 min at 70–90°C. Seed viability loss occurs at higher treatment temperatures and longer exposure durations. Manual scarification (i.e. nicking) is an alternative method to breaking physical dormancy.

Germination: High germination (*c.* 80%) achieved at 25°C after treating for 10–30 secs in wet heat at 90–95°C.

Corchorus aff. *lasiocarpus*

Common names: Newman Corchorus.

Indigenous names: None recorded.

Description: Small shrub to 60 cm. Stellate hairy. Leaves ovate, pale green. Flowers yellow, petals large. Fruit densely stellate hairy. Seeds brown.

Key features for identification: The petals are much larger than related species.

Similar species: Similar to *C. lasiocarpus* but with larger flowers.

Distribution: Only known from near Newman.

Habitat: On a rocky hill after fire.

Flowering period: May to August.

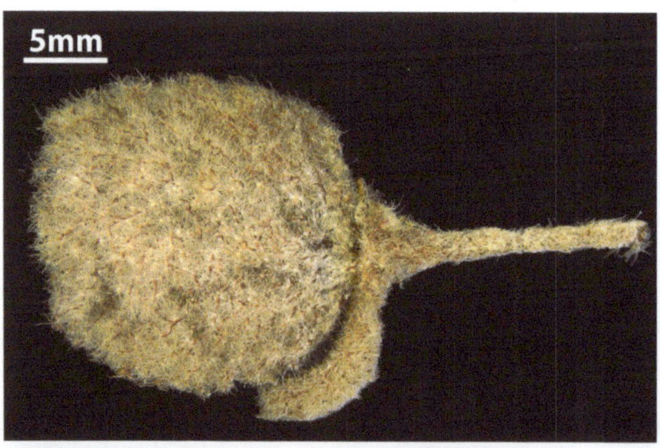

Corchorus aff. *lasiocarpus*

Fruit/Seed collection

Fruit/Seed maturity: September.

Appearance of mature fruit/seed: Densely covered in hairs, the cylindrical to ovoid capsules split into 4–5 segments at maturity. Optimum time to collect capsules is when the apex of the capsules first split. Each capsule contains many seeds (*c.* 10–30).

Cleaning requirements

Cleaning techniques: Capsules need to be manually opened or crushed over a sieve using a rubber bung to separate the seeds from the capsule. Vacuum separation assists in removing the lighter hairs and heavier capsule debris.

Purity and viability: The contribution of the capsule and dense hairs results in low (<15%) purity. Capsules and seeds are prone to insect predation. Once cleaned to pure seeds, viability is generally high (>90%).

Seeds per gram: 1422.

Germination requirements

Dormancy type: Physical.

Embryo type: Spatulate. **Embryo:seed ratio:** 1.22.

Pre-treatments: Wet heat for 1–2 min at 70–90°C. Seed viability loss occurs at higher treatment temperatures and longer exposure durations.

Germination: >85%. Unimpeded across a wide temperature range (20–35°C) once treated to alleviate physical dormancy. Germination is decreased at lower incubation temperatures (10–15°C).

Gossypium australe

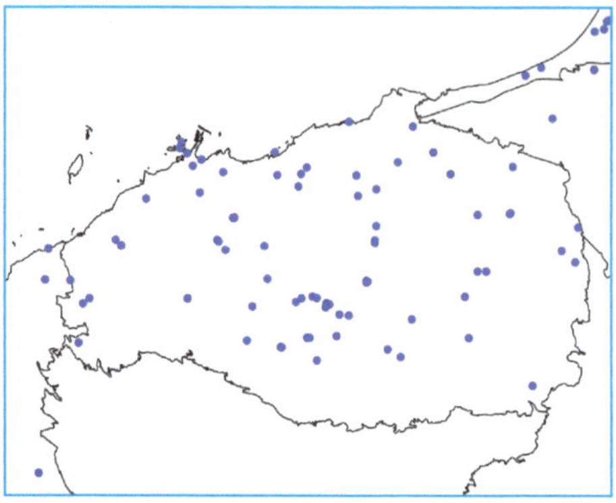

Common names: Native cotton, Australian desert rose.

Indigenous names: None recorded.

Description: Shrub to 3 m high, with a felted indumentum of short very dense stellate hairs. Stipules subulate, 3–7 mm long, caducous. Leaves: petiole 10–30 mm long; blade elliptic to obovate or broadly ovate, 32–100 mm long, 15–85 mm wide, usually unlobed, densely stellate-hairy above and below, nectary gland red and conspicuous, base obtuse, apex obtuse to truncate. Flowers solitary or paired, often on short axillary shoots; pedicel stout, articulate towards base. Epicalyx bracteoles linear to subulate, 5–15 mm long, stellate-hairy; nectary glands conspicuous. Calyx expanded distally, 10–16 mm long, somewhat ribbed towards base, with prominent oil glands distally; lobes broadly triangular at the base but acuminate and linear in upper part, 5–11 mm long, 1–2 mm wide, with only 1 prominent vein. Petals mauve to pink and dark red at the base, 25–40 mm long. Capsule ovoid-ellipsoid to obovoid-ellipsoid, 10–20 mm long, 8–12 mm diam., coriaceous, 5-celled, densely and minutely stellate-hairy, acuminate. Seeds with more or less straight dense spreading hairs 5–6 mm long.

Key features for identification: Dull-green, entire leaves; long, slender epicalyx lobes.

Similar species: *G. robinsonii* has lobed leaves.

Distribution: From Wyndham south to Carnarvon, inland to the Great Sandy Desert. Also occurs in N.T. and Qld.

Habitat: Occurs in savannah woodland, pindan vegetation, shrubland or grassland, often on sandy soils.

Flowering period: April to October.

Gossypium australe

Fruit/Seed collection

Fruit/Seed maturity: September to November.

Appearance of mature fruit/seed: The whole fruiting body (a capsule) turns from dark green to brown. At maturity, the individual capsule compartments begin to separate and split along a central line releasing individual seeds. The capsule remains intact. If collected at the right time, seeds will be bursting from the capsule due to the dense cotton-like hairs on the seed coat.

Cleaning requirements

Cleaning techniques: If the capsule compartments remain intact, seeds can be easily separated using a suitably sized sieve. If the capsule compartments are broken down, using vacuum separation will assist in removing capsule debris.

Purity and viability: Viability is generally high (>95%), but seeds can be prone to insect predation.

Seeds per gram: 80 (range 63–85).

Germination requirements

Dormancy type: Physical (or potentially combinational [physical + physiological]).

Embryo type: Folded. **Embryo:seed ratio:** Not determined.

Pre-treatments: To date, seeds of this species have only been tested with wet heat ranging between 40°C and 100°C for 1–5 mins and appear to be unresponsive to these treatments. Alternatives to alleviate physical dormancy include immersion in 98% sulphuric acid or mechanically scarifying (i.e. nicking) the seed coat.

Germination: Maximum seed germination achieved on water at 25°C after wet heat treatment at 95°C is 30–40%.

Notes: A large fraction of the ungerminated seeds comprises seeds that remain physically dormant (i.e. heat treatment did not alleviate physical dormancy), seeds that imbibed water but potentially were still physiologically dormant (i.e. have combinational dormancy), or seeds that lost viability after the heat treatments were applied. Further optimisation of heat treatments are required for many species from the Malvaceae.

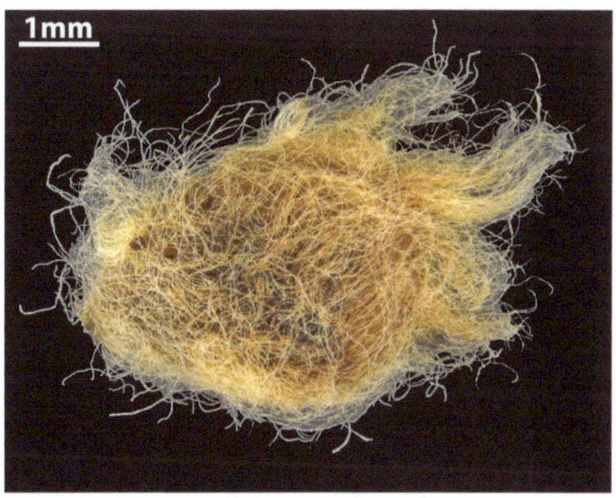

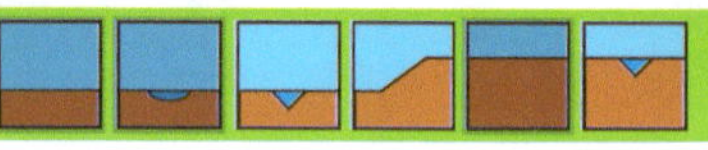

Gossypium robinsonii

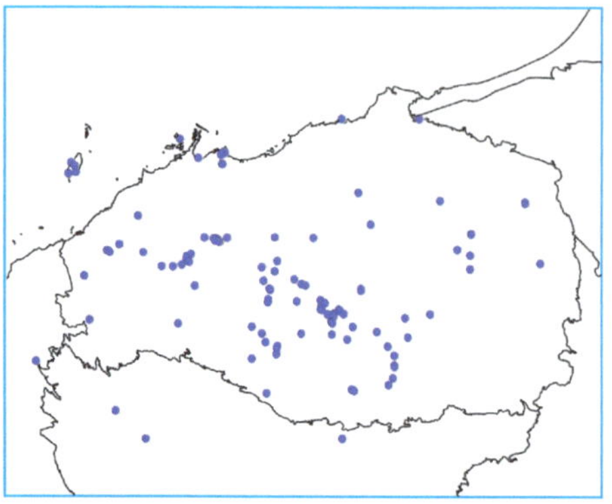

Common names: Wild cotton.

Indigenous names: Wathawayi (Yindjibarndi & Ngarluma).

Description: Upright shrub to 3 m with open foliage, glandular-punctate. Leaves palmate, 3–5-lobed; lobes acuminate, entire, 50–110 mm long, 60–180 mm wide; petioles 40–70 mm long. Foliar nectaries 3–5, median to subapical, 1–2 mm long, 0.5–1.0 mm diam., red. Bracts narrowly triangular to cuneate or ovate, entire. Flowers solitary or paired, often on short axillary shoots; pedicel stout, articulate towards base. Epicalyx bracteoles ovate to narrowly triangular, 15–25 mm long, glabrous. Calyx lobes acuminate, 2–5 mm long. Petals mauve to pink and dark red at the base, 40–70 mm long. Capsule globose, to 20 mm long, glabrous, 4- or 5-celled. Seeds 4–6 mm long; seed hairs whitish. Cotyledons slightly obcordate.

Key features for identification: Leaves deeply 3- to 5-lobed, lobes acuminate; foliar nectaries 3–5, prominent, red, near apices of leaf lobes.

Similar species: *G. australe* has entire leaves.

Distribution: Almost restricted to the Pilbara region where it is widespread.

Habitat: Usually on rocky soils, often along drainage lines.

Flowering period: July to November.

Photographs by Murray Fagg.
Australian National Botanic Garden Collection.

Gossypium robinsonii

Fruit/Seed collection

Fruit/Seed maturity: October to November.

Appearance of mature fruit/seed: The whole fruiting body (a capsule) turns from dark green to brown. At maturity, the individual capsule compartments begin to separate and split along a central line releasing individual seeds. The capsule remains intact.

Cleaning requirements

Cleaning techniques: If the capsule compartments remain intact, seeds can be easily separated using a suitably sized sieve. If the capsule compartments are broken down, using vacuum separation will assist in removing capsule debris.

Purity and viability: Viability is generally high (>95%), but seeds can be prone to insect predation.

Seeds per gram: 84 (range 71–84).

Germination requirements

Dormancy type: Physical (or potentially combinational [physical + physiological]).

Embryo type: Folded. **Embryo:seed ratio:** Not determined.

Pre-treatments: Wet heat for 1–2 mins at 80–95°C. Higher treatment temperatures and longer exposure durations result in loss of seed viability.

Germination: 65–75% once heat treated. Germination has only been tested at 25°C.

Notes: A small fraction of ungerminated seeds remains after heat treatment. This is potentially due to the heat treatment not alleviating physical dormancy in all seeds, or it may indicate that there is some level of physiological dormancy (i.e. seeds have combinational dormancy).

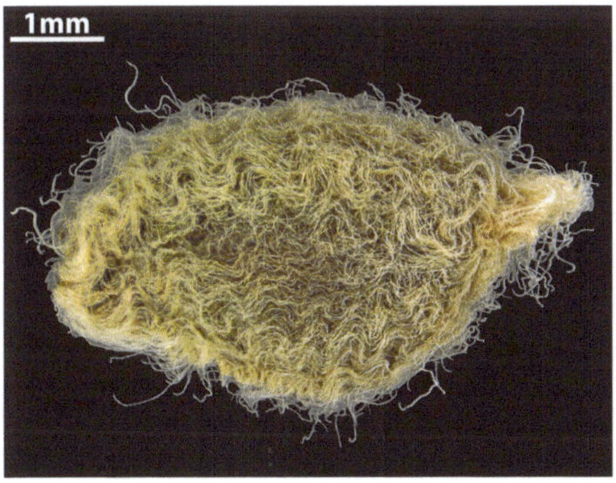

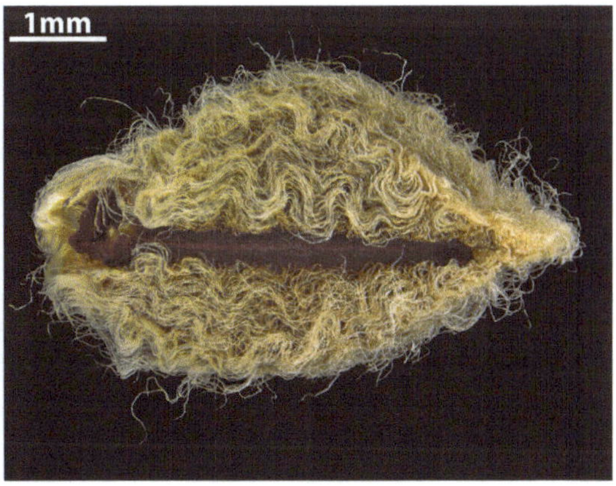

Hibiscus haynaldii

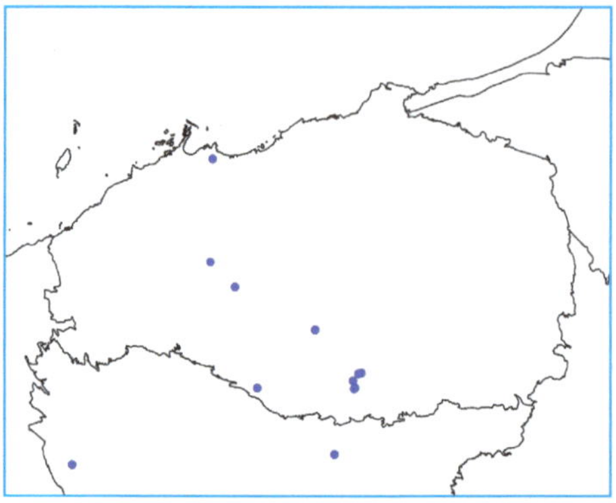

Common names: Haynald's hibiscus.

Indigenous names: None recorded.

Description: Slender, erect annual, biennial or perennial, 0.8–2.7 m high, densely stellate hairy. Leaves 50–80 mm long, 25–50 mm wide, cordate, entire, lanceolate to ovate, margins slightly serrate, densely rusty-brown stellate hairy; stipules linear-filiform, short. Epicalyx lobes 6–8, ovate in bud, elongating to oblong-linear in fruit, 1/3–1/2 the calyx length; calyx lobes 20–25 mm long, thin, acuminate. Flowers pale blue-purple, centre deep red-purple; staminal column glabrous; filaments longer than the anthers; styles free towards the summit; stigmas capitate; ovary glabrous, with 5–6 ovules in each cell. Fruit a brown, dehiscent capsule. Seeds kidney-shaped, covered in dense, golden-brown hairs.

Key features for identification: The combination of short, broad epicalyx lobes and long, narrow calyx lobes is distinctive.

Similar species: There are a number of similar species in the area. *H. leptothecus* is a variable species which usually has much narrower leaves and relatively shorter sepals. *H. sturtii* differs in having very short epicalyx lobes. *H. coatesii* differs in having longer epicalyx lobes and much longer stellate hairs on the calyx. *H. gardneri* has a very sparse indumentum, with dark green leaves.

Distribution: Restricted to the southern Pilbara and adjacent regions, uncommon and poorly known.

Habitat: Red or pebbly sand, loam. Creeklines, rocky crevices and gullies on the side of mesas.

Flowering period: Sporadic, from May to August.

Hibiscus baynaldii

Fruit/Seed collection

Fruit/Seed maturity: July (observations limited).

Appearance of mature fruit/seed: The whole fruiting body (a capsule) turns from green to yellow/light brown. At maturity, the individual capsule compartments begin to separate and split along a central line releasing individual seeds. The capsule remains intact.

Cleaning requirements

Cleaning techniques: If the capsule compartments remain intact, seeds can be easily separated using a suitably sized sieve. If the capsule compartments are broken down, using vacuum separation will assist in removing capsule debris.

Purity and viability: Viability is generally high (>95%), but seeds can be prone to insect predation.

Seeds per gram: 153.

Germination requirements

Dormancy type: Combinational (physical + physiological).

Embryo type: Folded. **Embryo:seed ratio:** 1.66.

Pre-treatments: Wet heat for 2 mins at 70–90°C. Higher treatment temperatures and longer exposure durations result in loss of seed viability. Alternatives to alleviate physical dormancy include immersion in 98% sulphuric acid for 8 hours. Additional treatment with GA_3 after heat treating promotes higher germination.

Germination: Maximum seed germination achieved on water after alleviation of physical dormancy is 40–45% at 20–30°C. Treatment with GA_3 increases germination to 60–85% at 15–35°C.

Notes: Identification of combinational dormancy in this species means additional dormancy break pre-treatments such as dry after-ripening or wet/dry cycling may be required to overcome the physiological component of dormancy, avoiding the need for the additional GA_3 treatment[9].

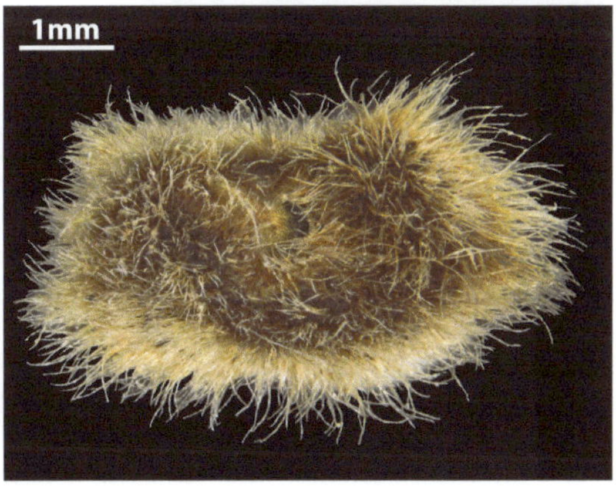

Sida echinocarpa

Common names: Hook-seeded sida.

Indigenous names: None recorded.

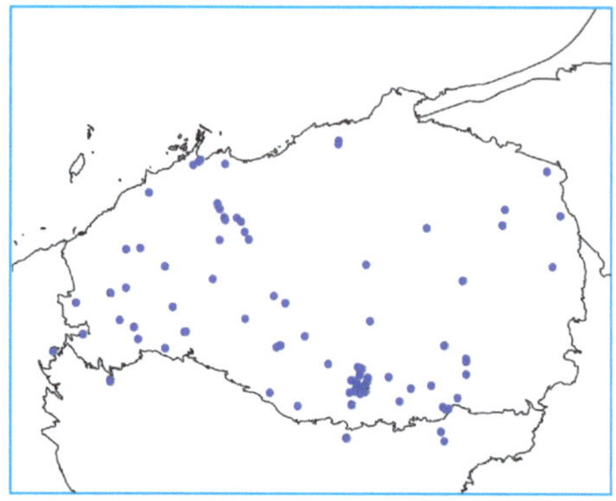

Description: Decumbent or upright, perennial subshrub, 0.3–1.3 m tall, 0.3–1 m wide; indumentum dense, greenish-white, a mixture of sessile and stalked stellate hairs. Leaves alternate, petiolate, blade ovate, 20–65 mm long, 5–30 mm wide, base truncate, margins crenulate, apex rounded or obtuse; green to rust green, indumentum dense; petiole 5–40 mm long. Stipules narrowly linear, 7.5–10 mm long, 0.2 mm wide. Inflorescence a terminal raceme. Pedicel 10–25 mm, densely stellate hairy. Calyx cup-shaped in open flower, not membranous, persistent, enlarging and far exceeding not enclosing fruit, lobes 5–7 mm long, 3–3.5 mm wide in flower, 10–14 mm long in fruit, ovate, united in lower half, faintly ribbed; densely stellate-hairy. Corolla yellow, exceeding or similar length to calyx, 16–20 mm diam.; petals obovate, 8–10 mm long, 6–7 mm wide, ciliate on claw, non-ciliate above. Stamens 14–20. Styles 6–11, longer than stamens; stigma discoid. Schizocarp 6.5–7.5 mm diam., transversely elliptic, glabrous or sparsely puberulent, echinate. Mericarps 10–13, 1.5–2 mm high, shallowly to deeply grooved apically. Seed 1.8–2 mm high, glabrous, surface smooth, red-brown. Funicle obvious.

Key features for identification: Broad sepals; mericarps 10–13 per fruit, spined.

Similar species: *S. arsiniata* has shorter calyx lobes.

Distribution: Widespread in the Pilbara. Also in N.T.

Habitat: Red sand, stony loam, ironstone hills and flats.

Flowering period: July to October.

Sida echinocarpa

Fruit/Seed collection

Fruit/Seed maturity: September to November.

Appearance of mature fruit/seed: The whole fruiting unit is arranged in a doughnut-shaped ring of numerous individual spiny fruits (10–15 segments) surrounded by a persistent calyx. When mature, the drooping fruiting unit turns from green to pale yellow and the individual fruits turn brown and fall easily from the plant.

Cleaning requirements

Cleaning techniques: Collections typically contain both the persistent calyx and the fruiting unit. These can either be separated using a suitably sized sieve or by using vacuum separation. To clean to pure seeds, the outer spiny fruit wall can be removed using a rubber bung and sieve and the fruit debris can be removed using vacuum separation.

Purity and viability: Viability high (>95%).

Seeds per gram: Unprocessed fruits = 167. Seeds = 493.

Germination requirements

Dormancy type: Combinational (physical + physiological).

Embryo type: Folded. **Embryo:seed ratio:** 1.90.

Pre-treatments: The effectiveness of heat treatments to overcome physical dormancy is variable. Wet heat for 2 mins at 80–90°C, and dry heat for 5–30 mins at 80–90°C, have rendered 40–50% of seeds permeable to water. By immersing seeds in 98% sulphuric acid for 1 hour permeability increases to 100%.

Germination: Once treated to remove physical dormancy, germination of fresh seeds on water is variable (*c.* 10–30%) after incubation at 25°C. Maximum germination of stored seeds (2–3 months) on water after heat treatment is 80–100% across a wide temperature range (10–35°C). Exposure to GA_3 or smoke does not increase germination of stored seeds.

Notes: Treatments to overcome the physiological dormancy component of dormancy have not been fully investigated. However, dry after-ripening is likely to assist in promoting germination once seeds are rendered permeable to water[9].

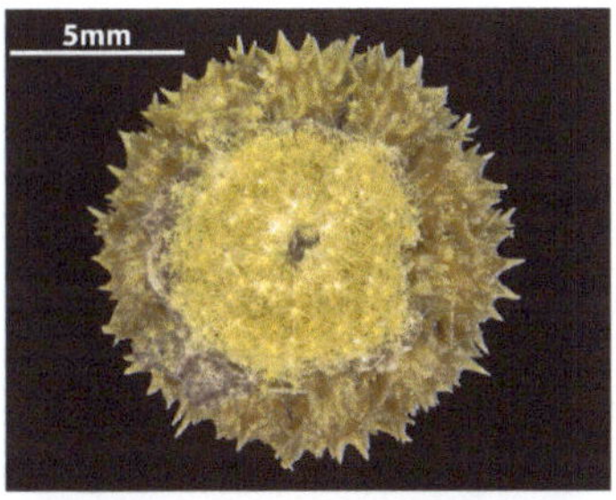

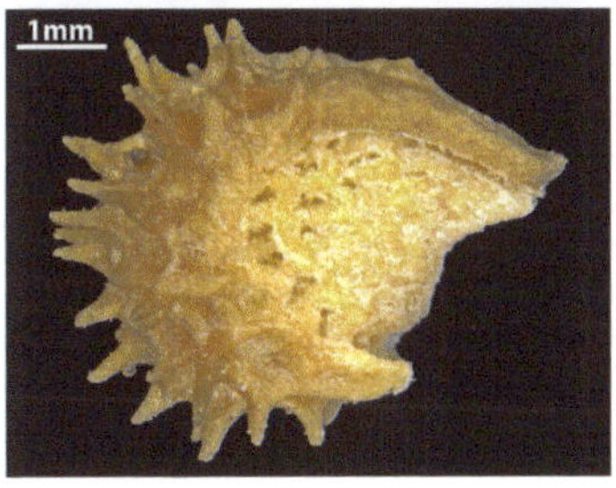

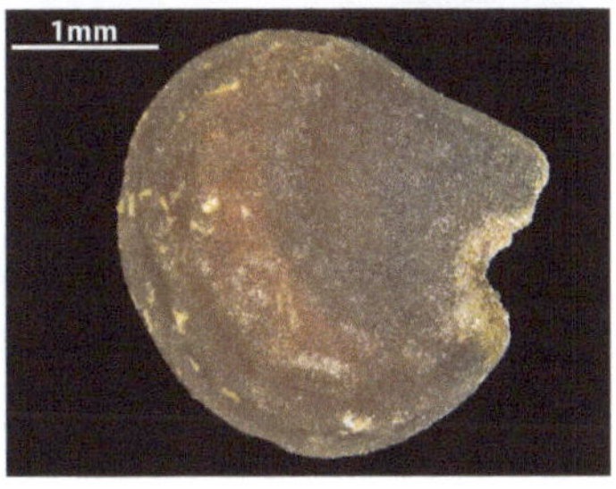

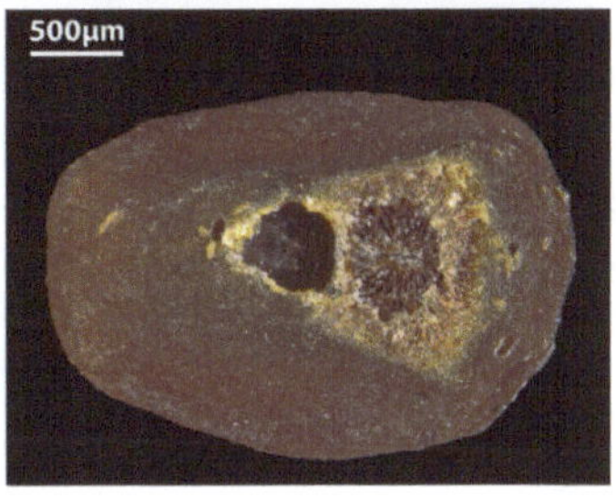

Eucalyptus gamophylla

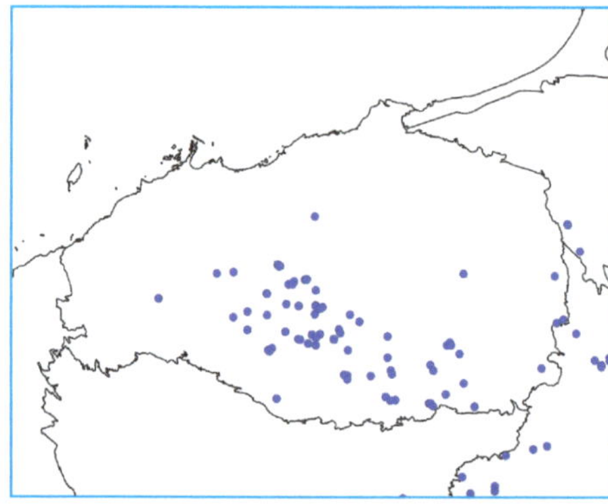

Common names: Twin-leaf mallee, Blue mallee.

Indigenous names: None recorded.

Description: A small mallee, occasionally almost prostrate, 1.5–7 m high, bark smooth or rough, white or grey to brown. Leaves and stems pruinose, bluish green. Adult leaves opposite or subopposite, lanceolate, 60–80 mm long, 10–15 mm wide, glaucous, thick; petioles 2–5 mm long; veins distinct. Mostly juvenile or intermediate leaves are seen on mature plants, opposite-decussate, each pair is fused at the base around the stem, 30–80 mm long, 20–50 mm wide. Conflorescence terminal or axillary panicles; buds 3–7, clavate or obpyriform (club-shaped or pear-shaped, respectively), mostly green, 4–6 mm long, 2–3 mm wide; operculum patelliform, shorter than hypanthium; peduncles 4–8 mm long; pedicels 1–4 mm long, occasionally sessile; flowers cream. Fruit tapering to base, sometimes pruinose, 6–10 mm long, 5–7 mm wide; 3 or 4-valved, broadly deltoid, inserted just below rim.

Key features for identification: Pruinose, opposite decussate leaves, fused at the base.

Similar species: It could possibly be confused with *Corymbia aspera*, but that species has hairy leaves that are not fused at the base and papery-walled fruit.

Distribution: Widespread in the southern Pilbara region. Also in N.T. and Qld.

Habitat: On red sand, sandplains and dunes or on stony hills, usually with *Triodia*.

Flowering period: October to March.

Eucalyptus gamophylla

Fruit/Seed collection

Fruit/Seed maturity: All months.

Appearance of mature fruit/seed: Seeds are canopy stored and are held within woody capsules. The semi-cylindrical to conical-shaped capsules turn from green to light brown (and/or pruinose) when mature and can be collected at any time of the year. Closed capsule valves that are dark brown or grey provide a good indication that seeds within the capsule are mature. Sectioning a capsule with secateurs to confirm the capsules are dry and 'woody' is recommended. Capsules appear to last 2–5 years in the canopy.

Cleaning requirements

Cleaning techniques: Within a few days of collection, capsules will dry out and the valves will open releasing the seeds and chaff. Given the large size of the seed in comparison to the chaff, seeds can be easily separated using a suitably sized sieve.

Purity and viability: Chaff contributes up to 80% of collected material, leading to low seed purity. Once the chaff is removed, viability of seed is generally high (>90%).

Seeds per gram: 376.

Germination requirements

Dormancy type: Non-dormant.

Embryo type: Folded. **Embryo:seed ratio:** 1.00.

Pre-treatments: No pre-treatment requirements.

Germination: 100%. Unimpeded across a wide temperature range (10–35°C).

Notes: A high proportion of chaff in the collected material can contribute to low seed purity values. It is recommended to take into consideration the ratio of seed:non-seed material when acquiring large quantities.

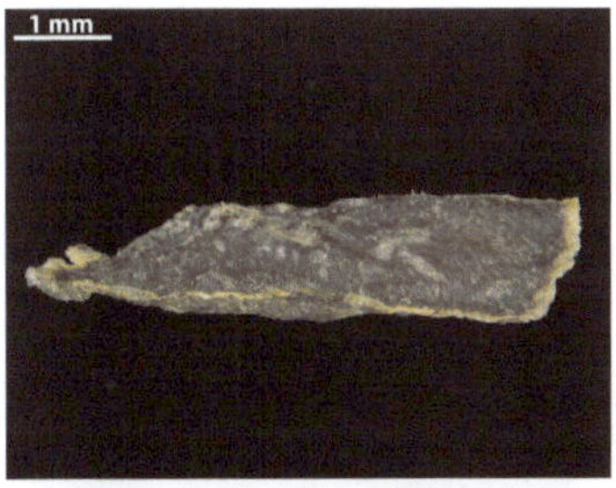

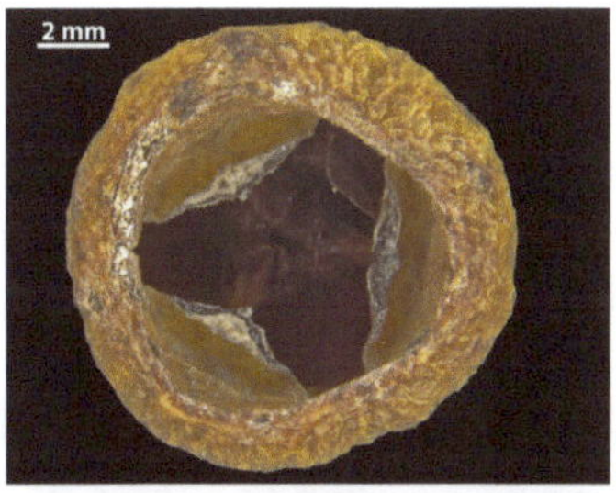

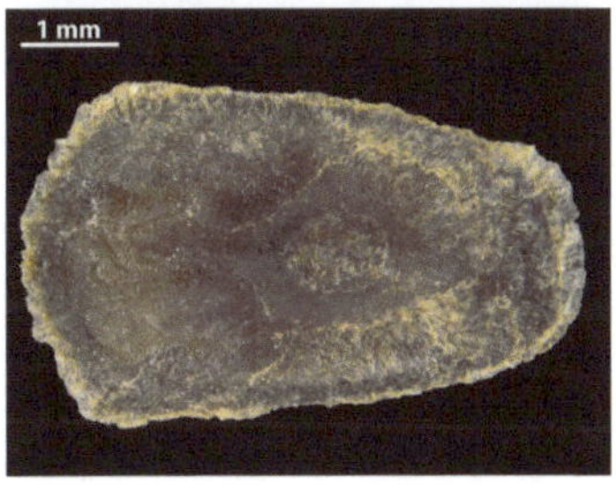

Eucalyptus leucophloia subsp. *leucophloia*

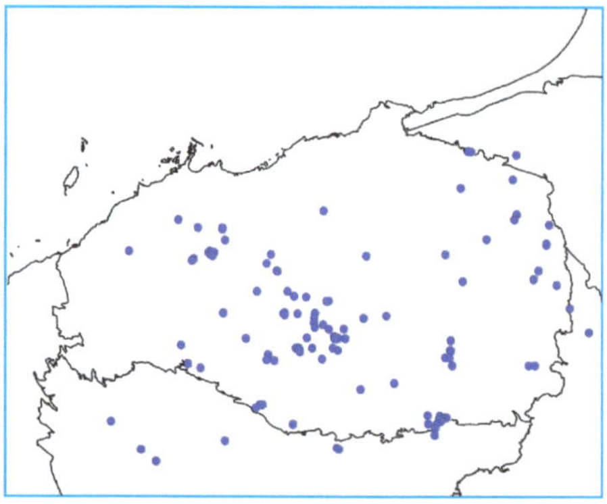

Common names: Snappy gum, Migum.

Indigenous names: Majgan (Yindjibarndi); Malygan (Ngarluma).

Description: Small tree or mallee, 2.5–10 m high, bark smooth, powdery, white, or pinkish when newly exposed. Juvenile leaves subopposite, ovate, subglaucous. Adult leaves alternate, lanceolate, subglaucous to pruinose; petiole 8–15 mm long. Umbels 7–11-flowered; peduncle slender, terete, 4–10 mm long; pedicels terete, 2–3 mm long. Buds ovoid, glaucous; operculum hemispherical, 2–3 mm long, 3–4 mm wide. Flowers white to cream. Fruit small, woody, hemispherical, 5–8 mm long, 4–6 mm wide; disc narrow, descending; valves 3, exserted.

Key features for identification: White powdery bark throughout; alternate, pruinose leaves; small pruinose fruit with exserted valves.

Similar species: Unlikely to be confused with other species in the Pilbara region.

Distribution: Widespread in the Pilbara region, but not extending far beyond the region.

Habitat: Skeletal soils over sandstone or quartz on stony hills, in gullies or on plains.

Flowering period: March to August.

Eucalyptus leucophloia subsp. *leucophloia*

Fruit/Seed collection

Fruit/Seed maturity: All months (but see notes section).

Appearance of mature fruit/seed: Seeds are canopy stored and are held within woody capsules. The cup-shaped, spherical capsules turn from lime green to light brown (and/or pruinose) when mature and can be collected at any time of the year. Closed capsule valves that are dark brown provide a good indication that seeds within the capsule are mature. Sectioning a capsule with secateurs to confirm the capsules are dry and 'woody' is recommended. Retention of capsules appears to be short (<1 year).

Cleaning requirements

Cleaning techniques: Within a few days of collection, capsules will dry out and the valves will open releasing the seeds and chaff. Seeds can be separated from the majority of chaff using a suitably sized sieve. Final vacuum separation will aid in removing the chaff debris.

Purity and viability: Chaff contributes up to 80% of collected material, leading to low seed purity. Once the chaff is removed, viability of seed is generally high (>90%).

Seeds per gram: 2100.

Germination requirements

Dormancy type: Non-dormant.

Embryo type: Folded. **Embryo:seed ratio:** 1.00.

Pre-treatments: No pre-treatment requirements.

Germination: 100%. Unimpeded across a wide temperature range (10–35°C).

Notes: A high proportion of chaff in the collected material can contribute to low seed purity. It is recommended to take into consideration the ratio of seed:non-seed material when acquiring large quantities.

This species appears to be moderately serotinous whereby seed retention in woody capsules is shorter than typical *Eucalyptus* species (<1–2 years) and seed production appears to mainly occur after above-average rainfall seasons. Therefore, seed production and capsule retention may be seasonally dependent.

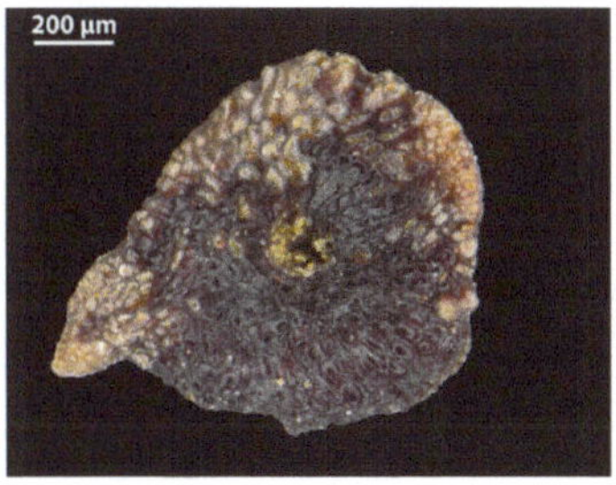

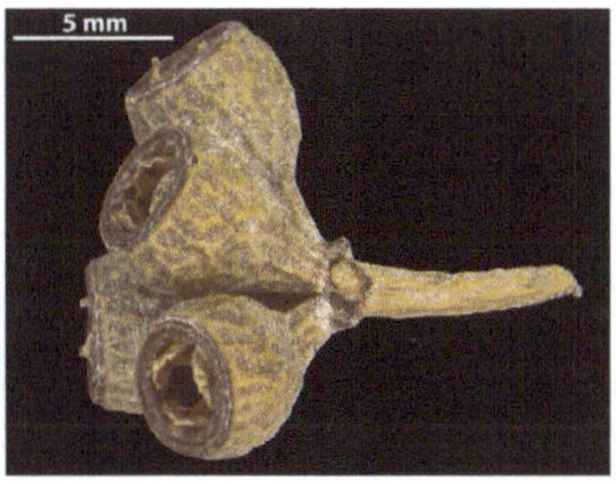

Melaleuca glomerata

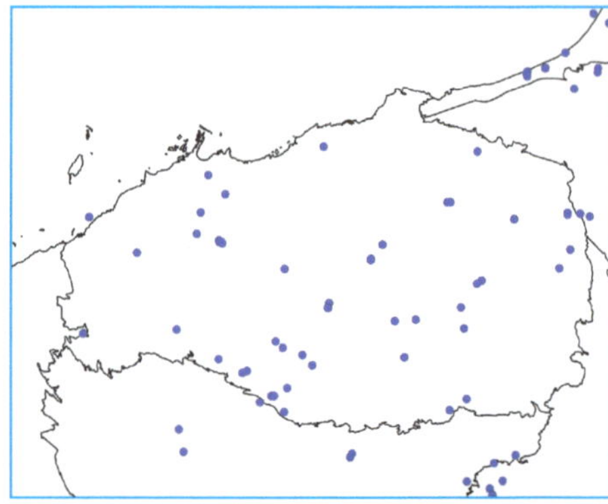

Common names: Desert paperbark, Inland paperbark.

Indigenous names: Gurliwirn (Yindjibarndi); Gurlibirn (Ngarluma).

Description: Shrub or tree, 1–5 m high, with a dense indumentum of medium to long appressed hairs on young stems and leaves; bark chartaceous. Leaves alternate, attenuate at base to a poorly defined petiole, glaucous, linear-obovate, 12–80 mm long, 1.5–6 mm wide, sometimes glabrescent, with only the midvein prominent, apiculate to acute. Inflorescence of globular heads in the upper leaf axils; axillary peduncles or short peduncle-like branchlets 3–7 mm long, with short spreading hairs; heads *c.* 10 mm in diameter, commonly 9–16-flowered, the axis growing out in fruiting stage. Flowers pale yellow or cream. Floral tube *c.* 1 mm long, densely hairy outside with short spreading hairs, glabrous inside. Sepals petal-like, 0.4–0.5 mm long, densely hairy outside with short spreading hairs inside, obtuse. Petals *c.* 1.5 mm long, ciliate, persistent in flower. Staminal bundles *c.* 4.5 mm long; 1 mm long; filaments *c.* 6, in 1 row. Capsule *c.* 1.5 mm long, 2–2.5 mm diam.

Key features for identification: Flowers and fruit in globular clusters on stems.

Similar species: *M. xerophila* has very small leaves. All other Pilbara species have flowers and fruits in elongate spikes.

Distribution: From the southern Kimberley to near Warburton Mission and the Gascoyne River. Also occurs in N.T. and S.A.

Habitat: On shallow depressions and sandy flats.

Flowering period: June to August.

Melaleuca glomerata

Fruit/Seed collection

Fruit/Seed maturity: All months (but see notes section).

Appearance of mature fruit/seed: Seeds are canopy stored and are held within woody capsules. Many individual capsules make up an agglomerated fruiting body that is spherical in shape. The capsules turn light brown when mature and can be collected at any time of the year (if present). Closed capsule valves that are dark brown provide a good indication that seeds within the capsule are mature. Sectioning a capsule with secateurs to confirm the capsules are dry and 'woody' is recommended. Retention of capsules appears to be short (<1 year).

Cleaning requirements

Cleaning techniques: Within a few days of collection, capsules will dry out and the valves will open releasing the seeds and chaff. Seeds can be separated from the majority of chaff using a suitably sized sieve. Final vacuum separation will aid in removing the chaff debris.

Purity and viability: Chaff contributes up to 75% of collected material, leading to low seed purity. Once the chaff is removed, viability of seed is generally high (>90%).

Seeds per gram: 11000.

Germination requirements

Dormancy type: Non-dormant.

Embryo type: Linear-straight. **Embryo:seed ratio:** 1.00.

Pre-treatments: No pre-treatment requirements..

Germination: >85%. Unimpeded across a wide temperature range (15–35°C). Germination is reduced to 20–65% at 10°C.

Notes: A high proportion of chaff in the collected material can contribute to low seed purity. It is recommended to take into consideration the ratio of seed:non-seed material when acquiring large quantities. This species appears to be moderately serotinous whereby seed retention in woody capsules is shorter than typical *Melaleuca* species (<1 year) and seed production appears to mainly occur after above-average rainfall seasons. Therefore, seed production and capsule retention may be seasonally dependent.

Boerhavia coccinea

Common names: Tar vine.

Indigenous names: None recorded.

Description: Annual or perennial herb, prostrate and spreading to 2 m in diameter or upright to 1 m high; stems usually densely glandular-hairy; hairs a mixture of type 1 (thin, *c.* 1 mm long with a small apical gland) and 2 (stout, *c.* 0.5 mm long with a large round apical gland), predominantly the latter; nodes not widely spaced. Lower leaves: petiole up to 50 mm long; blade narrowly ovate to broadly ovate or elliptic, largest up to 50 mm long, 48 mm wide, glandular-hairy, paler and prominently veined below, usually undulate, obtuse or acute; upper leaves sessile and bract-like. Inflorescence axillary and terminal, usually compound but not a repeatedly branched cyme, sometimes simple; basic unit a glomerule of 3–10(–14) flowers; peduncle stout, up to 30 mm long, glandular hairy; flowers pink; pedicels 0–2.5 mm long. Calyx: proximal part glandular hairy; distal part campanulate, 1.5–3.5 mm long, glandular hairy. Stamens 2 or 3; anthers included. Style usually exserted. Fruit fusiform, 3–4.5 mm long, 1–1.5 mm wide, glandular-hairy; ribs 5, narrow, obtuse; furrows narrow to wide, sometimes white.

Key features for identification: Mixed type 1 and type 2 hairs on stems; flowers in a tight, terminal cluster.

Similar species: *Boerhavia* species are often difficult to distinguish and further studies are required on how many species occur in the Pilbara.

Distribution: Widespread from the Kimberley Region, throughout the Pilbara, south to Merredin and east to Warburton Range. Extends to N.T., S.A., Qld and N.S.W. Extends to the Pacific Islands and Central America.

Habitat: Most common on sandy plains.

Flowering period: February to August.

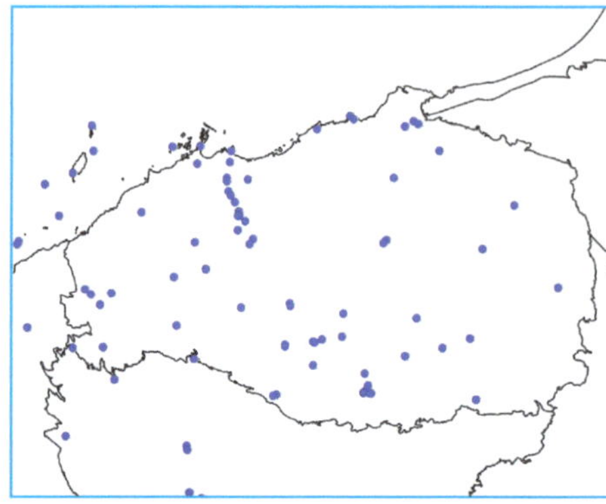

Boerhavia coccinea

Fruit/Seed collection

Fruit/Seed maturity: March to June.

Appearance of mature fruit/seed: The small teardrop-shaped fruits turn from light green, to reddish to dark brown when mature. Fruits fall from the plants easily when mature.

Cleaning requirements

Cleaning techniques: The single-seeded indehiscent fruit is typically collected pure and therefore requires minimal cleaning. Removal of the woody endocarp has not been attempted.

Purity and viability: Purity and viability are generally high (100% and >85%, respectively).

Seeds per gram: 367 (indehiscent fruits, range 367–438).

Germination requirements

Dormancy type: Physiological.

Embryo type: Peripheral. **Embryo:seed ratio:** 1.85.

Pre-treatments: Dormancy alleviation treatments unknown.

Germination: Maximum germination (of seeds within the woody fruit) with no pre-treatment is 15–20% at 20–25°C. Treatment with smoke increases germination to 40–45% at 20–25°C.

Notes: Additional treatments such as physical nicking of the fruit coat may assist in weakening the woody endocarp. Dry after-ripening and exposure to smoke would be an alternative option. When wet, exudates from the glandular hairs on the fruit mix with water to form a thick, gelatinous coating around the fruit which presumably assists with germination through water retention. This species, and others in the genus, also produces occasional geocarpic fruits as a 'bet-hedging' strategy against destruction or removal of fruit remaining on the surface.

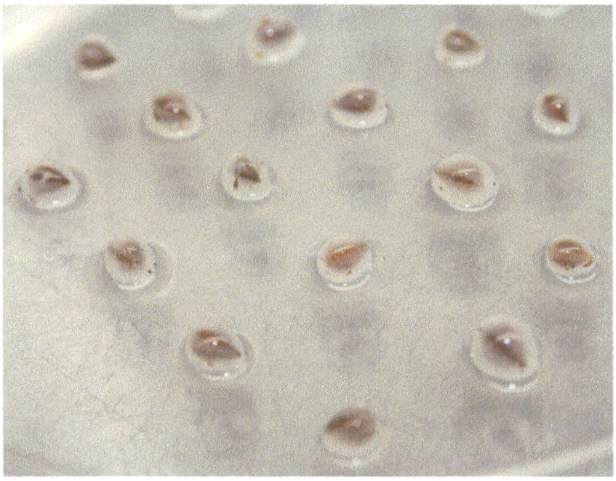

Aristida contorta

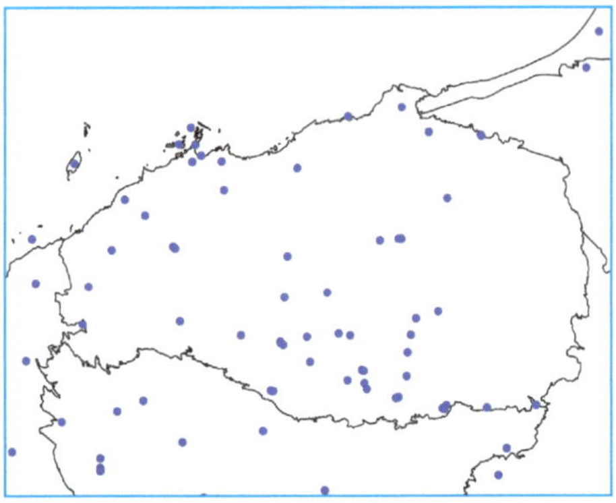

Common names: Bunched kerosene grass, Windgrass.

Indigenous names: Jilyilygura (Thalanyji); Bawuny (Yindjibarndi); Thambirrgura (Ngarluma).

Description: Annual or short-lived perennial, compactly tufted, 0.1–0.4 m high. Leaves: sheath shorter than internode; ligule 0.3–0.5 mm long; blade 30–100 mm long, 1 mm wide, glabrous or hairy, flexuous at maturity. Inflorescence 90–260 mm long, 40–90 mm diam., contracted or interrupted; branches to 25 mm long, naked at base. Glumes normal, not inverse; lower glume 7.4–14.2 mm long, 1-veined, glabrous, entire, awnless; upper glume 12–30 mm long, 1-veined, glabrous, awnless. Lemma brownish when mature, 4.9–7 mm long, shorter than glumes, convolute, muriculate towards apex, with a distinct articulation; column distinct, 7–26 mm long, scabridulous; awns equal or subequal, erect: median 18–95 mm long; laterals 16–92 mm long. Caryopsis 3–4.5 mm long, terete, less than twice hilum length.

Key features for identification: Usually annual; lemma 4.9–7 mm long, shorter than glumes; column 7–26 mm long.

Similar species: *A. holathera* has straight (not wavy) culms, longer lemmas (>7 mm long) and mature plants become purplish.

Distribution: Common and widespread in most of W.A. including the Pilbara region and occurring in all mainland states.

Habitat: Grows on red, white or yellow sand, red or brown clay or loam.

Flowering period: February to June.

Aristida contorta

Fruit/Seed collection

Fruit/Seed maturity: March to April.

Appearance of mature fruit/seed: When immature, the inflorescence is tightly packed along the main stem and green-brown in colour. During maturation, the inflorescence head opens up to display the three prominent awns of the individual florets. From a distance, the inflorescence appears golden-red/brown at maturity. Mature florets will detach easily from the plant when collected by hand.

Cleaning requirements

Cleaning techniques: The splayed nature of the three awns on each individual floret makes bulk collections hard to handle. Successful removal of the awns has been achieved through rubbing the florets between a thin, ribbed rubber mat or sieve. Care must be taken not to damage seeds contained within the florets. The elongated nature of the floret unit makes it prone to fracturing during cleaning. Isolating seeds from the floret has never been achieved.

Purity and viability: High (>90%).

Seeds per gram: Filled florets = 707 (unable to clean to pure seeds).

Germination requirements

Dormancy type: Physiological.

Embryo type: Lateral. **Embryo:seed ratio:** 0.57.

Pre-treatments: Dormancy alleviation treatments unknown.

Germination: Maximum germination of seeds (within florets) with no pre-treatment is 30–40% at 25–35°C. Treatment with GA_3 increases germination to 60% at 25°C.

Notes: Additional dormancy break pre-treatments such as dry after-ripening and wet/dry cycling may assist in promoting greater germination.

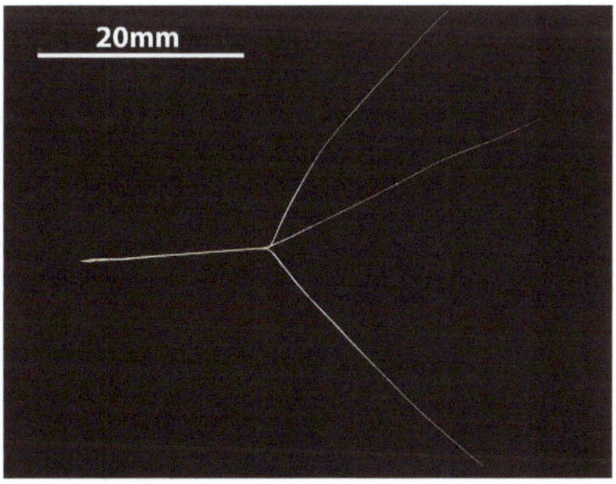

Aristida inaequiglumis

Common names: Unequal threeawn, Feathertop threeawn, Feathertop wiregrass.

Indigenous names: None recorded.

Description: Perennial, loosely tufted, 0.3–1.5 m high. Leaves: sheath longer than internode; ligule *c.* 1 mm long; blade to 300 mm long, 2.3–4.5 mm wide, curling at maturity. Inflorescence 130–400 mm long, 15–40 mm wide, contracted or interrupted; branches to 80 mm long, bearing spikelets from base. Glumes inverse; lower glume 9–18.5 mm long, 1(–3)-veined, glabrous, notched or emarginate, aristulate; upper glume 6.4–13 mm long, 1-veined, glabrous, aristulate. Lemma 3.7–9 mm long, shorter than glumes, involute with a furrow on the adaxial surface, smooth or scabrous towards apex, with a pseudo-articulation, rarely disarticulating, margins of lemma furrow smooth or with 1-several rows of tubercles; awns subequal: median 14–46 mm long; laterals 11–40 mm long. Caryopsis 3.1–7 mm long, furrowed, about twice hilum length.

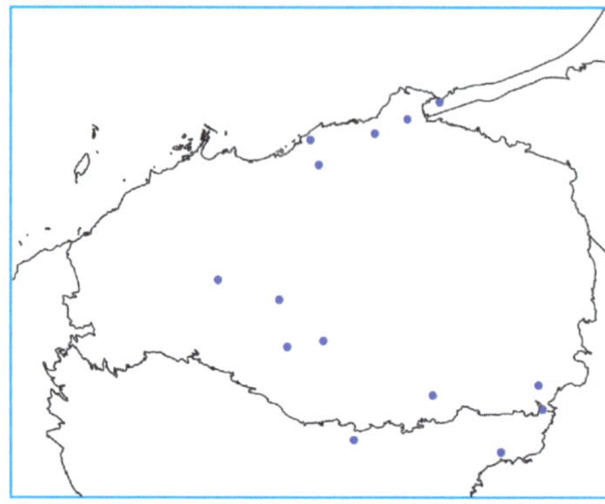

Key features for identification: Perennial; leaves with a distinct pseudoarticulation, curling at maturity; inverse glumes differing in length by >3 mm.

Similar species: Allied to *A. pruinosa* in having inverse glumes differing in length by >3 mm. It differs by being less robust and less pruinose; a distinct pseudoarticulation; and leaf blades which curl up at maturity. The glumes of *A. ingrata* differ by <3 mm.

Distribution: A fairly widespread species from arid to tropical areas in the Kimberley and Pilbara regions. Also recorded from arid regions of N.T., S.A. and Qld.

Habitat: Red, sandy loam, red earths, yellow sandy clay.

Flowering period: January to June.

Aristida inaequiglumis

Fruit/Seed collection

Fruit/Seed maturity: March to July.

Appearance of mature fruit/seed: When immature, the elongated, dense inflorescence is tightly packed along the main stem with individual florets green in colour. During maturation, the inflorescence head opens up to display the three prominent awns of the individual florets, and appears almost brush-like. From a distance the inflorescence appears silver at maturity. Mature florets will detach easily from the plant when collected by hand.

Cleaning requirements

Cleaning techniques: The splayed nature of the three awns on each individual floret makes bulk collections hard to handle. Successful removal of the awns has been achieved through rubbing the florets between a thin, ribbed rubber mat or sieve. Care must be taken not to damage seeds contained within the florets. The elongated nature of the floret unit makes it prone to fracturing during cleaning. Isolating seeds from the floret has never been achieved.

Purity and viability: High (>90%).

Seeds per gram: Filled florets = 684 (unable to clean to pure seeds).

Germination requirements

Dormancy type: Non-dormant.

Embryo type: Lateral. **Embryo:seed ratio:** 0.53.

Pre-treatments: No pre-treatment requirements.

Germination: 80-100% (seeds within florets). Unimpeded across a wide temperature range (10–35°C).

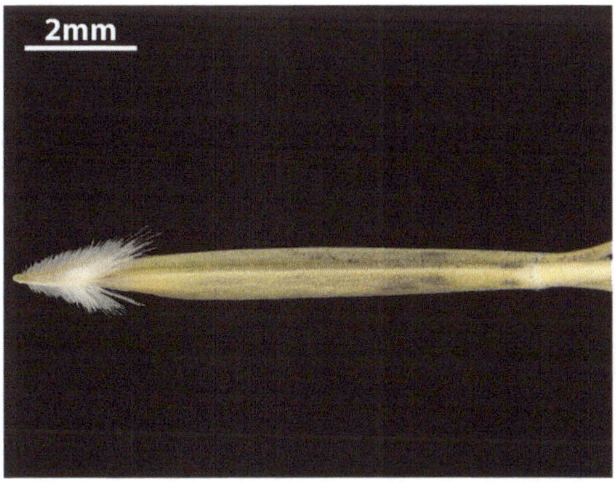

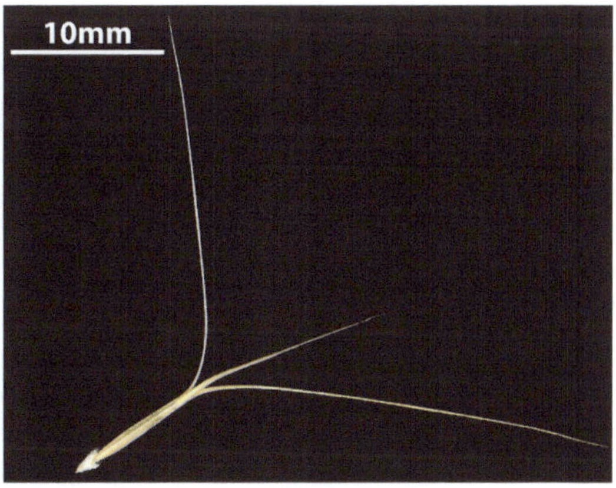

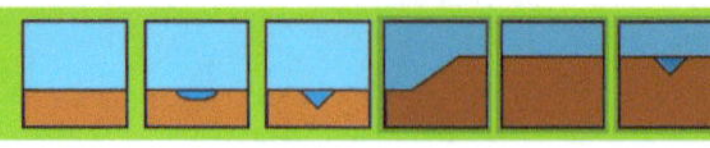

Aristida latifolia

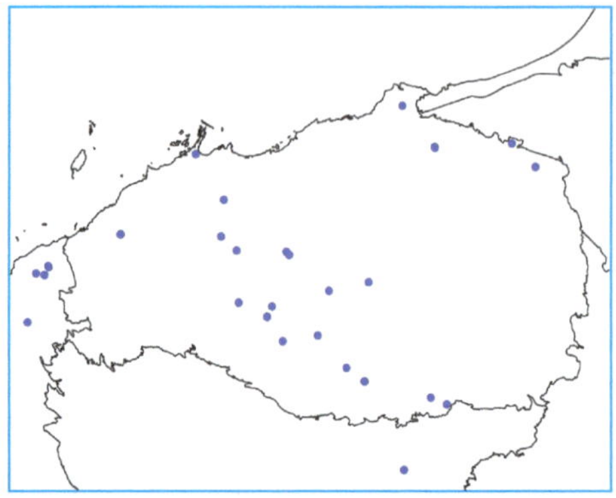

Common names: Feathertop wiregrass.

Indigenous names: None recorded.

Description: Perennial, loosely tufted, 0.2–1.3 m high. Leaves: sheath longer than internode; ligule 0.5–1 mm long; blade to 700 mm long, 4.5 mm wide. Inflorescence 80–690 mm long, 30–35 mm diam., spiciform, continuous and dense or interrupted; branches to 100 mm long, bearing spikelets from base or naked at base. Glumes more or less equal; lower glume 7–13 mm long, 1-veined, entire, with an awn to 4 mm long; upper glume 8–17 mm long, 1-veined, with an awn to 4 mm long. Lemma 8–11 mm long, shorter than glumes, convolute, without a furrow on the adaxial surface, muricate towards apex but sometimes sparsely so or smooth, without an articulation; column poorly developed or distinct with 1–9 spirals, 1–9 mm long; awns unequal: median 17 –50 mm long; laterals 15–46 mm long. Caryopsis 5–7.5 mm long, more or less twice hilum length.

Key features for identification: Perennial, sometimes pruinose; leaf blade to 4.5 mm wide, curling when dry; Lemma 8–9 mm long, column 1–9 mm long.

Similar species: Very distinctive for its broad leaves. *A. holathera* var. *latifolia* differs in having a lemma 8–13.5 mm long with a column 18–57 mm long.

Distribution: Widespread in the Kimberley and Pilbara regions and in all mainland states except Vic.

Habitat: Often on cracking clay soils.

Flowering period: April to June.

Aristida latifolia

Fruit/Seed collection

Fruit/Seed maturity: June.

Appearance of mature fruit/seed: When immature, the elongated, dense inflorescence is tightly packed along the main stem with individual florets green in colour. During maturation, the inflorescence head opens up to display the three prominent awns of the individual florets, and appears almost brush-like. From a distance the inflorescence appears silver at maturity. Mature florets will detach easily from the plant when collected by hand.

Cleaning requirements

Cleaning techniques: The splayed nature of the three awns on each individual floret makes bulk collections hard to handle. Successful removal of the awns has been achieved through rubbing the florets between a thin, ribbed rubber mat or sieve. Care must be taken not to damage seeds contained within the florets. The elongated nature of the floret unit makes it prone to fracturing during cleaning. Isolating seeds from the floret has never been achieved.

Purity and viability: High (>90%).

Seeds per gram: Filled florets = 894 (unable to clean to pure seeds).

Germination requirements

Dormancy type: Physiological.

Embryo type: Lateral. **Embryo:seed ratio:** 0.47.

Pre-treatments: Dormancy alleviation treatments unknown.

Germination: Maximum germination of seeds (within florets) with no pre-treatment is 60–75% at 20–25°C. Treatment with GA_3, and to a lesser extent smoke, increases germination to 70–95%, predominantly at lower temperatures (10–25°C).

Notes: Additional dormancy break pre-treatments such as dry after-ripening and wet/dry cycling may assist in promoting 100% germination.

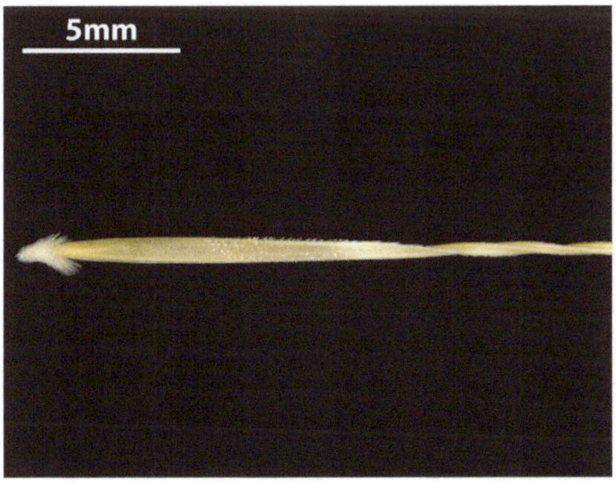

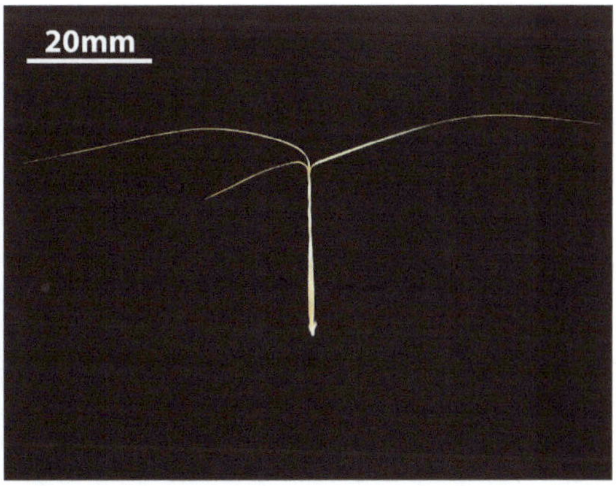

Cymbopogon ambiguus

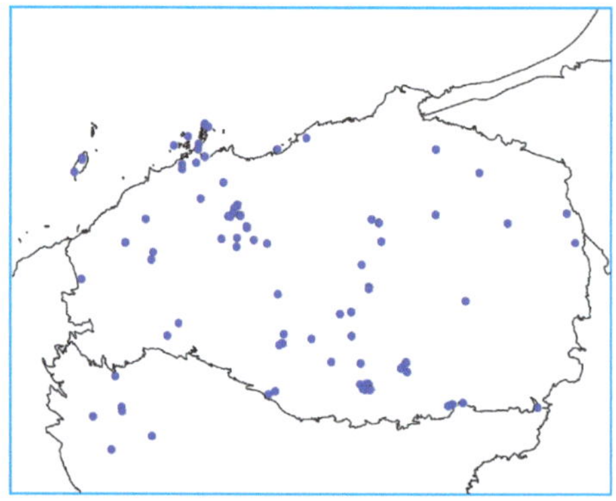

Common names: Scentgrass, Lemon grass, Scented oilgrass.

Indigenous names: Mararrany (Thalanyji); Malhanggaa (Yindjibarndi); Mathangura (Ngarluma).

Description: Perennial, culms 0.3–1 m high. Ligule an eciliate membrane, 2–5 mm long. Leaf blades 150–500 mm long, 1–5 mm wide, narrowed to base, flat. Inflorescence 150–400 mm long, 2–5 mm diam., 4–5-noded, branches mostly exceeding internodes; spathes very narrowly elliptic, 15–30 mm long; peduncle glabrous, one-seventh to two-fifths as long as spathe; racemes 15–20 mm long, not reflexed; internodes and pedicels flattened or concave on inner side, slightly convex dorsally, 2–3 mm long, densely villous except lower part of back, nearly equal; apex dilated, cup-shaped, crenulate. Sessile spikelet narrowly oblong to narrowly elliptic-oblong, 4–6 mm long, 0.7–1 mm diam.; longer callus hairs up to 1.3 mm long. Glumes: lower glume 5.5–7 mm long, 2-keeled from middle, with 2–5 intra-carinal veins, flat dorsally in upper half, concave to deeply furrowed towards base; upper glume 3–5-veined. Lower floret: lemma narrowly oblong, two-thirds as long as lower glume. Upper floret: lemma lobed to one-quarter to half its length, with an awn 12–20 mm long. Pedicellate spikelet male or sterile, ovate to linear, 3–5 mm long.

Key features for identification: Plant usually <1.5 m tall with 4–8-noded inflorescences; lower glume of sessile spikelet 5.5–7 mm long; longer callus hairs to 1.3 mm long.

Similar species: *C. procerus* is 1.5–2.5 m tall with 7–11-noded inflorescences; lower glume of sessile spikelet not >4.5 mm long.

Distribution: Widespread in the Kimberley and Pilbara regions. Occurs in all mainland states except Vic.

Habitat: Grows in rocky red loam, sand, ironstone, granite, limestone, sandstone and creeks.

Flowering period: November to June.

Cymbopogon ambiguus

Fruit/Seed collection

Fruit/Seed maturity: March to June.

Appearance of mature fruit/seed: Spikelets and individual florets have distinct white hairs. When immature, the glumes are green and tightly packed around each flowering unit. Once mature, the glumes brown off and open up exposing the very fluffy and cotton-like spikelets and florets. Whole spikelets or individual florets detach easily from the plant when hand collected.

Cleaning requirements

Cleaning techniques: Per floret, low seed set can result in low purity. Attempts to clean to 100% filled florets using vacuum separation have been unsuccessful. Pure seeds can be obtained by rubbing the florets through a suitably sized sieve. The lighter empty floret debris can then be removed by vacuum separation.

Purity and viability: Purity can be low (<15%) due to the large fraction of empty floret material. Once cleaned to seed, viability is high (85%).

Seeds per gram: 846 (seeds).

Germination requirements

Dormancy type: Non-dormant (tested on seeds).

Embryo type: Lateral. **Embryo:seed ratio:** 0.49.

Pre-treatments: Clean to pure seed.

Germination: Maximum seed germination achieved with no chemical pre-treatment is 80–90% over a broad range of temperatures (15–35°C). No additional benefit is evident after exposure to GA_3 or smoke.

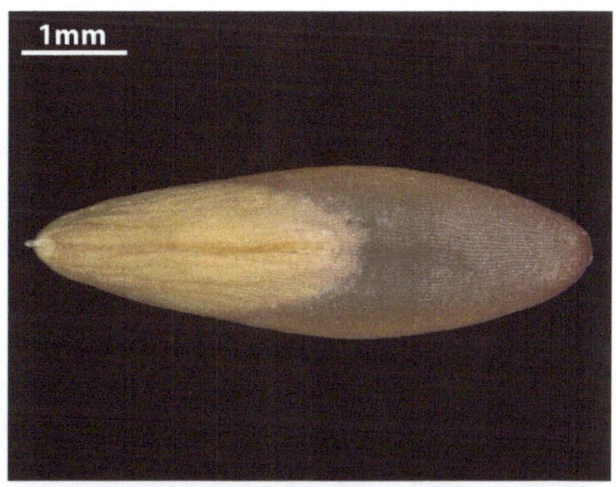

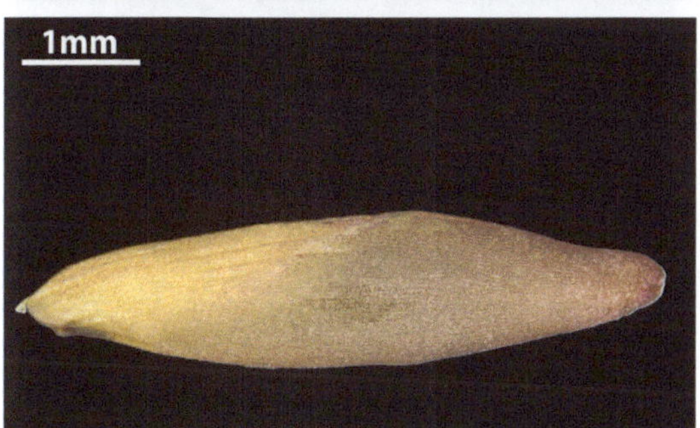

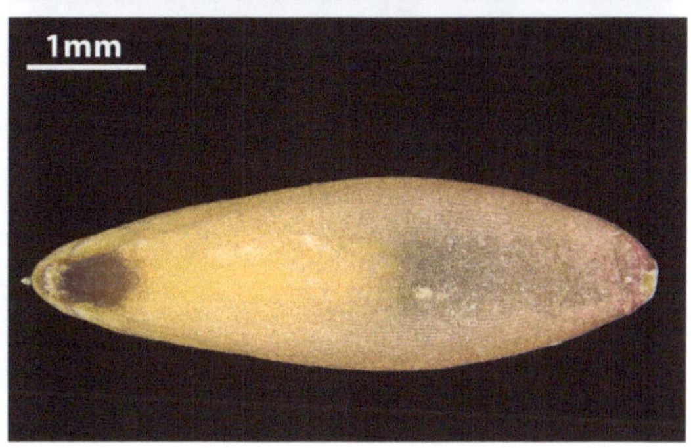

Cymbopogon obtectus

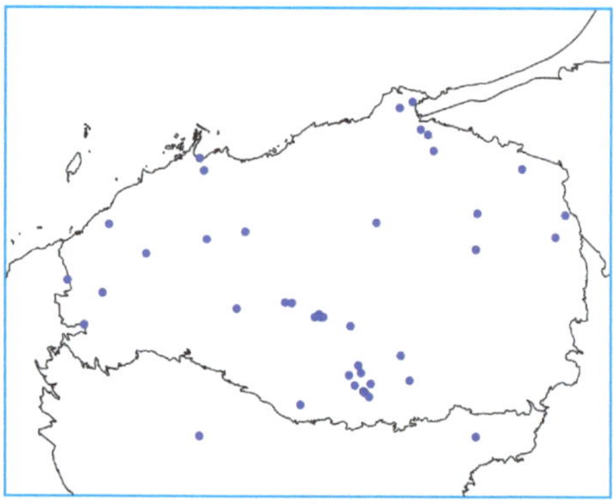

Common names: Silky heads.

Indigenous names: None recorded.

Description: Perennial, culms 0.5–1 m tall, 1–2-noded; mid-culm nodes glabrous or pubescent. Ligule an eciliate membrane, 3–5 mm long. Leaf blades flat or conduplicate or revolute, 150–350 mm long, 1–3 mm wide. Inflorescence compound; central inflorescence axis 70–250 mm long; rachis fragile at the nodes. Primary spikelet sessile; companion spikelet pedicellate; basal sterile spikelets 2, well-developed (lower raceme). Companion spikelet developed, comprising 2 unequal glumes (muticous) without lemmas or containing empty lemmas, 3–5 mm long. Fertile spikelets 2-flowered, the lower floret barren (rarely male), the upper fertile, comprising 1 basal sterile floret, 1 fertile floret, without rachilla extension, lanceolate or elliptic or oblong or obovate, dorsally compressed, 5–6.4 mm long. Glumes dissimilar, firmer than fertile lemma; lower glume lanceolate, chartaceous, keeled, 2-keeled, wingless, 0–5 -nerved; upper glume lanceolate, 1-keeled, 5–7-nerved. Basal sterile floret 1, barren. Lemma of lower sterile floret hyaline, 2–3-nerved; fertile lemma 3.6–4.5 mm long, without keel, 1–3-nerved; apex lobed, awned, 1-awned; median (principal) awn from a sinus, 6–8 mm long overall, without a column or with a straight or slightly twisted column. Palea absent. Anthers 3.

Key features for identification: Leaf sheaths 2–3 mm wide, not curling at maturity; median awn 6–8 mm long longer callus hairs 3.5–5 mm long.

Similar species: *C. bombycinus* differs by leaf sheaths to 6 mm wide, curling at maturity; median awn 6–21 mm long.

Distribution: Widespread in the Pilbara and across most of mainland Australia.

Habitat: Brown or red loam, red sand or granite.

Flowering period: All year round.

Cymbopogon obtectus

Fruit/Seed collection

Fruit/Seed maturity: March to June.

Appearance of mature fruit/seed: Spikelets and individual florets have distinct white hairs. When immature, the glumes are green and tightly packed around each flowering unit. Once mature, the glumes brown off and open up exposing the very fluffy and cotton-like spikelets and florets. Whole spikelets or individual florets detach easily from the plant when hand collected.

Cleaning requirements

Cleaning techniques: Per floret, low seed set can result in low purity. Attempts to clean to 100% filled florets using vacuum separation have been unsuccessful. Pure seeds can be obtained by rubbing the florets through a suitably sized sieve. The lighter empty floret debris can then be removed by vacuum separation.

Purity and viability: Purity can be very low (<5%) due to the large fraction of empty floret material. Once cleaned to seed, viability is high (100%).

Seeds per gram: 868 (seeds).

Germination requirements

Dormancy type: Non-dormant (tested on seeds).

Embryo type: Lateral. **Embryo:seed ratio:** 0.62.

Pre-treatments: Clean to pure seed.

Germination: Maximum seed germination achieved with no chemical pre-treatment is 90–95% over temperatures of 20–35°C. Exposure to GA_3 or smoke can improve germination by 15–20% at 15°C.

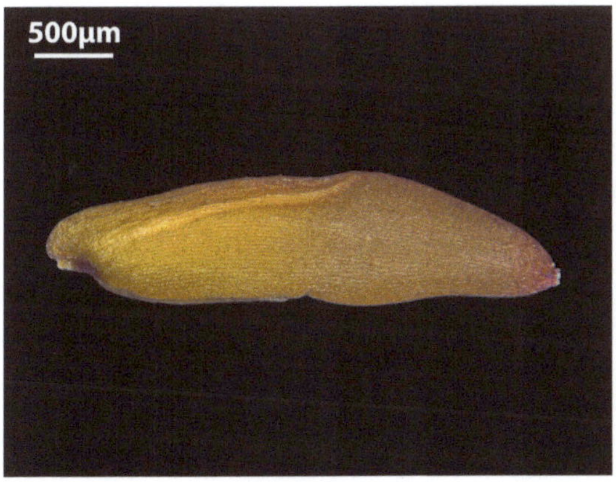

Cynodon convergens

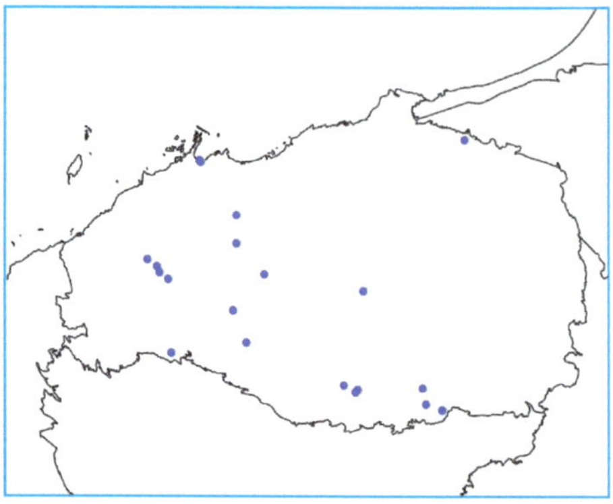

Common names: Common native couch, Kimberley couch, Spider grass.

Indigenous names: None recorded.

Description: Annual or short-lived perennial, 0.23–0.5 m high, usually stoloniferous. Leaves: sheath usually less than half as long as the internodes, bearded or pilose on the auricles; blade 35–60 mm long, subacute. Spikes 2–4, 30–70 mm long, divergent or rarely divaricate, stiff, with a broadly winged rachis. Spikelets sessile; fertile spikelets 1-flowered, comprising 1 fertile floret, without rachilla extension, lanceolate or oblong, laterally compressed, 3–5 mm long. Glumes oblong, 3–5 mm long, coriaceous, 1-keeled, broadly winged on keel, 1-nerved. Fertile lemma 2.2–2.8 mm long, glabrous and smooth between the veins, ciliate or hairy on the veins for their entire length, keeled, 3-nerved; apex muticous or mucronate. Palea 2-nerved, with contiguous keels, the keels ciliate or hairy, similar to the veins of the lemma. Caryopsis laterally flattened, narrowly obovoid, 1.7–2 mm long.

Key features for identification: The strongly laterally flattened caryopsis and palea with contiguous keels distinguish it from all other species in Australia.

Similar species: *C. prostratus* differs by its prostrate habit, pseudo-stolons and short, spatheate, often enclosed spikes.

Distribution: Widespread from the north Kimberley south to the Pilbara. Also occurs in N.T., Qld and N.S.W.

Habitat: In a range of habitats including claypans, river banks, levees and floodplains, irrigated cotton fields, gorges and sandplains, in sandy to clay soils, and in association with monsoon vine thickets, pindan, and open woodland.

Flowering period: January to June.

Cynodon convergens

Fruit/Seed collection

Fruit/Seed maturity: March to April.

Appearance of mature fruit/seed: Flowering spikes (2–4) splay away from the main stem and are green when immature. Once mature, the whole flowering spike turns light brown. Mature spikes and/or individual florets fall easily from the plant when touched.

Cleaning requirements

Cleaning techniques: Collections should be relatively absent of plant debris, with occasional flower stems present. Stem debris can be removed quickly via sieving. Pure seeds can be obtained by passing floret material through a sieve, using a rubber bung. Floret debris can be removed via vacuum separation.

Purity and viability: Purity can be low (<15%) due to the large fraction of empty floret material. Once cleaned to seed, viability is high (100%).

Seeds per gram: 6440 (seeds).

Germination requirements

Dormancy type: Physiological.

Embryo type: Lateral. **Embryo:seed ratio:** 0.55.

Pre-treatments: Clean to pure seed. Seeds moderately response to GA_3. Alternative dormancy alleviation treatments unknown.

Germination: Maximum seed germination achieved with no chemical pre-treatment is 10–20% at 25–35°C. Treatment with GA_3 increases germination to 40–70% at 25–35°C.

Notes: Additional dormancy break pre-treatments such as dry after-ripening or wet/dry cycling may assist in overcoming the physiological component of dormancy.

Enneapogon caerulescens

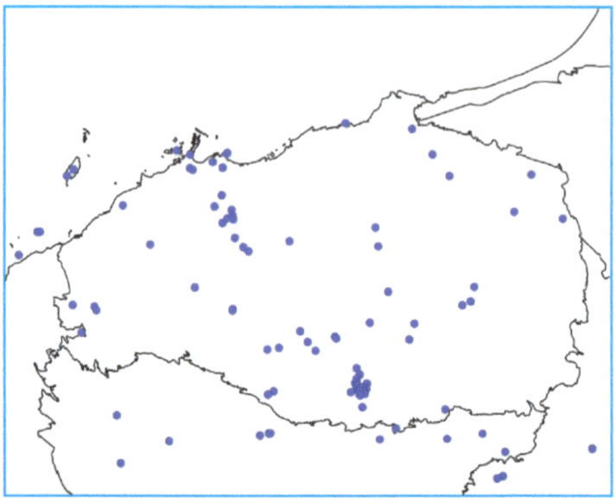

Common names: Limestone grass.

Indigenous names: None recorded.

Description: Annual or perennial, culms 0.1–0.4 m tall; lateral branches simple or sparsely branched. Ligule a fringe of hairs. Leaf-blades flat or involute, 1 mm wide. Inflorescence solid, a panicle, oblong or ovate, 10–20 mm long, 1–1.5 cm wide. Spikelets pedicellate; fertile spikelets 1 or more flowered, with 1–2 fertile floret(s), with diminished florets at the apex, oblong, laterally compressed, 2.7–9.2 mm long. Glumes similar, thinner than fertile lemma; lower glume ovate, membranous, without keels, 5–7-nerved; surface indumented; upper glume ovate, 2.7–9.2 mm long, membranous, without keels, 5–7(–11)-nerved; surface scabrous, indumented; apex entire or erose. Fertile lemma 1.6–2.8 mm long, without keel, 9–13-nerved; surface indumented; apex dentate or lobed, 9-awned; median (principal) awn 3–10 mm long overall; lateral lemma awns present. Lodicules present. Anthers 3. Caryopsis 1.2–2 mm long.

Key features for identification: Axillary panicles and cleistogenes; unequal spikelets in upper and lower parts of the panicle; relatively long hairs on lobes and palea keels; body of the basal lemma strongly ribbed, short relative to its lobes, and usually covered by long, dorsal hairs.

Similar species: *E. avenaceus* and *E. polyphyllus* differ by having spikelets in panicle all similar in length; axillary panicles and spikelets absent; panicles longer than wide; basal lemma nerves flattened or slightly ribbed, obscure or slightly prominent.

Distribution: Widespread in the arid zone of W.A., southern N.T. and S.A.

Habitat: White, brown or red sand, clay, sandstone.

Flowering period: March to July.

Enneapogon caerulescens

Fruit/Seed collection

Fruit/Seed maturity: March to April.

Appearance of mature fruit/seed: Immature inflorescences are green during flowering and tightly packed at the apex of stems. Once mature, ovoid-shaped inflorescences open up and are white (occasionally purple-tinged) in colour. Whole inflorescences or individual florets detach easily from the stems when hand collected.

Cleaning requirements

Cleaning techniques: Seeds can be removed from the florets by rubbing through a suitably sized sieve. The lighter floret debris can then be removed by vacuum separation.

Purity and viability: Purity not assessed in this species. Low purity (20–40%) in other species has been assessed, with many florets not containing any seed. Once cleaned to pure seeds, viability is high (>95%).

Seeds per gram: 2976 (seeds).

Germination requirements

Dormancy type: Physiological.

Embryo type: Lateral. **Embryo:seed ratio:** 0.71.

Pre-treatments: Clean to pure seed. Exposure to GA_3 in fresh seeds increases germination slightly (10–15%) at higher temperatures. Other dormancy alleviation treatments unknown.

Germination: Maximum seed germination achieved with no pre-treatment is 30–60% at 20–35°C. Treatment with GA_3 increases germination to 40–65% at 20–35°C.

Notes: Additional dormancy break pre-treatments such as dry after-ripening or wet/dry cycling may assist in improving germination.

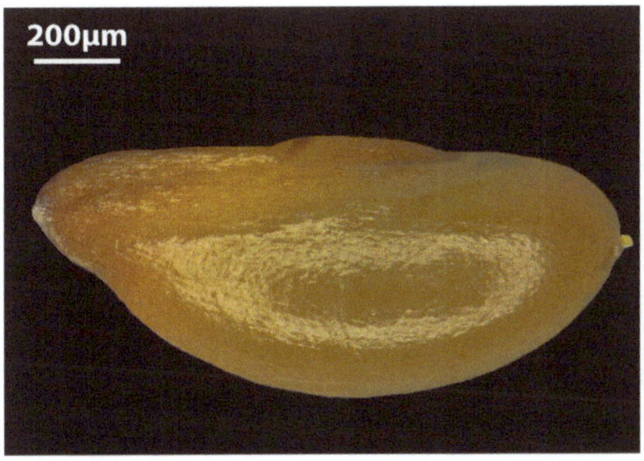

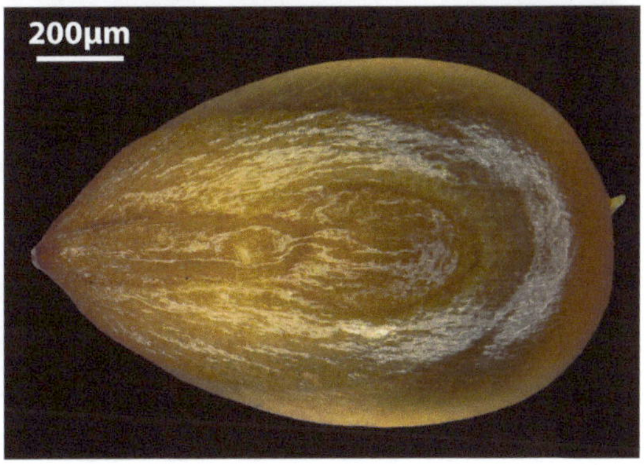

Enneapogon polyphyllus

Common names: Leafy nineawn.

Indigenous names: None recorded.

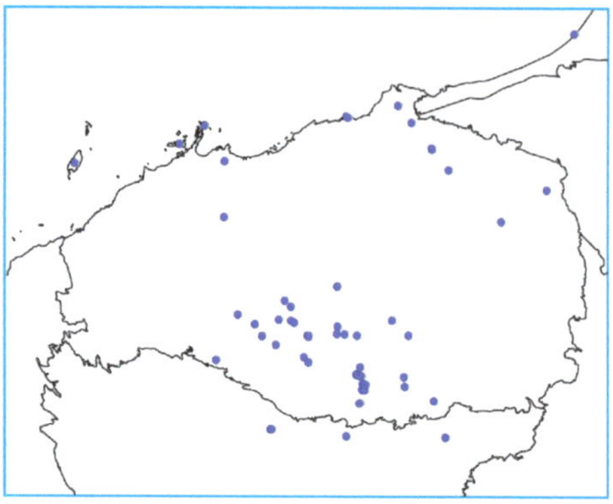

Description: Annual or perennial, 0.03–0.5 m high. Panicle linear to elliptic, 40–90 mm long, 10–20 mm diam. including awns, spiciform; axillary inflorescences absent. Spikelets all similar in length; florets 5 or 6, 1 fertile. Glumes 4–9 mm long: lower glume 9–13-veined; upper glume 5–7-veined. Lowest floret: lemma 2–3 mm long and similar in width, cartilaginous to indurate, with slightly ribbed veins, the dorsal hairs exceeding the base of the awns, awns 4–7 mm long, the 2 outer awns wider in their lower part than above; palea membranous, as long as the body of its lemma or shorter, glabrous on the body in the upper one-third to half and sparsely hairy below or entirely glabrous, sparsely hairy on the flaps above and glabrous below. Upper florets: palea of second floret with broad flaps, almost circular, usually hairy; florets 3–6 reduced.

Key features for identification: Distinguished by the structure of its distally crowded sterile florets.

Similar species: *E. intermedius* is closely allied, but more robust and always perennial, with a differently shaped spikelet and 2 fertile florets.

Distribution: Widespread in the Kimberley and Pilbara regions. Occurs in all mainland states except Vic.

Habitat: Grows in pindan and savannah communities in sandy and clayey soils, and in disturbed sites such as homestead grounds and gravelly quarry hills.

Flowering period: January–July.

Enneapogon polyphyllus

Fruit/Seed collection

Fruit/Seed maturity: March to June.

Appearance of mature fruit/seed: Immature inflorescences are green during flowering and tightly packed at the apex of stems. Once mature, the inflorescences open up to an elongated and linear arrangement, and are white in colour. Whole inflorescences or individual florets detach easily from the stems when hand collected.

Cleaning requirements

Cleaning techniques: Seeds can be removed from the florets by rubbing through a suitably sized sieve. The lighter floret debris can then be removed by vacuum separation.

Purity and viability: Low purity (<20%) with many florets not containing a seed. Once cleaned to pure seeds, viability is high (>95%).

Seeds per gram: 1492–2097 (seeds).

Germination requirements

Dormancy type: Physiological.

Embryo type: Lateral. **Embryo:seed ratio:** 0.84.

Pre-treatments: Clean to pure seed. Exposure of fresh seeds to GA_3 increases germination by 30–50% at higher temperatures. Other dormancy alleviation treatments unknown.

Germination: Maximum seed germination achieved with no pre-treatment is 10–25% at 20–35°C. Treatment with GA_3 increases germination to 20–75% at 20–35°C.

Notes: Additional dormancy break pre-treatments such as dry after-ripening or wet/dry cycling may assist in improving germination.

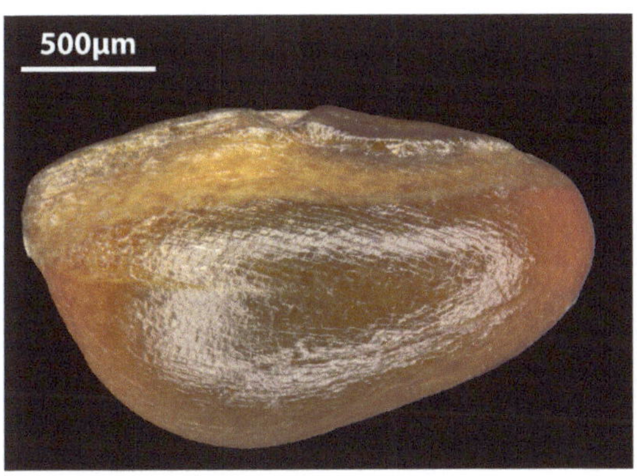

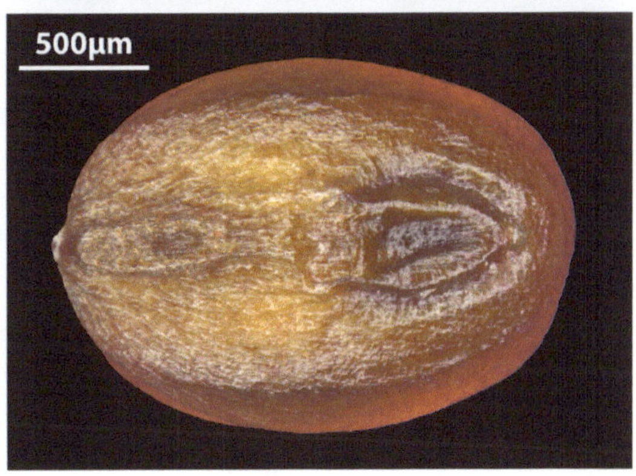

Enneapogon robustissimus

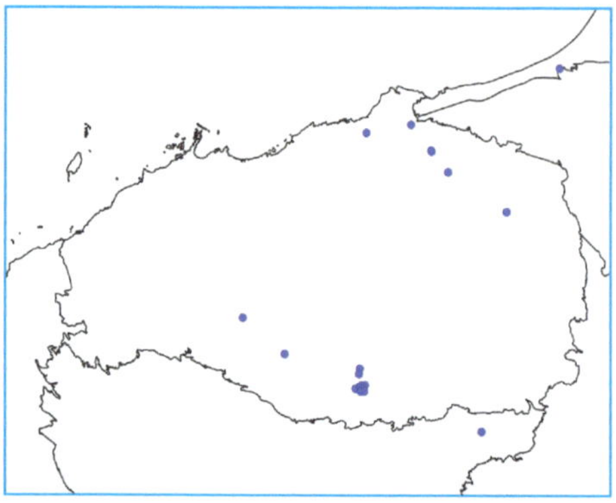

Common names: Robust nineawn.

Indigenous names: None recorded.

Description: Perennial, culms 0.3–1.1 m tall, 7-noded; lateral branches simple or sparsely branched. Leaf-sheaths glabrous on surface or hairy. Ligule a fringe of hairs. Leaf-blades flat or involute, 2–5 mm wide; surface scabridulous. Inflorescence solid, a linear, lanceolate, ovate or pyramidal panicle, 15–40 mm long, 10–25 mm wide. Spikelets pedicellate; fertile spikelets 1 or more flowered, with 1 fertile floret and diminished florets at the apex, oblong, laterally compressed, 3–5.6 mm long. Glumes similar, thinner than fertile lemma; lower glume ovate, membranous, without keels, 3–9-nerved; surface glabrous or indumented; apex mucronate; upper glume ovate, 3–5.6 mm long, membranous, without keels, 3–9-nerved; surface asperulous or scabrous, glabrous or indumented; apex erose, muticous or mucronate. Fertile lemma 2–2.6 mm long, without keel, 9–11-nerved; surface indumented; apex dentate or lobed, awned, 9-awned; median (principal) awn 2.3–6 mm long overall; lateral lemma awns present. Palea apex muticous or awned. Lodicules present. Anthers 3. Caryopsis 1.3–1.6 mm long.

Key features for identification: Inflorescence capitate, linear to ovate; dorsal lemma hairs shorter than or subequal to base of awns; axillary spikelets or panicles usually present.

Similar species: *E. lindleyanus* differs in the palea apex being glabrous and the awns are 2 nerved at base.

Distribution: Scattered in the Pilbara and northern deserts of W.A. Also in N.T., S.A. and Qld.

Habitat: Red sandy loam, alluvium, gravel, undulating sandplains, coastal areas, river banks and creeklines.

Flowering period: April to August.

Enneapogon robustissimus

Fruit/Seed collection

Fruit/Seed maturity: May to June.

Appearance of mature fruit/seed: Immature inflorescences are green during flowering and tightly packed at the apex of stems. Once mature, the ovoid to spherical inflorescences open up and are white in colour. Whole inflorescences or individual florets detach easily from the stems when hand collected.

Cleaning requirements

Cleaning techniques: Seeds can be removed from the florets by rubbing through a suitably sized sieve. The lighter floret debris can then be removed by vacuum separation.

Purity and viability: Purity is *c.* 40% with many florets not containing a seed. Once cleaned to pure seeds, viability is high (>95%).

Seeds per gram: 2269 (seeds).

Germination requirements

Dormancy type: Non-dormant.

Embryo type: Lateral. **Embryo:seed ratio:** 0.76.

Pre-treatments: Clean to pure seeds. Germination of seeds retained in the floret has not been tested.

Germination: Maximum seed germination achieved with no pre-treatment is 50–80% at 15–35°C (peak at 25°C). Treatment with GA_3 and smoke increases germination (50–100%) at the same temperatures, with GA_3 producing higher germination (>70%) at all temperatures.

Notes: Additional dormancy break pre-treatments such as dry after-ripening or wet/dry cycling may assist in improving germination without the need for GA_3 and smoke.

Eragrostis dielsii

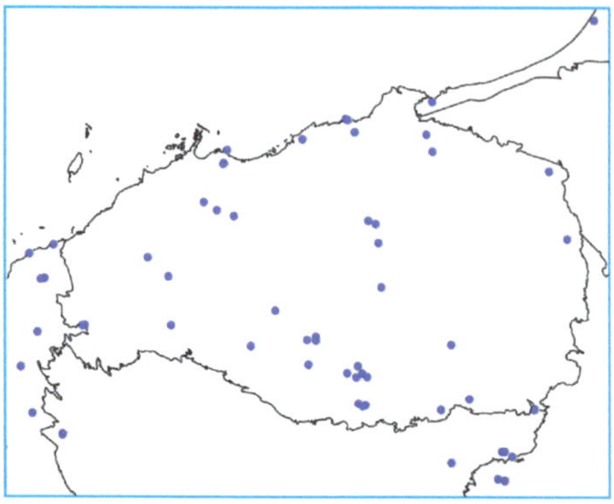

Common names: Mallee lovegrass.

Indigenous names: None recorded.

Description: Annual or sometimes short-lived perennial, 0.03–0.55 m high, eglandular; when perennial the plants slightly thickened and sparsely hairy at the base. Leaf blades to 3.5 mm wide, at first flat, soon becoming conduplicate or involute. Panicle 20–170 mm long, 10–70 mm diam., axils glabrous or sparsely hairy. Spikelets sessile or subsessile, sometimes biconvex, 4.5–50 mm long, 0.7–1.7mm diam.; rachilla persistent or finally disarticulating; florets 21–81. Glumes persistent, obtuse, sometimes notched; both glumes 1.3–1.8 mm long. Lemmas 1.5–2.5 mm long, sometimes 4 or 5-veined, persistent, firmly membranous to indurate with scarious margins prominent at the apex, smooth or scabrous-granular and often scabrous-tuberculed near apex, obtuse or subacute, sometimes shortly 2-lobed; lateral veins *c.* midway between the margin and midvein, each lateral often comprising 2 adjacent veins. Paleas subequal in length to their lemmas, smooth to scabridulous on the keels, obtuse or truncate. Anthers 3, 0.3–0.5 mm long. Caryopsis longer than wide, strongly compressed dorsiventrally.

Key features for identification: Spikelets sessile or subsessile, clustered; panicle often reduced to raceme or spike and spiciform; spikelets biconvex or terete, linear; rachilla persistent or subpersistent, compressed; lemmas and paleas often granular-scabrous; keels and flaps of palea thickened; caryopsis flattened; florets closely imbricate.

Similar species: *E. lanicaulis* is perennial. *E. pergracilis* has lemmas 1.5–1.6 mm long and spikelet 0.75–1 mm wide.

Distribution: Common throughout the Pilbara region. Occurs in all mainland states.

Habitat: On plains in loams and in heavy-textured soils.

Flowering period: April to September.

Eragrostis dielsii

Fruit/Seed collection

Fruit/Seed maturity: July.

Appearance of mature fruit/seed: The predominantly prostrate culms result in inflorescences radiating away from the centre of the plant. Spikelets are upright and turn yellow at maturity. Spikelets and/or individual florets detach easily when mature.

Cleaning requirements

Cleaning techniques: Seeds can be removed from the florets by rubbing through a suitably sized sieve. The lighter floret debris can then be removed by vacuum separation.

Purity and viability: Low purity (*c.* 15%) with many florets not containing a seed. Once cleaned to pure seeds, viability is high (>95%).

Seeds per gram: 22000 (seeds).

Germination requirements

Dormancy type: Physiological.

Embryo type: Lateral. **Embryo:seed ratio:** 0.60.

Pre-treatments: Clean to pure seed. Other dormancy alleviation treatments unsuccessful (e.g. dry after-ripening after 6 months storage).

Germination: Maximum seed germination achieved at 10–35°C is 5% (tested on water, GA_3, and smoke).

Notes: The low germination of fresh seeds and lack of response to dry after-ripening suggests seeds have a deeper form of physiological dormancy. Additional dormancy alleviation techniques need to be investigated.

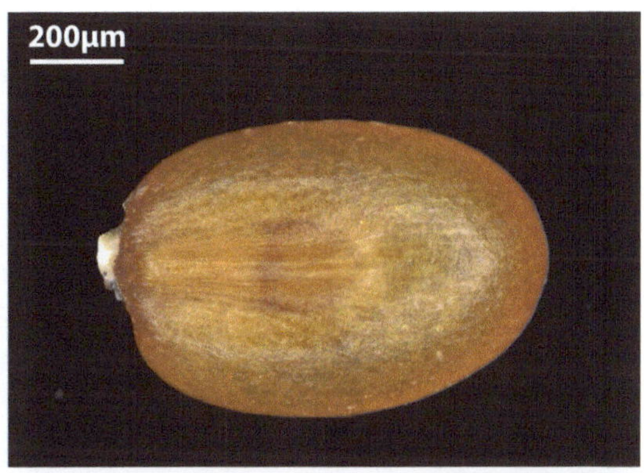

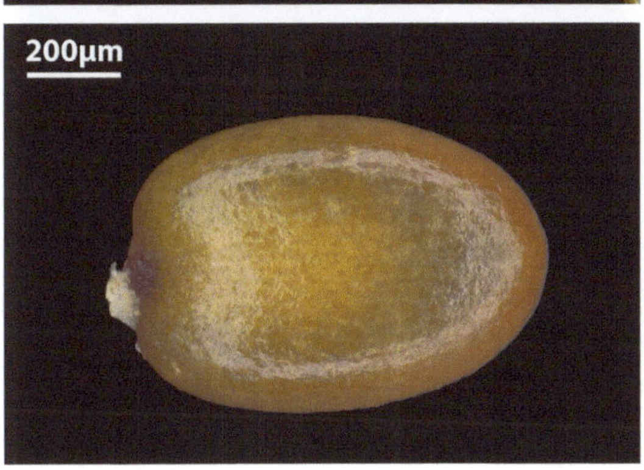

Eragrostis eriopoda

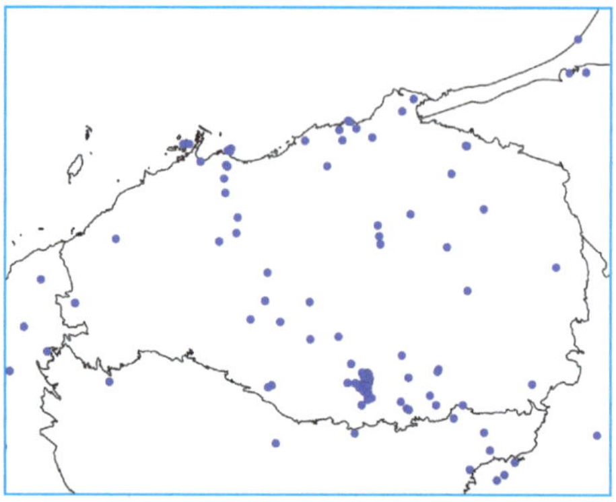

Common names: Woollybutt grass, Naked woollybutt.

Indigenous names: None recorded.

Description: Perennial, 0.2–0.6 m high, culms erect, eglandular, often forming tussocks 150–220 mm wide, with thickened or bulbous, hirsute or woolly base, often with a knotty woolly rhizome and with hirsute cataphylls. Leaf blades *c.* 10 mm long, tightly convolute or involute, often stiffly setaceous. Panicle 90–200 mm long, 10–75 mm diam., with glabrous axils. Spikelets 5–30 mm long, 1.5–3.5 mm diam., strongly compressed laterally; rachilla persistent or finally disarticulating; florets 8–62. Glumes 1–3-veined, 1.1–2.5 mm long, obtuse. Lemmas 2–3 mm long, scabrous-papillose near apex or smooth or granular, obtuse; lateral veins *c.* midway between margin and midvein. Paleas subequal in length to their lemmas, minutely scabridulous on the keels, obtuse to truncate; keel as long as palea itself. Anthers usually 3, 0.8–1.3 mm long. Caryopsis longer than wide, laterally compressed and quadrangular or obtusely trigonous.

Key features for identification: Bulbous woolly root base; leaf sheaths short relative to internodes, tight on culms, abruptly wider than blades, with ciliate margins; 1–3-nerved glumes; stipitate caryopsis; closely imbricate florets usually falling entire; pouched paleas.

Similar species: *E. desertorum* has spikelets *c.* 1 mm wide.

Distribution: From the southern Kimberley south to the Pilbara region. Occurs in all mainland states except Vic, also in New Guinea.

Habitat: Grows in pindan, white to red sands and loams.

Flowering period: March to July.

Eragrostis eriopoda

Fruit/Seed collection

Fruit/Seed maturity: March to June.

Appearance of mature fruit/seed: Inflorescences are upright and produced in large numbers from the tussock. Immature spikelets and florets appear green (with an occasional purple-tinge). At maturity, the elongated spikelets and florets detach from the inflorescence easily when collected by hand.

Cleaning requirements

Cleaning techniques: Seeds can be removed from the florets by rubbing through a suitably sized sieve. The lighter floret debris can then be removed by vacuum separation.

Purity and viability: Low purity (*c.* 15%) with many florets not containing a seed. Once cleaned to pure seeds, viability is high (>95%).

Seeds per gram: 7996 (seeds).

Germination requirements

Dormancy type: Physiological.

Embryo type: Lateral. **Embryo:seed ratio:** 0.43.

Pre-treatments: Clean to pure seed. Other dormancy alleviation treatments unsuccessful (e.g. dry after-ripening after 6 months storage).

Germination: Maximum seed germination achieved at 10–35°C is 5% (tested on water, GA_3 and smoke).

Notes: The low germination of fresh seeds and lack of response to dry after-ripening suggests seeds have a deeper form of physiological dormancy. Additional dormancy alleviation techniques need to be investigated.

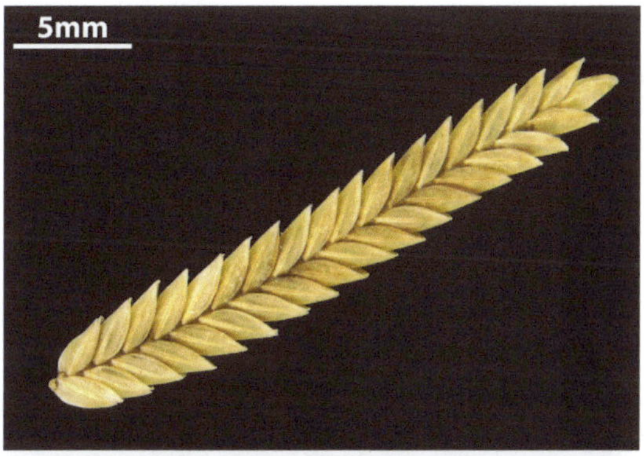

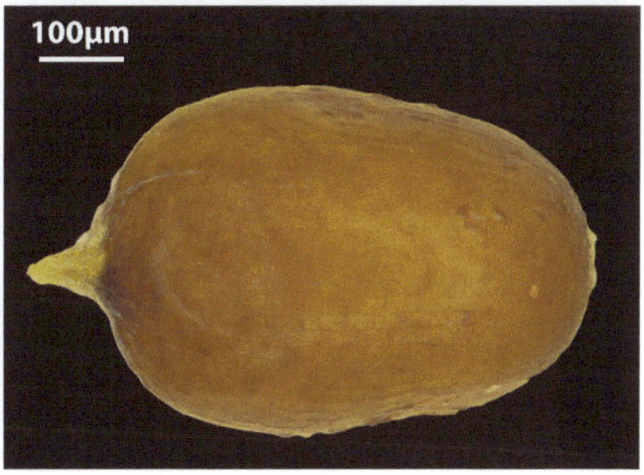

Eragrostis tenellula

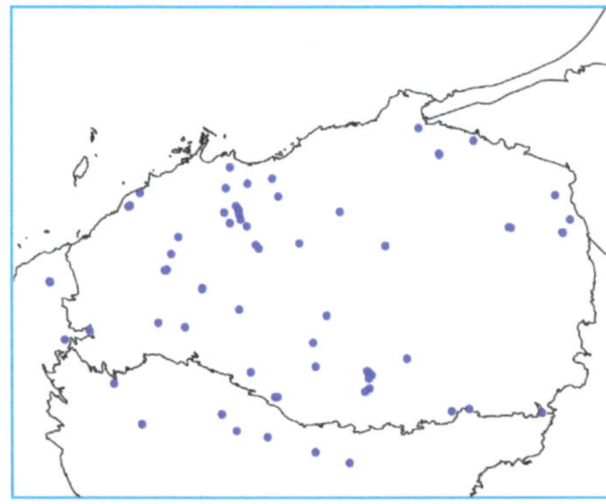

Common names: Delicate lovegrass.

Indigenous names: None recorded.

Description: Annual, 0.06–0.2 m high, eglandular. Leaves: ligule membranous; blade to 300 mm long, 6 mm wide, flat or loosely convolute or involute. Panicle 80–150 mm long, 15–50 mm diam., with glabrous axils. Spikelets 2–4.7 mm long, 0.9–2 mm diam.; rachilla jointed and fragile, readily disarticulating downwards; florets 3–12. Glumes persistent; lower glume 0.5–1.1 mm long; upper glume 0.7–1.3 mm long. Lemmas 0.7–1 mm long, smooth, obtuse; lateral veins closer to the margin than the midvein. Paleas subequal in length to their lemmas, acute; keels distinctly shorter than the palea itself and dividing the apex into 3 acute lobes, mostly smooth; flaps in their lower part wider than the palea body. Anthers 2, *c.* 0.2 mm long. Caryopsis longer than wide, not compressed.

Key features for identification: Panicle decompound, occupying 1/2–3/4 of plant; ligule membranous; florets loose, falling entire; rachilla fragile, straight, the internodes cupular at apex; glumes, lemmas and paleas hyaline; palea 3-lobed, keels relatively short, mostly smooth; anthers two, minute; caryopsis subulate, shiny.

Similar species: *E. tenella* has distinctly ciliate palea margins.

Distribution: Widespread throughout the Kimberley and Pilbara regions. In all mainland states and New Guinea.

Habitat: Often associated with watercourses, plains and gentle slopes, on a wide variety of soils including saline areas, black cracking clays, sands, lateritic or basaltic loams.

Flowering period: March to August.

Eragrostis tenellula

Fruit/Seed collection

Fruit/Seed maturity: July.

Appearance of mature fruit/seed: Mature spikelets can be collected by hand when they detach readily from the plant. Given a large population size, the small stature of the plant would allow for removal of the main flowering stem for bulk collections.

Cleaning requirements

Cleaning techniques: Seeds can be removed from the spikelets and individual florets by rubbing through a suitably sized sieve. The lighter floret debris can then be removed by vacuum separation.

Purity and viability: Low purity (*c.* 25%) with many florets not containing a seed. Once cleaned to pure seeds, viability is high (>95%).

Seeds per gram: 46800 (seeds).

Germination requirements

Dormancy type: Physiological.

Embryo type: Lateral. **Embryo:seed ratio:** 0.45.

Pre-treatments: Clean to pure seed. Other dormancy alleviation treatments unsuccessful (e.g. dry after-ripening after 6 months storage).

Germination: Maximum seed germination achieved with no pre-treatment is 2%. Exposure to GA_3 improved germination to 20% at 25°C.

Notes: The low germination of fresh seeds and lack of response to dry after-ripening suggests seeds have a deeper form of physiological dormancy. Additional dormancy alleviation techniques need to be investigated.

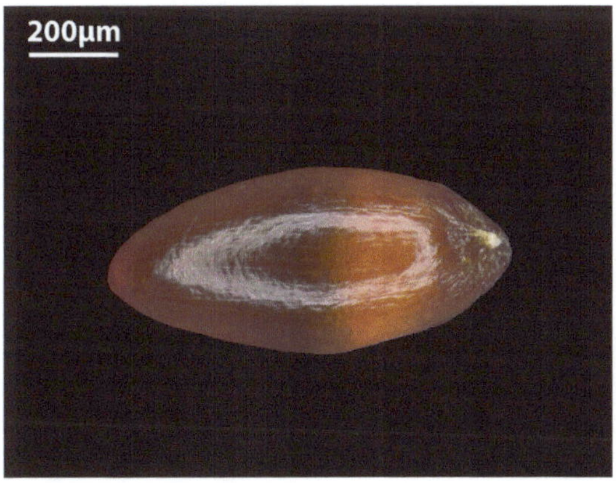

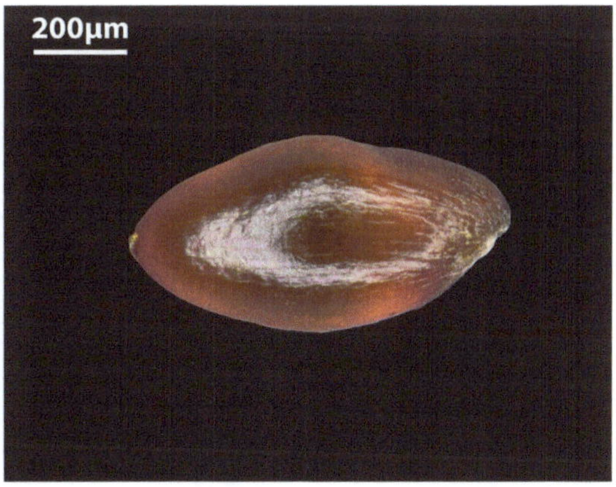

Eriachne pulchella subsp. *dominii*

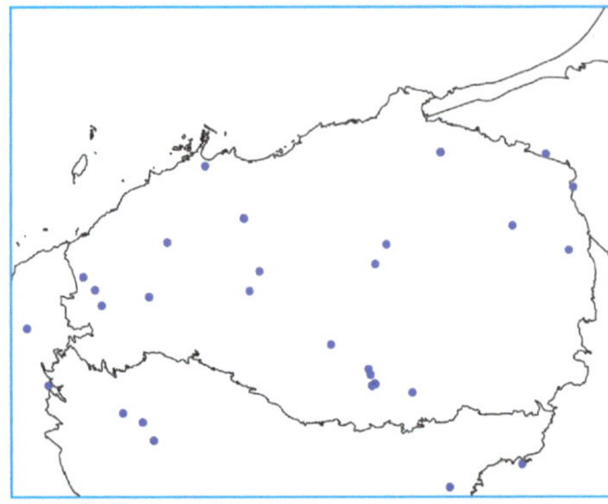

Common names: Pretty wanderrie grass.

Indigenous names: None recorded.

Description: Annual or ephemeral, culms stature slender to delicate, 0.02–0.3 m tall, 1–2-noded; mid-culm nodes glabrous; lateral branches simple or sparsely branched; leaves mostly basal. Ligule a fringe of hairs. Leaf blades to 200 mm long, 1–3.5 mm wide; surface scabrous, hairy. Inflorescence a compound panicle, linear, 10–50 mm long, 5–10 mm wide. Spikelets pedicellate; fertile spikelets 2-flowered, both fertile, without rachilla extension, ovate, laterally compressed, 2–5 mm long. Glumes similar, thinner than fertile lemma; lower glume ovate, scarious, without keels, 5–7-nerved, surface glabrous, apex muticous or mucronate; upper glume ovate, 2–5 mm long, scarious, without keels, 5–7-nerved, surface smooth, glabrous, apex muticous or mucronate. Fertile lemma 3–4.8 mm long, without keel, 5–7-nerved, surface hairy, apex muticous or mucronate. Palea apex entire or dentate. Caryopsis 1.3–1.5 mm long.

Key features for identification: Small annual. Inflorescence compact. Lemma not awned.

Similar species: There are two subspecies. Subsp. *pulchella* has the inflorescence densely spicate, up to 3.5 cm long when mature while subsp. *dominii* has the inflorescence loosely spicate, more than 3.5 cm long when mature.

Distribution: Widespread in the arid zone of Western Australia. Also in N.T. and Qld.

Habitat: Common in skeletal often pebbly or stony soils on hillslopes, outcrops and ridges of sandstone, quartzite and laterite; recorded also from sandplains, reddish-brown loams and from disturbed ground.

Flowering period: April to July.

Eriachne pulchella subsp. *dominii*

Fruit/Seed collection

Fruit/Seed maturity: April to May.

Appearance of mature fruit/seed: The few erect stems of this plant are largely made up of the elongated inflorescences. At maturity, the florets can be collected by hand when they detach readily from the plant.

Cleaning requirements

Cleaning techniques: Seeds can be removed from the spikelets and individual florets by rubbing through a suitably sized sieve. The lighter floret debris can then be removed by vacuum separation.

Purity and viability: Purity is *c.* 40% with many florets not containing a seed. Once cleaned to pure seeds, viability is very high (100%).

Seeds per gram: 3108 (seeds).

Germination requirements

Dormancy type: Non-dormant.

Embryo type: Lateral. **Embryo:seed ratio:** 0.33.

Pre-treatments: Clean to pure seed. A distinct germination pattern in response to temperature exists at >25°C.

Germination: Maximum seed germination achieved with no pre-treatment is 70–85% (at 25–35°C). Treatment with GA_3 or smoke increased germination to >85% at 25–35°C.

Notes: Additional dormancy break pre-treatments such as dry after-ripening or wet/dry cycling may assist in improving germination across a wider temperature range.

Neurachne muelleri

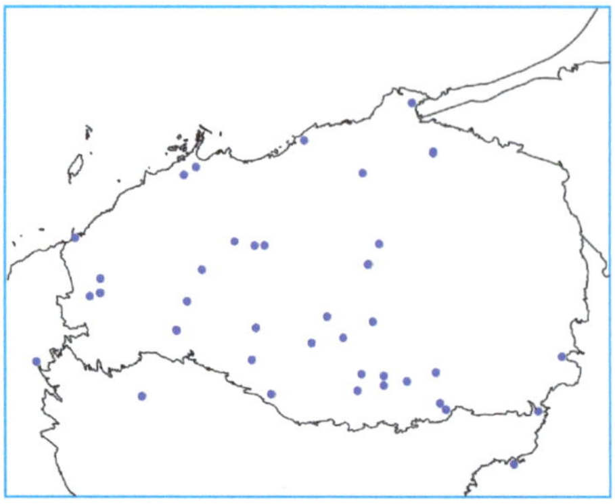

Common names: Northern mulga grass.

Indigenous names: None recorded.

Description: Tufted or stoloniferous, up to 0.45 m high. Leaves: sheath almost glabrous to hairy but usually with tubercles or tubercle-based hairs; blade flat or loosely inrolled, 2–4 mm wide, sparsely to rather densely hairy with in addition conspicuous tubercles or tubercle-based hairs. Inflorescence a dense, cylindric or ovoid uninterrupted raceme, not secund, *c.* 30–35 mm long, on a short to long peduncle. Spikelets 7.4–12 mm long; pedicels up to 0.7 mm long; callus at base of glumes with long spreading stiff hairs. Glumes subequal or the lower a little shorter, at least the upper as long as the spikelet; lower glume prominently veined dorsally, especially in the upper two-thirds, lateral veins bearing rather sparse stiff tubercle-based hairs, acuminate to somewhat awned; upper glume smooth dorsally and indurate in the lower two-thirds, somewhat prominently veined above, with very dense submarginal bands of long stiff hairs in the middle one-third, elsewhere glabrous. Lower floret: lemma prominently veined, inner pair of lateral veins becoming indurate, scabrous dorsally, acute; palea small, membranous. Upper floret: lemma smooth and glabrous on back, ciliate on upper margins, acute; palea similar.

Key features for identification: Creeping stolons. Lower glume membranous and flat on the back; upper glume convex in the lower half and much hardened, smooth and inconspicuously nerved.

Similar species: Readily distinguished from other species by the dissimilar glumes.

Distribution: Through the southern Kimberley, Pilbara and interior, extending south to the latitude of Shark Bay. Also in N. T., S.A. and Qld.

Habitat: In *Triodia* grassland with *Eucalyptus* and *Acacia*.

Flowering period: March to October.

Neurachne muelleri

Fruit/Seed collection

Fruit/Seed maturity: March to May.

Appearance of mature fruit/seed: Immature inflorescences are green (with an occasional purple-black tinge) during flowering and tightly packed at the apex of stems. Once mature, the elongated, almost cylindrical, inflorescences open up and are silvery-white in colour. Whole inflorescences or individual florets detach easily from the stems when hand collected.

Cleaning requirements

Cleaning techniques: Seeds can be removed from the florets by rubbing through a suitably sized sieve. The lighter floret debris can then be removed by vacuum separation. Care must be taken during the sieving process as the embryo is prone to damage.

Purity and viability: Low purity (<20%) with many florets not containing a seed. Once cleaned to pure seeds, viability is high (>90%).

Seeds per gram: 509.

Germination requirements

Dormancy type: Physiological.

Embryo type: Lateral. **Embryo:seed ratio:** 0.59.

Pre-treatments: After cleaning to seeds, germination is increased significantly with GA_3 exposure. The response to dry after-ripening is unknown.

Germination: Maximum seed germination achieved with no chemical pre-treatment is 10% at 25°C. Treatment with GA_3 increases germination to 60–80% at 15–35°C.

Notes: Additional dormancy break pre-treatments such as dry after-ripening or wet/dry cycling may assist in improving germination without the need for GA_3 (or smoke).

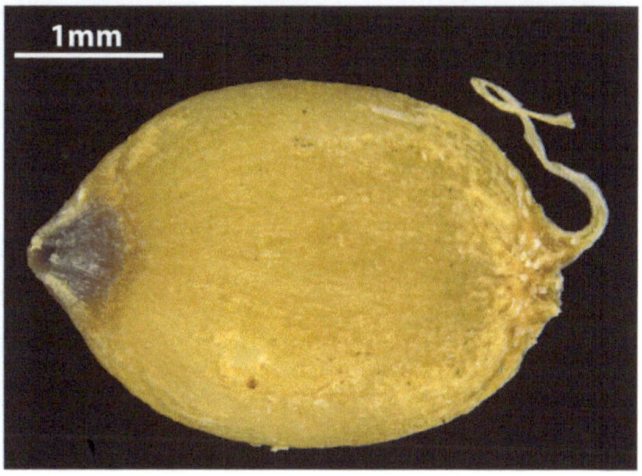

Triodia angusta

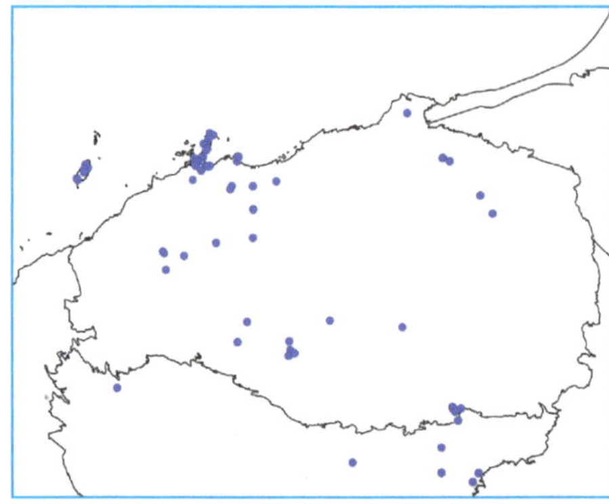

Common names: Slender spinifex.

Indigenous names: None recorded.

Description: Perennial, stolons present, culms 0.5–2 m tall; mid-culm internodes glabrous; leaf-sheath auricles absent. Ligule a fringe of hairs. Leaf blades straight or flexuous, aciculate, conduplicate, 80–200 mm long, 1.2–1.8 mm wide. Inflorescence a compound panicle, central inflorescence axis 100–410 mm long, appressed lateral racemes 20–30 mm long. Spikelets pedicellate; fertile spikelets many-flowered, with 3–6 fertile florets with diminished florets at the apex, cuneate, laterally compressed or terete, 4–12 mm long. Glumes similar, thinner than fertile lemma; lower glume lanceolate, membranous or scarious, without keels, 1-nerved, surface glabrous, apex mucronate; upper glume lanceolate, 3–5 mm long, hyaline or scarious, without keels, 1–3-nerved, surface glabrous, apex entire, mucronate. Fertile lemma 2.5–4 mm long, without keel or keeled, 3-nerved, surface glabrous or hairy; apex dentate or lobed. Anthers 3.

Key features for identification: Spiciform racemose inflorescence with secund, biseriate spikelets; unequal mucronate glumes *c.* 1/2 as long as the spikelet; shortly 3-lobed, almost glabrous, 3-nerved lemmas; long scabrous-pubescent paleas; stoloniferous, often bluish habit.

Similar species: Allied to *T. secunda*, but lacks the fringed auricles and hirsute lemmas of that species.

Distribution: Relatively widespread in the Pilbara region.

Habitat: On the banks of rivers and creeks, margins of salt flats, cracking clay plains and flats, beach sands, sand, quartzite and sandstone.

Flowering period: All seasons, mostly late summer.

Triodia angusta

Fruit/Seed collection

Fruit/Seed maturity: March to April.

Appearance of mature fruit/seed: Many spikelets containing individual florets make up the inflorescence. If present, a single seed is enclosed in a floret. Low seed fill rates are common among *Triodia* species. When mature, the whole inflorescence will turn straw yellow. Whole spikelets or individual florets will fall from the inflorescence when ripe. Floret size is much smaller than other *Triodia* species so collections by weight contain many more florets. Lobes of the florets are also reduced, to almost absent. Collection methods range from hand collection to machine harvesting using modified brush harvesters.

Cleaning requirements

Cleaning techniques: Vacuum separation can vastly increase the proportion of filled florets as the lighter florets that do not contain seeds can be isolated. To clean to pure seed, filled florets can be passed through sieves using a rubber bung. Rubbing florets across a ribbed rubber mat is an alternative method. Repeated vacuum separation and sieving will ensure all seeds are removed and separated from the florets. When cleaning to seed, damage to the embryo can easily occur. Periodically checking for damage will ensure embryo damage is minimised.

Purity and viability: Purity or floret fill rates can be highly variable, but seed viability is high (>90%).

Seeds per gram: Filled florets = 1450. Seeds = 1965.

Germination requirements

Dormancy type: Physiological. **Embryo type:** Lateral. **Embryo:seed ratio:** 0.34.

Pre-treatments: Dry after-ripening for at least 12 months promotes germination of smoke-treated florets. Cleaning to pure seeds and smoke treatment promotes germination after 3–12 months storage.

Germination: Maximum germination of after-ripened florets (seeds retained in the floret) for 12 months on water is <5% (tested at 25°C). Treatment of florets after-ripened for 12 months with smoke increases germination to 50–60%. Maximum germination of pure seeds (extracted from florets) after-ripened for 12 months is 20% when tested on water, increasing to 70–80% following treatment with smoke.

Notes: Cleaning to pure seeds and treating with smoke appears to be the optimal treatment for germination. Yet, caution must be taken as seeds that are retained in the floret may have an advantage during establishment in the soil. Additional dormancy break pre-treatments such as wet/dry cycling may assist in promoting germination in shorter periods of storage time (<3 months).

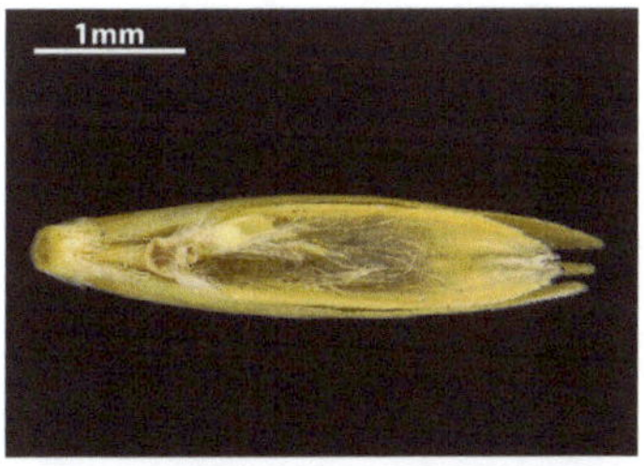

Triodia basedowii

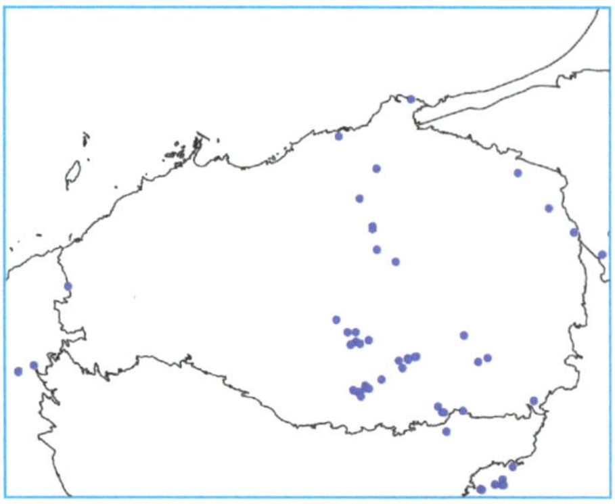

Common names: Basedow's spinifex, Lobed spinifex.

Indigenous names: None recorded.

Description: Perennial, stolons present, culms 0.15–1.3 m tall; leaf sheaths hairy, auricles absent. Ligule a fringe of hairs. Leaf blades straight, aciculate, conduplicate, 50–250 mm long, 0.7 mm wide. Inflorescence compound, a linear panicle, 60–130 mm long, 10–15 mm wide. Spikelets pedicellate; fertile spikelets many-flowered, with 5–6 fertile florets with diminished florets at the apex, oblong, laterally compressed, 7–18 mm long. Glumes similar, thinner than fertile lemma; lower glume oblong, scarious, without keels, 7–13-nerved, apex muticous or mucronate; upper glume oblong, 6–10 mm long, scarious, without keels, 7–13-nerved, apex entire, muticous or mucronate. Fertile lemma 5.5–9 mm long, without keel or keeled, 9–15-nerved, surface hairy, apex lobed. Anthers 3.

Key features for identification: The deeply lobed, bitextured lemma; bitextured mostly glabrous, long palea; many-nerved scarious usually glabrous glumes long relative to spikelet; linear usually short and dense panicle, and usually woolly indumentum are diagnostic features.

Similar species: Related to *T. lanigera* which has much longer hairs on the ligule and *T.* sp. Shovelanna Hill which differs in its typically smaller habit with shorter leaves and in its fewer-branched inflorescence.

Distribution: Widespread in the eastern Pilbara. Also in N.T., S.A., N.S.W. and Qld.

Habitat: On flat to gently sloping and undulating sandplains, dunefields, dune swales and sandhills in deep, red or reddish-yellow, siliceous sands and sandy loams.

Flowering period: All seasons, but most common a month after heavy summer rainfall.

Triodia basedowii

Fruit/Seed collection

Fruit/Seed maturity: March to April.

Appearance of mature fruit/seed: Many spikelets containing individual florets make up the inflorescence. If present, a single seed is enclosed in a floret. Low seed fill rates are common among *Triodia* species. When mature, the whole inflorescence will turn straw yellow. Whole spikelets or individual florets will fall from the inflorescence when ripe. Collection methods range from hand collection to machine harvesting using modified brush harvesters.

Cleaning requirements

Cleaning techniques: Vacuum separation can vastly increase the proportion of filled florets as the lighter florets that do not contain seeds can be isolated. To clean to pure seed, filled florets can be passed through sieves using a rubber bung. Rubbing florets across a ribbed rubber mat is an alternative method. Repeated vacuum separation and sieving will ensure all seeds are removed and separated from the florets. When cleaning to seed, damage to the embryo can easily occur. Periodically checking for damage will ensure embryo damage is minimised.

Purity and viability: Purity or floret fill rates can be highly variable, but are typically 30–40%. But, once seeds are produced, viability is high (>90%).

Seeds per gram: Filled florets = 419. Seeds = 755.

Germination requirements

Dormancy type: Physiological. **Embryo type:** Lateral. **Embryo:seed ratio:** 0.54.

Pre-treatments: Cleaning to pure seeds, after-ripening seeds for >12 months, and treating with smoke appears to be the optimal treatment. Little germination is achieved if seeds are retained in the florets.

Germination: Maximum germination of after-ripened florets (seeds retained in the floret) for 12 months on water or smoke is <5% (tested at 25°C). Maximum germination of pure seeds (extracted from florets) is *c.* 40–50% when after-ripened for 12 months and tested on water. When treated with smoke, no benefit of after-ripening is observed for cleaned seeds, with maximum germination of *c.* 60–65% occurring immediately after cleaning.

Notes: Cleaning to pure seeds and treating with smoke appears to be the optimal treatment for germination. Germination of seeds within florets appears to be more complex in this species and additional dormancy break pre-treatments such as wet/dry cycling need to be developed.

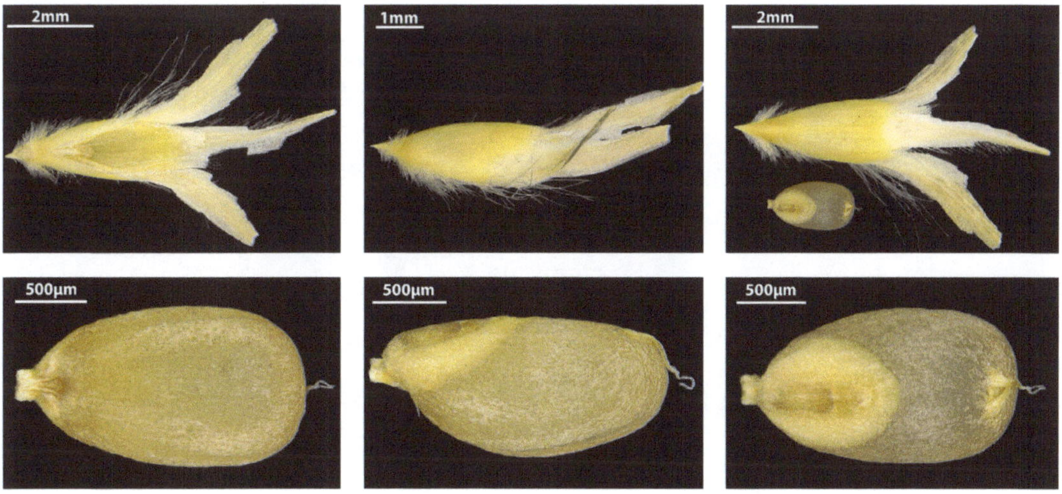

Triodia brizoides

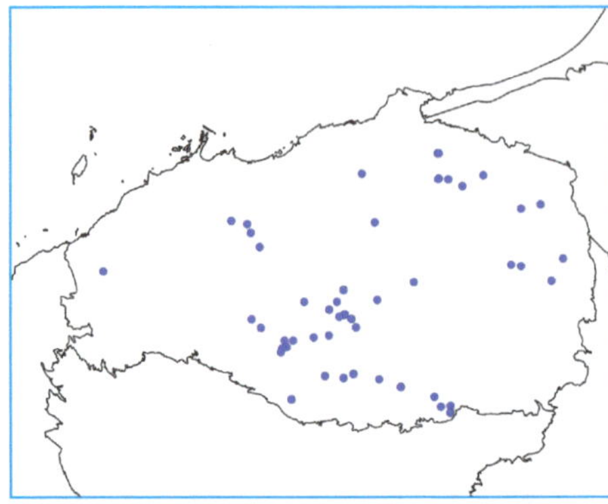

Common names: Oat-flowered spinifex.

Indigenous names: None recorded.

Description: Perennial, stolons present, culms 0.3–1.5 m tall; mid-culm internodes glabrous; leaf sheaths glabrous on surface, auricles absent. Ligule a fringe of hairs. Leaf blades straight, aciculate, conduplicate, 30–200 mm long, 0.6–1 mm wide. Inflorescence a compound panicle, linear or elliptic, effuse, 50–170 mm long. Spikelets pedicellate. Fertile spikelets many-flowered, with 7–10 fertile florets, with diminished florets at the apex, oblong, laterally compressed, 8–22 mm long. Glumes similar, thinner than fertile lemma; lower glume oblong, scarious or cartilaginous or indurate, without keels, 3–5-nerved, surface glabrous, apex muticous or mucronate; upper glume oblong, 4–10 mm long, scarious or cartilaginous or indurate, without keels, 3–5-nerved, surface glabrous, apex entire, muticous or mucronate. Fertile lemma 4.5–7 mm long, without keel or keeled, 9-nerved, surface hairy; apex dentate or lobed. Anthers 3.

Key features for identification: Hard leaf anatomy; leaves bluish green; hairy ligules; broad, pendant spikelets.

Similar species: *T. longiceps* has short hairs on the ligule. *T. wiseana* has the lemma hirsute only in the lower 1/2 and lobes 1/4–1/2 as long as the lemma.

Distribution: Widespread in the northern arid zone of W.A. Also in N.T. and Qld.

Habitat: In skeletal, rocky or stony soils on hills, ranges, ridges, slopes and outcrops with quartzite, sandstone, laterite and limestone, shale, basalt and metamorphic rocks, occasionally on sandy plains, along rivers or on clays.

Flowering period: All seasons, but most common a month after heavy summer rainfall.

Triodia brizoides

Fruit/Seed collection

Fruit/Seed maturity: (March to) April.

Appearance of mature fruit/seed: Many spikelets containing individual florets make up the inflorescence. If present, a single seed is enclosed in a floret. Low seed fill rates are common among *Triodia* species. When mature, the whole inflorescence will turn straw yellow. Whole spikelets or individual florets will fall from the inflorescence when ripe. Florets are distinctly hairy on the floret margins which turn golden brown when mature. Collection methods range from hand collection to machine harvesting using modified brush harvesters.

Cleaning requirements

Cleaning techniques: Vacuum separation can vastly increase the proportion of filled florets as the empty florets can be isolated. To clean to pure seed, filled florets can be passed through sieves using a rubber bung. Rubbing florets across a ribbed rubber mat is an alternative method. Repeated vacuum separation and sieving will ensure all seeds are removed and separated from the florets. When cleaning to seed, damage to the embryo can easily occur. Periodic checks will ensure embryo damage is minimised.

Purity and viability: Purity or floret fill rates in this species have been observed to be very low (<10%). But, once seeds are produced, viability is high (>90%).

Seeds per gram: Filled florets = 500. Seeds = 1058.

Germination requirements

Dormancy type: Physiological. **Embryo type:** Lateral. **Embryo:seed ratio:** 0.52.

Pre-treatments: Cleaning to pure seeds and treating with smoke appears to be the optimal treatment. Little germination is achieved if seeds are retained in the florets and after-ripened.

Germination: Maximum germination of after-ripened florets (seeds retained in the floret) for 12 months on water or smoke is <40% (tested at 25°C). Maximum germination of pure seeds (extracted from florets) is *c.* 30–50% when after-ripened for 3–12 months and tested on water. When treated with smoke, no benefit of after-ripening is observed for cleaned seeds, with maximum germination of *c.* 80–90% occurring immediately after cleaning or within 6 weeks of after-ripening treatments.

Notes: Cleaning to pure seeds and treating with smoke appears to be the optimal treatment for germination. Germination of seeds within florets appears to be more complex in this species and additional dormancy break pre-treatments such as wet/dry cycling need to be developed.

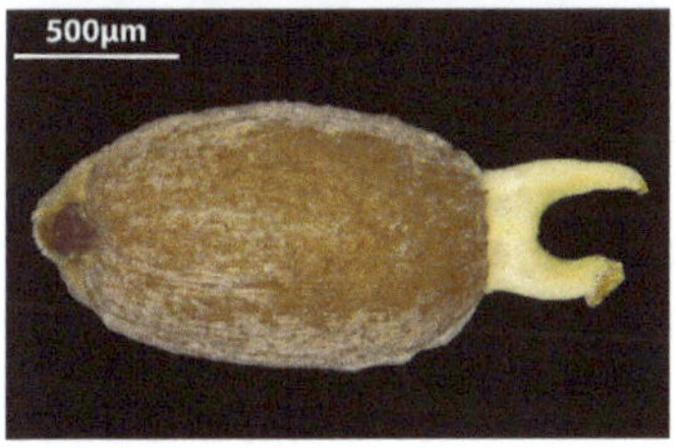

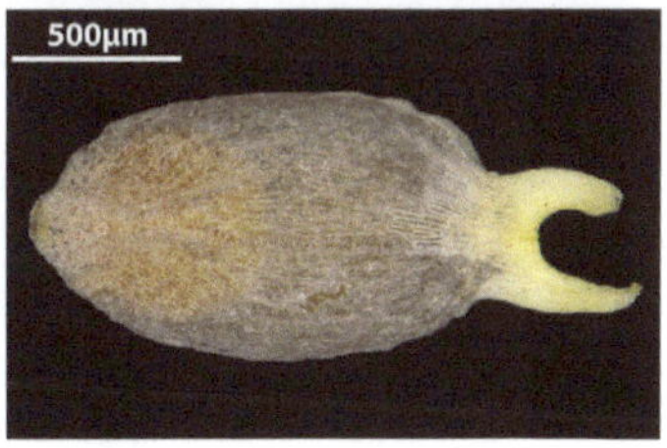

Triodia epactia

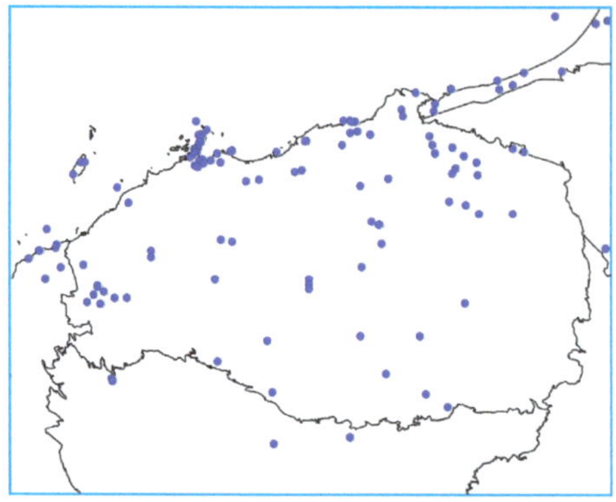

Common names: Sandplain spinifex, Soft spinifex.

Indigenous names: Ngajarri (Thalanyji); Yalhi (Yindjibarndi); Mina (Ngarluma).

Description: Perennial, stolons absent or present, culms 0.6–2 m tall; lateral branches fastigiate; leaf sheaths smooth, glabrous on surface. Ligule a fringe of hairs, 0.5–1.3 mm long. Leaf blades straight or curved or curled or flexuous, flat or conduplicate, 150–350 mm long, 0.7–2 mm wide, surface smooth. Inflorescence a compound panicle, linear or lanceolate or oblong, dense, 100–510 mm long, 15–35 mm wide. Spikelets pedicellate; fertile spikelets many-flowered, with 5–9 fertile florets with diminished florets at the apex, lanceolate, laterally compressed or terete, 9–12 mm long. Glumes similar, thinner than fertile lemma; lower glume lanceolate or oblong, scarious or cartilaginous or indurate, 1-keeled, 3–5(–7)-nerved, surface glabrous, apex muticous or mucronate; upper glume lanceolate or oblong, 4.5–11 mm long, scarious or cartilaginous or indurate, 1-keeled, 3–5(–7)-nerved, surface asperulous, glabrous, apex entire, muticous or mucronate. Fertile lemma 4–10 mm long, without keel or keeled, 9–15 -nerved; surface hairy; apex lobed, muticous. Lodicules present. Anthers 3.

Key features for identification: Slender panicle; dense spikelets; bitextured, broadly winged lemmas; prominent lemma lobes terminated with a bristle, subequal to lemma body in length; bitextured, beaked paleas, keels winged.

Similar species: *T. pungens* has a non-bitextured lemma with smaller winged lobes and sparser indumentum.

Distribution: Widespread in the Pilbara, southern Kimberley and Great Sandy Desert Regions. Also in N.T.

Habitat: On sandplains, sandhills, coastal dunes, on granite, basalt, quartzite, sandstone, shale, and laterite.

Flowering period: All seasons, mostly late summer.

Triodia epactia

Fruit/Seed collection

Fruit/Seed maturity: February to March.

Appearance of mature fruit/seed: Many spikelets containing individual florets make up the inflorescence. If present, a single seed is enclosed in a floret. Low seed fill rates are common among *Triodia* species. When mature, the whole inflorescence will turn straw yellow. Whole spikelets or individual florets will fall when ripe. The floret lobes are distinctly pointy and curl outwards away from the spikelet when mature. Collection methods range from hand collection to machine harvesting using modified brush harvesters.

Cleaning requirements

Cleaning techniques: Vacuum separation can vastly increase the proportion of filled florets as the empty florets can be isolated. To clean to pure seed, filled florets can be passed through sieves using a rubber bung. Rubbing florets across a ribbed rubber mat is an alternative method. Repeated vacuum separation and sieving will ensure all seeds are removed and separated from the florets. When cleaning to seed, damage to the embryo can easily occur. Periodic checks will ensure embryo damage is minimised.

Purity and viability: Purity or floret fill rates in this species have been observed to be very low (10–20%). But, once seeds are produced, viability is high (>90%).

Seeds per gram: Filled florets = 520. Seeds = 1113.

Germination requirements

Dormancy type: Physiological. **Embryo type:** Lateral. **Embryo:seed ratio:** 0.47.

Pre-treatments: Dry after-ripening for at least 6–24 months promotes germination of seeds within florets (smoke treated florets germinate marginally better). Alternatively, cleaning to pure seeds and treating with smoke removes the need for any dry after-ripening treatments.

Germination: Maximum germination of after-ripened florets (seeds retained in the floret) for 6–12 months on water or on smoke is 50–60% (tested at 25°C). After 24 months, germination for both water and smoke treated florets is *c.* 70%. Maximum germination of pure seeds (extracted from florets) is *c.* 45% when tested on water and *c.* 65% when tested on smoke. No benefit of after-ripening is observed for cleaned seeds, with germination gradually decreasing during dry after-ripening.

Notes: After-ripening florets for 6–24 months or cleaning to pure seeds and treating with smoke appears to be the optimal treatment for germination. Additional dormancy break pre-treatments such as wet/dry cycling may assist in promoting germination of florets in shorter periods of storage time (<3 months).

Triodia pungens

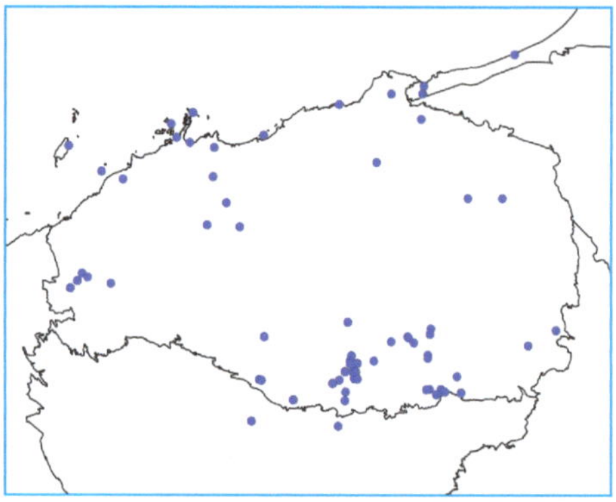

Common names: Soft spinifex, Gummy spinifex.

Indigenous names: Yalhi (Yindjibarndi); Mina (Ngarluma).

Description: Perennial grass with stolons, culms 0.2–2.3 m high, resinous. Leaf-blades flexuous or tortuous, aciculate, flat or conduplicate, to 340 mm long, 0.8–1.2 mm wide. Inflorescence a panicle, linear or pyramidal, 140–270 mm long, 10–25 mm wide, contracted about primary branches. Spikelets pedicellate, green to purple. Fertile spikelets many flowered, lanceolate, laterally compressed or terete, 7–12 mm long. Glumes similar, thinner than fertile lemma. Lower glume oblong, scarious or cartilaginous or indurate, without keels, 3–5-nerved, apex muticous or mucronate. Upper glume oblong, 3–10 mm long, scarious or cartilaginous or indurate, without keels, 3–5-nerved, apex entire, muticous or mucronate. Fertile lemma 5–8 mm long, 9-nerved, apex entire or lobed, muticous. Anthers 3.

Key features for identification: Resinous foliage; linear panicles; usually closely flowered spikelets; sparsely hairy, usually deeply lobed lemmas.

Similar species: Similar to many species of *Triodia* in general appearance. Most often confused with *T. epactia* which has longer spikelets, usually 9–12 mm long.

Distribution: As currently defined, very widespread from the Pilbara region to the Gulf of Carpentaria. This species complex is to be broken up and a new name will probably be required for Pilbara populations. Some of the points mapped here will apply to *T. epactia*.

Habitat: Grows on a wide range of soils, particularly in gravelly or rocky areas, but also on plains and flats.

Flowering period: Recorded all year round, usually 4–6 weeks following rainfall, peaking February to April.

Triodia pungens

Fruit/Seed collection

Fruit/Seed maturity: March to April.

Appearance of mature fruit/seed: Many spikelets containing individual florets make up the inflorescence. If present, a single seed is enclosed in a floret. Low seed fill rates are common among *Triodia* species. When mature, the whole inflorescence will turn straw yellow. Whole spikelets or individual florets will fall from the inflorescence when ripe. Collection methods range from hand collection to machine harvesting using modified brush harvesters.

Cleaning requirements

Cleaning techniques: Vacuum separation can vastly increase the proportion of filled florets as lighter florets lacking seeds can be isolated. To clean to pure seed, filled florets can be passed through sieves using a rubber bung or florets can be rubbed across a ribbed rubber mat. Repeated vacuum separation and sieving will ensure all seeds are removed and separated from the florets. When cleaning to seed, damage to the embryo can easily occur. Periodic checks will ensure embryo damage is minimised.

Purity and viability: Purity or floret fill rates can be highly variable. But, once seeds are produced, viability is high (>90%).

Seeds per gram: Filled florets = 470–600. Seeds = 800.

Germination requirements

Dormancy type: Physiological. **Embryo type:** Lateral. **Embryo:seed ratio:** 0.45.

Pre-treatments: Dry after-ripening for at least 6–12 months promotes germination of smoke-treated florets or cleaning to pure seeds and treating with smoke promotes earlier germination.

Germination: Maximum germination of after-ripened seeds enclosed in florets is 30% (tested at 25°C). Treatment of after-ripened seeds in florets with smoke increases germination to 60–70%. Maximum germination of pure seeds (extracted from florets) after dry after-ripening is 80%, increasing to 90% following treatment with smoke.

Notes: Cleaning to pure seeds and treating with smoke appears to be the optimal treatment for germination. Yet, caution must be taken as seeds that are retained in the floret may have an advantage during establishment in the soil. Additional dormancy break pre-treatments such as wet/dry cycling may assist in promoting germination in shorter periods of storage time (<3 months).

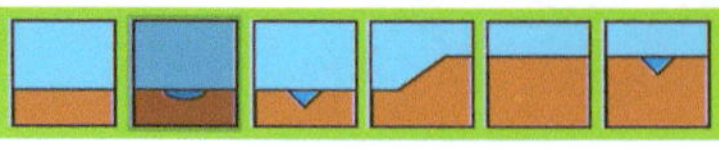

Triodia wiseana

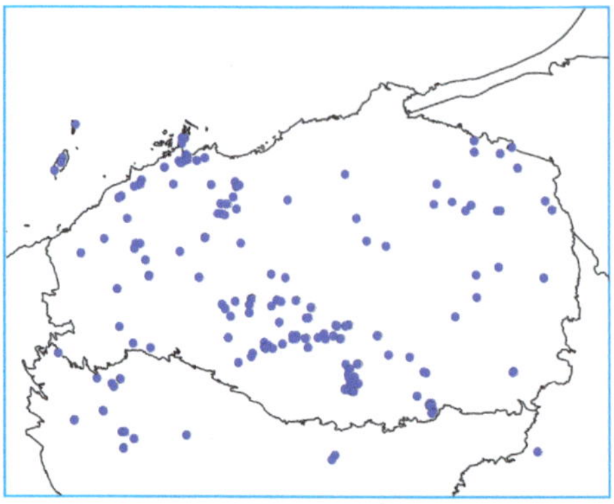

Common names: Wise's spinifex, Limestone spinifex, Hard spinifex.

Indigenous names: Baru (Yindjibarndi, Ngarluma).

Description: Perennial with or without stolons, culms 0.3–1.2 m high, not resinous. Leaf-sheaths glabrous on surface or hairy. Ligule a fringe of long straight stiff hairs. Leaf-blades curved or flexuous, aciculate, conduplicate, 45–300 mm long, 0.8–1.3 mm wide. Inflorescence a dense panicle, linear or oblong, 90–350 mm long, 15–30 mm wide. Spikelets sessile or pedicellate, green to purple. Fertile spikelets many flowered, lanceolate or oblong, laterally compressed or terete, 7–19 mm long. Glumes similar, thinner than fertile lemma; lower glume lanceolate or oblong, scarious or cartilaginous or indurate, 1-keeled, 3–5-nerved, surface glabrous, apex muticous or mucronate; upper glume lanceolate or oblong, 4.5–9 mm long, scarious or cartilaginous or indurate, 1-keeled, 3–5-nerved, surface smooth or asperulous, glabrous, apex muticous or mucronate. Fertile lemma 4–7.5 mm long, without keel, 9–10-nerved, surface glabrous or indumented, apex deeply and narrowly lobed, muticous. Anthers 3.

Key features for identification: Sparsely hairy lemmas, with usually deep subulate lobes; strongly 3-nerved glumes; long leaves; indumentum of long, straight, stiff hairs on orifice, which extend along margins of blades in tubercular-based clusters.

Similar species: Similar to *T. brizoides* which has woolly tangled hairs on the ligule and lower leaf blade.

Distribution: Widespread in the Pilbara region and adjacent areas.

Habitat: Grows on a variety of shallow and skeletal soils, alluvium, stony clay loam, coarse sandy soils, in gullies, on hills, ranges, outcrops, sandplains and creek beds.

Flowering period: October to May.

Triodia wiseana

Fruit/Seed collection

Fruit/Seed maturity: February to April.

Appearance of mature fruit/seed: Many spikelets containing individual florets make up the inflorescence. If present, a single seed is enclosed in a floret. Low fill rates are common among *Triodia* species. When mature, the whole inflorescence will turn straw yellow. Whole spikelets or individual florets will fall from the inflorescence when ripe. Collection methods range from hand collection to machine harvesting using modified brush harvesters.

Cleaning requirements

Cleaning techniques: Vacuum separation can vastly increase the proportion of filled florets as the lighter florets that do not contain seeds can be isolated. To clean to pure seed, filled florets can be passed through sieves using a rubber bung. Rubbing florets across a ribbed rubber mat is an alternative method. Repeated vacuum separation and sieving will ensure all seeds are removed and separated from the florets. When cleaning to seed, damage to the embryo can easily occur. Periodically checking for damage will ensure embryo damage is minimised.

Purity and viability: Purity or floret fill rates can be highly variable. But, once seeds are produced, viability is high (>90%).

Seeds per gram: Filled florets = 800–1000. Seeds = 1200–1300.

Germination requirements

Dormancy type: Physiological. **Embryo type:** Lateral. **Embryo:seed ratio:** 0.40.

Pre-treatments: Dry after-ripening for at least 12–24 months promotes germination of smoke-treated florets or treating pure seeds with smoke promotes germination after shorter storage periods.

Germination: Maximum germination of after-ripening seeds enclosed in florets is 70%, increasing to 80% if treated with smoke (tested at 25°C). Maximum germination of pure seeds (extracted from florets) after dry after-ripening is 70%, increasing to 80% if treated with smoke.

Notes: Cleaning to seeds and treating with smoke appears to be the optimal treatment. Yet, caution must be taken as seeds that are retained in the floret may have an advantage during establishment in the soil. Significant viability loss has been observed during after-ripening of seeds. Additional dormancy break pre-treatments such as wet/dry cycling may assist in promoting germination in <3 months.

Triodia sp. Shovelanna Hill (S. van Leeuwen 3835)

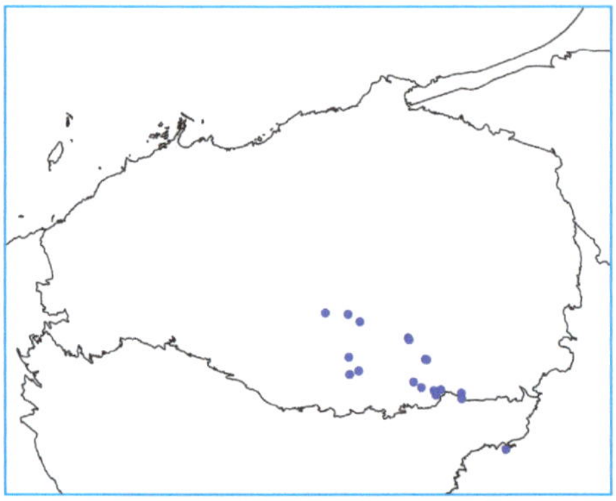

Common names: Hamersley Range spinifex.

Indigenous names: None recorded.

Description: Perennial, often with stolons, 0.15–0.5 m tall, bluish hummocks. Flowering culms 0.5–1 m high. Leaf anatomy type 'hard' (not resinous), but often with small resin droplets on sheaths. Leaf sheaths glabrous; orifice with straight, tangled or woolly hairs, not extending onto the blade. Ligule a fringe of straight hairs 0.5–1 mm long. Leaf blades 45–90 mm long, rolled, sharply pointed. Inflorescence a racemose panicle or raceme, 34–72 mm long, few branches with >1 spikelet on them, 5–9 spikelets per inflorescence; pedicels 0.6–1.1 cm long. Spikelets 9–12 mm long, with 5–8 florets. Lower glumes 5–7 mm long with 6–12 prominent nerves; apex acute to acuminate. Lemma prominently three-lobed; indurated body 2–3 mm long, pubescent; midlobe narrowly elliptic to oblong, 3–4 mm long, glabrous, with 5 nerves, apex acute to acuminate; lateral lobes broadly ovate to elliptic, 2.5–3.2 mm long, glabrous, with 5 nerves, apex rounded to acute. Palea 3–3.5 mm long, 1.25 mm wide, glabrous. Anthers 3.

Key features for identification: Hummocks small, bluish. Lower glumes with 6–12 prominent nerves.

Similar species: Closely related to *T. basedowii* and *T. lanigera*, from which it differs in its typically smaller habit with shorter leaves and in its fewer-branched inflorescence. It is recognisably distinct from other 'hard' species in its many-nerved glumes.

Distribution: Restricted to the Hamersley Ranges in the southern Pilbara region.

Habitat: Associated with hilltops and slopes, in skeletal, rocky or gravelly soil as well as in stony or sandy loam, also occurring on lower slopes of hills and on valley floors.

Flowering period: February to April.

Triodia sp. Shovelanna Hill (S. van Leeuwen 3835)

Fruit/Seed collection

Fruit/Seed maturity: March to April.

Appearance of mature fruit/seed: Many spikelets containing individual florets make up the inflorescence. If present, a single seed is enclosed in a floret. Low fill rates are common among *Triodia* species. When mature, the whole inflorescence will turn straw yellow. This species occasionally has a strong purple tinge colouration on the stems and glumes. Whole spikelets or individual florets will fall from the inflorescence when ripe. Collection methods range from hand collection to machine harvesting using modified brush harvesters.

Cleaning requirements

Cleaning techniques: Vacuum separation can vastly increase the proportion of filled florets as lighter florets lacking seeds can be isolated. To clean to pure seed, filled florets can be passed through sieves using a rubber bung or florets can be rubbed across a ribbed rubber mat. Repeated vacuum separation and sieving will ensure all seeds are removed and separated from the florets. When cleaning to seed, damage to the embryo can easily occur. Periodic checks will ensure embryo damage is minimised.

Purity and viability: Purity or floret fill rates can be highly variable. But, once seeds are produced, viability is high (>90%).

Seeds per gram: Filled florets = 370–460. Seeds = 660–800.

Germination requirements

Dormancy type: Physiological. **Embryo type:** Lateral. **Embryo:seed ratio:** 0.55.

Pre-treatments: Cleaning to pure seeds, after-ripening seeds for 12–24 months, and treating with smoke appears to be the optimal treatment. Little germination is achieved if seeds are retained in florets.

Germination: Maximum germination of after-ripening seeds enclosed in florets is 10%, increasing to 20% if treated with smoke. Maximum germination of pure seeds (extracted from florets) after dry after-ripening is 50%, increasing to 60% if treated with smoke.

Notes: Cleaning to seeds and treating with smoke appears to be the optimal treatment. Yet, caution must be taken as seeds that are retained in the floret may have an advantage during establishment in the soil. Additional dormancy break pre-treatments such as wet/dry cycling may assist in promoting germination in shorter periods of time (<3 months).

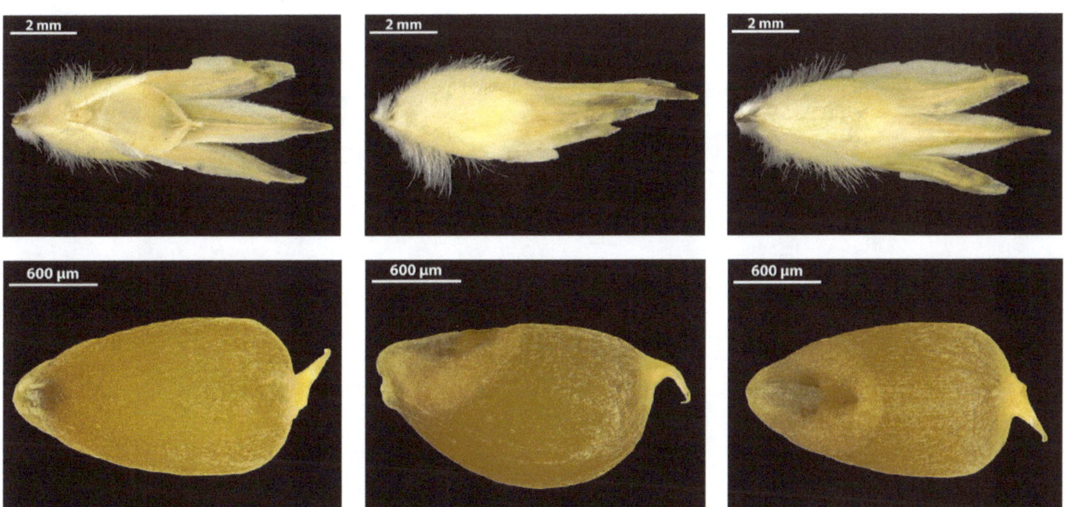

Grasses

Calandrinia schistorhiza

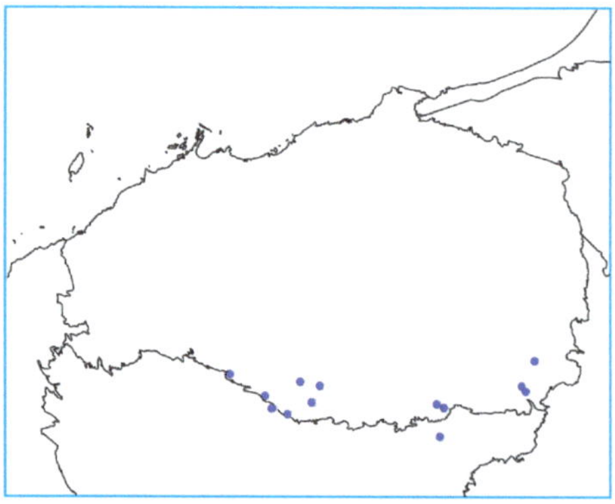

Common names: Spoon-leaf parakeelya.

Indigenous names: None recorded.

Description: Perennial herb with a tuberous tap-root, prostrate to decumbent, 30–140 mm tall, 40–300 mm wide, glabrous. Basal leaves fleshy, broadly spathulate, 15–700 mm long, 3–25 mm wide. 2–4 bract pairs per stem with a whorl of 3 bracts on the lowest node. Stem bracts spreading, scarious, triangular to broadly ovate, 3–7 mm long, apex sometimes recurved. Inflorescence a terminal solitary flower on an erect pedicel which may deflex in fruit. Sepals orbicular, 7–11.5 mm long, commonly spreading and crinkled when dry, often with a hyaline margin. Petals 5, normally bright to deep pink, 14–28 mm long, obovate to broadly obovate and often with a slightly crenulate and depressed apex. Stamens numerous; filaments free; anthers versatile. Ovary ellipsoid to subglobose. Stigmatic lobes 3, united onto a short style. Capsule ellipsoid to broadly ovoid, 8–11 mm long, 3–5 mm wide; valves 3, splitting fully to base. Seeds black, roughly kidney-shaped, strongly verrucose and sometimes tuberculate, 0.5–0.8 mm long.

Key features for identification: A medium to large plant with large flowers. Leaf length is extremely variable; the blade always broadly spathulate and petiole often strap-like.

Similar species: The tuber and sepals somewhat resemble those of *C. crispisepala*. The sepals sometimes have a hyaline margin similar to *C. balonensis*.

Distribution: Widespread from the Goldfields to the southern Pilbara regions.

Habitat: Usually grows in red sandy loam or clays often stony or rocky soils in a variety of communities.

Flowering period: June to September.

Photographs by Rob Davis.

Calandrinia schistorhiza

Fruit/Seed collection

Fruit/Seed maturity: October (limited observations).

Appearance of mature fruit/seed: Individual capsules are located at the apex of the flower stems. When mature, the capsules turn brown and split easily when collected. Each capsule contains numerous small kidney-shaped black seeds. If there is a large population, bulk collections can be made by removing multiple flowering stems at the same time.

Cleaning requirements

Cleaning techniques: Seeds can easily be separated from capsule debris by using a suitably sized sieve and vacuum separation.

Purity and viability: Viability high (>95%).

Seeds per gram: 9633.

Germination requirements

Dormancy type: Non-dormant.

Embryo type: Peripheral. **Embryo:seed ratio:** 1.37.

Pre-treatments: No pre-treatment requirements. However, exposure to GA_3 enhances germination of fresh seeds at lower incubation temperatures.

Germination: Maximum seed germination achieved with no pre-treatment is 80–90% at 20–35°C. Treatment with GA_3 increases germination to >70% at 10–15°C.

Notes: Additional dormancy break pre-treatments such as dry after-ripening may assist in increasing overall germination at lower incubation temperatures.

Portulaca oleracea

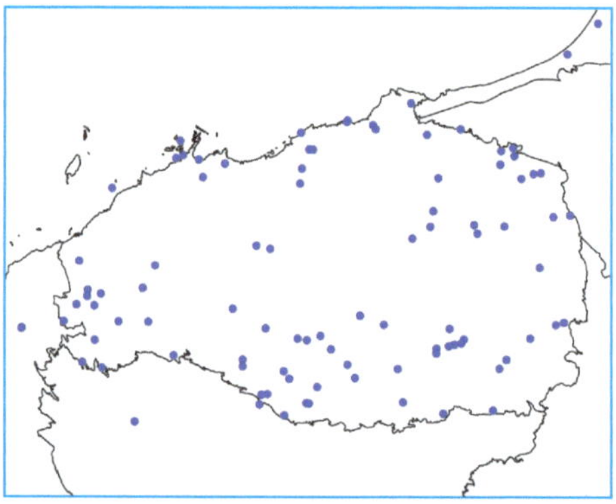

Common names: Pig weed, Purslane.

Indigenous names: Gudburru (Yindjibarndi); Gulburru (Ngarluma).

Description: Prostrate to decumbent annual herb, with stems to 250 mm long. Axillary hairs few, inconspicuous, to 1 mm long, confined to axil. Leaves alternate to subopposite, subsessile, narrowly obovate to obovate or cuneate, 10–25 mm long, 6–15 mm wide, truncate, obtuse or emarginate. Inflorescence a head of 2–30 sessile flowers. Sepals narrowly ovate to ovate, 3–6 mm long, keeled. Petals yellow, obovate, 4–7 mm long. Stamens 8–15. Stigmas 4–6 on a common style as long or slightly longer than the stigmatic arms. Capsule ovoid, 3–5 mm long; operculum conic, as long as the base. Seeds black, globular to more or less obovoid, 0.6–1 mm long, shiny to dull, with rounded or pointed tubercles, especially on the margin.

Key features for identification: Decumbent to spreading habit; succulent, flat leaves; yellow flowers; elongate operculum with compressed apex; black seeds with irregular tuberculae.

Similar species: *P. australis* has sepals that lack a keel or wing and a blunt operculum.

Distribution: Widespread in the Kimberley and Pilbara regions. A cosmopolitan species occurring as both a native and a weed in disturbed areas of all states of Australia. Occurs in temperate and tropical regions of the world.

Habitat: Clay loam and sand. Often in disturbed sites

Flowering period: April to May.

Note: The weedy form is illustrated here, which may be introduced to Australia.

Portulaca oleracea

Fruit/Seed collection

Fruit/Seed maturity: May to June.

Appearance of mature fruit/seed: Numerous small black seeds are contained in ovoid capsules at the apex of flower stems. When mature, the capsule turns light brown and the lid separates from the capsule base. A small collection window exists once the capsule is open.

Cleaning requirements

Cleaning techniques: Seeds should be collected in a relatively pure state. Capsule debris, if present, can easily be separated using a suitably sized sieve and vacuum separation.

Purity and viability: Viability high (>90%).

Seeds per gram: 12480.

Germination requirements

Dormancy type: Physiological.

Embryo type: Peripheral. **Embryo:seed ratio:** 2.00.

Pre-treatments: Exposure to GA_3 significantly enhances germination of fresh seeds.

Germination: Treatment with GA_3 results in >90% germination across the temperature range of 20–35°C. Germination at lower temperatures is <40%. Germination of fresh seeds does not occur on water or with smoke treatment.

Notes: A period of >3 months in dry storage increases germination of seeds on water indicating that dormancy loss occurs via dry after-ripening.

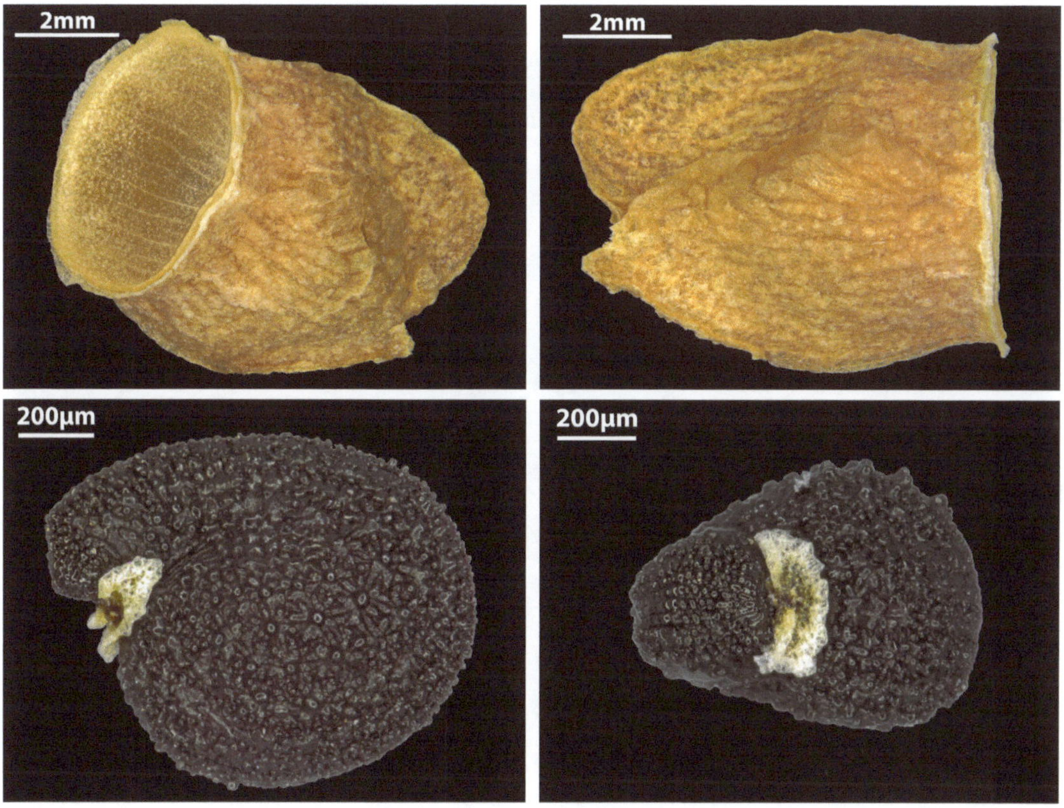

Grevillea pyramidalis subsp. *leucadendron*

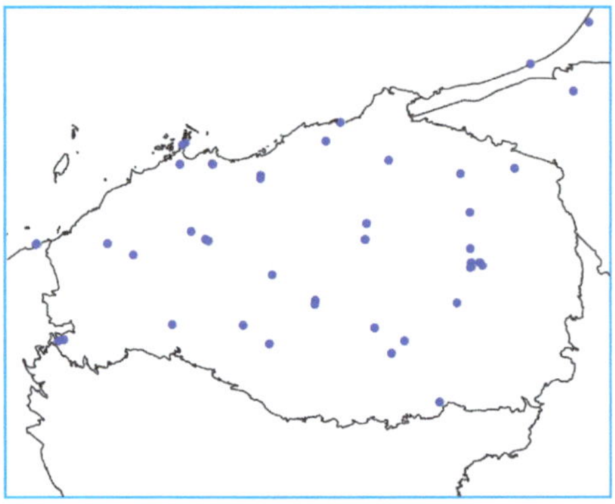

Common names: Caustic bush.

Indigenous names: Gura (Yindjibarndi); Jiingu (Ngarluma).

Description: Shrub or small tree, 2–6 m high. Leaves 150–300 mm long, blade dissected, subpinnatisect, lobes 50–290 mm long, 1–8 mm wide, margins flat, hairs straight. Flowers white, perianth 3–6 mm long, glabrous, lobes all free, pistil 5–7 mm long, stipitate; stipe 2–3 mm long, ovary glabrous, styles glabrous, white, pollen presenter conical. Fruit viscid, obovoid, glabrous, 18–23 mm long.

Key features for identification: Divided leaves with long, narrow lobes; viscid (caustic) fruit.

Similar species: There are two subspecies. Subsp. *leucadendron* has densely hairy adult leaves with lobes 1–8 mm wide. Subsp. *pyramidalis* has glabrous adult leaves with lobes 3–20 mm wide. *G. stenobotrya* has very narrow leaf lobes, 1–2 mm wide, and the fruit are only lightly viscid when young. *G. striata* has leaves that are not lobed.

Distribution: Widespread in the Pilbara region, extending to the Kimberley and N.T.

Habitat: Sandy and loam soils, usually on flats.

Flowering period: May to July.

Note: The sticky resin on the fruit and inflorescence can be highly caustic and can cause burns. If contact with skin occurs, wash the resin off with copious amounts of water.

Grevillea pyramidalis subsp. *leucadendron*

Fruit/Seed collection

Fruit/Seed maturity: October to December.

Appearance of mature fruit/seed: Two seeds are encased within a woody fruit that splits into two segments when mature. The best time to collect the fruit is when the incision line between the two fruit segments begins to split. During maturation, the woody fruit turns from lime green to dark brown. Bulk collections can be made easily as the woody fruits are held upright on stems, providing greater seed retention, and allowing whole stems to be upturned into buckets/bags.

Cleaning requirements

Cleaning techniques: Once the fruits have split open, seeds are easily separated from the woody fruit. Seeds can be separated from the woody segments by passing through a sieve.

Purity and viability: Viability high (>95%).

Seeds per gram: 23 (range 19–23).

Germination requirements

Dormancy type: Non-dormant.

Embryo type: Investing. **Embryo:seed ratio:** 1.00.

Pre-treatments: No pre-treatment requirements.

Germination: Maximum seed germination achieved with no pre-treatment is 90–100% at 20–35°C. Response to GA_3 and smoke unknown.

Grevillea refracta subsp. *refracta*

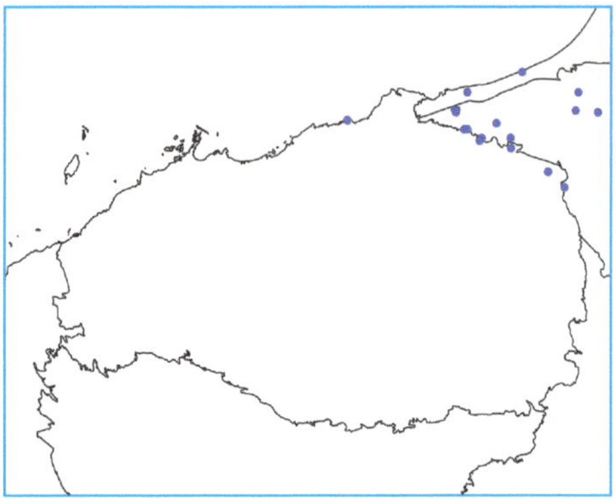

Common names: Silver-leaf grevillea, Silver oak.

Indigenous names: None recorded.

Description: Shrub or tree, 1.2–6 m high, with dense, fine, 2-branched hairs. Leaves simple or divided; petiole 10–50 mm long; blade silvery green on lower surface and pale to dark green above, narrowly obovate to linear, 55–155 mm long, 12–33 mm wide or with up to 30 leaflets and 90–155 mm long, with conspicuous veins, acute or attenuate; leaflets similar to simple blade in shape, 22–155 mm long, 3–30 mm wide. Inflorescence of axillary panicles of racemes; racemes several per panicle, 20–35 mm long, few–10-flowered. Pedicels 2.5–6 mm long. Calyx orange or red with yellowish apex, distally recurved to revolute, 9.5–11 mm long, densely white-hairy outside, glabrous inside except for a belt of hairs below middle, the hairs short and fine; claw 4–5 mm wide; limb 3–3.5 mm long. Gynoecium 18–21 mm long, glabrous; stipe 3–4 mm long; stigmatic disc lateral, broadly elliptic, 2–2.5 mm long. Follicle very dark brown to black, 25–30 mm long, 22–25 mm diam., thick-walled. Seeds: body 12–14 mm long, 9–11 mm wide; wing 4.5–6 mm wide.

Key features for identification: Silvery, simple or divided leaves; large, yellow to red flowers; large, thin-walled fruit.

Similar species: No similar species in the Pilbara.

Distribution: Widespread from the Kimberley Region to the northern Pilbara. Also occurs in N.T. and Qld.

Habitat: Occurs mainly in sand on sandstone outcrops, cliffs and plateaus.

Flowering period: Mainly April to September.

Grevillea refracta subsp. *refracta*

Fruit/Seed collection

Fruit/Seed maturity: September to October.

Appearance of mature fruit/seed: Two seeds are encased within a woody fruit that splits into two segments when mature. The best time to collect the fruit is when the incision line between the two fruit segments begins to split. During maturation, the woody fruit turns from lime green to dark brown. The collection window can occur prior to the fruit splitting provided the fruit has turned 'woody' and is internally dry.

Cleaning requirements

Cleaning techniques: Once the fruits have split open, seeds are easily separated from the woody fruit. Seeds can be separated from the woody segments by passing through a sieve.

Purity and viability: Viability high (100%).

Seeds per gram: 8 (range 8–19).

Germination requirements

Dormancy type: Non-dormant.

Embryo type: Investing. **Embryo:seed ratio:** 1.00.

Pre-treatments: No pre-treatment requirements.

Germination: Maximum seed germination achieved with no pre-treatment is 80% at 25°C. Germination decreases by at least 20% at 20°C and 30°C. The response to GA_3 and smoke is unknown.

Notes: Exposure to GA_3 and smoke and or additional dormancy break pre-treatments such as dry after-ripening or wet/dry cycling may assist in improving germination across a wider temperature range.

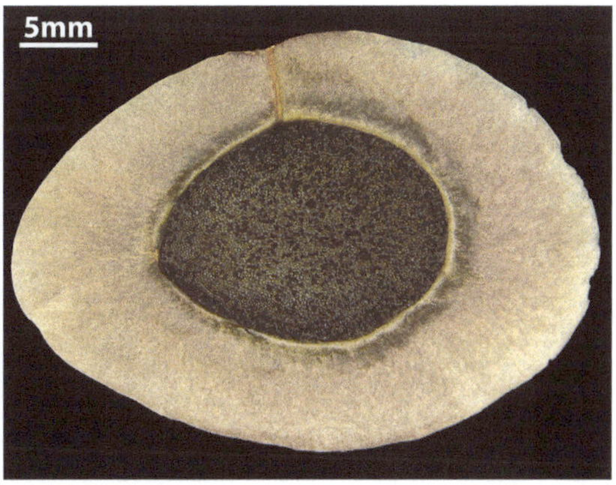

Grevillea stenobotrya

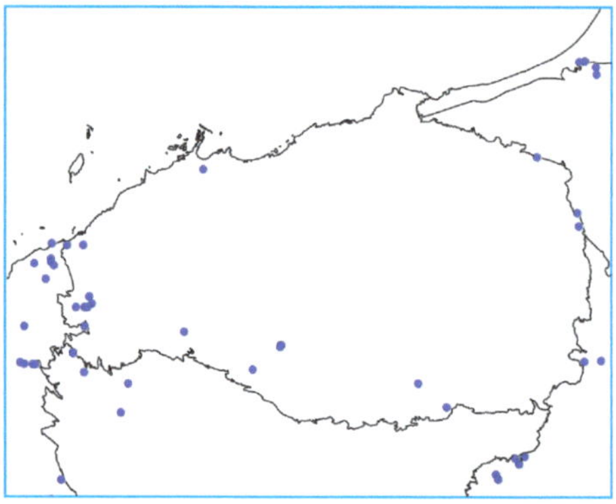

Common names: Rattle-pod grevillea.

Indigenous names: None recorded.

Description: Shrubs or small trees, 2–7 m high. Branchlets not glaucous. Leaves 60–280 mm long; blade 0.5–2.5 mm wide when simple, dissected or undissected, flat, linear, subpinnatisect; lobes 60–120 mm long, 1–2 mm wide; margins entire, revolute, enclosing the lower surface of the leaf blade, forming two grooves with the midvein. Hairs straight. Inflorescence terminal; a 5–12(–20)-branched raceme. Flowers green or white or cream, very irregular. Pedicel 2–5 mm long. Perianth 3–6 mm long, simple-hairy, 4-partite; lobes all free; stamens 4; pistil 6–10 mm long, stipitate; stipe 0.5–1.5 mm long; ovary glabrous; styles glabrous, white; pollen presenter oblique (sometimes almost lateral). Fruit viscid when young, oblong or ellipsoidal, glabrous, 13–16 mm long.

Key features for identification: Leaves tightly revolute, only midvein obvious above, completely hidden below.

Similar species: Similar to *G. parallela* which has the lower surface of the leaves with only partly revolute margins, the midvein being exposed; leaves with 1–7 parallel veins. *G. berryana* has a 2–6-branched inflorescence.

Distribution: Widespread but scattered in the Pilbara region and arid zone of W.A. Also in N.T., S.A., Qld and N.S.W.

Habitat: In woodland on sandhills.

Flowering period: May to December.

Grevillea stenobotrya

Fruit/Seed collection

Fruit/Seed maturity: October to November.

Appearance of mature fruit/seed: Two seeds are encased within a woody fruit that splits into two segments when mature. The best time to collect the fruit is when the incision line between the two fruit segments begins to split. During maturation, the woody fruit turns from lime green-yellow to dark brown. Bulk collections can be made easily as the woody fruits are held upright on stems, providing greater seed retention, and allowing whole stems to be upturned into buckets/bags.

Cleaning requirements

Cleaning techniques: Once the fruits have split open, seeds are easily separated from the woody fruit. Seeds can be separated from the woody segments by passing through a sieve.

Purity and viability: Viability moderately high (*c.* 80%).

Seeds per gram: 32.

Germination requirements

Dormancy type: Non-dormant.

Embryo type: Investing. **Embryo:seed ratio:** 1.00.

Pre-treatments: Exposure to GA_3 and smoke enhances germination of fresh seeds over a wider temperature when compared to testing on water alone.

Germination: Maximum seed germination achieved with no pre-treatment is 95% at 15°C. At 10°C and 20°C germination on water decreases to 40% and 80%, respectively. Germination is absent on water at >25°C. Treatment with GA_3 or smoke increases germination to >90% at 10–20°C, and increases germination to 50% at 25°C using smoke.

Notes: Additional dormancy break pre-treatments such as dry after-ripening or wet/dry cycling may assist in improving germination across a wider temperature range and remove the requirement to treat with GA_3 or smoke.

Grevillea wickhamii subsp. *hispidula*

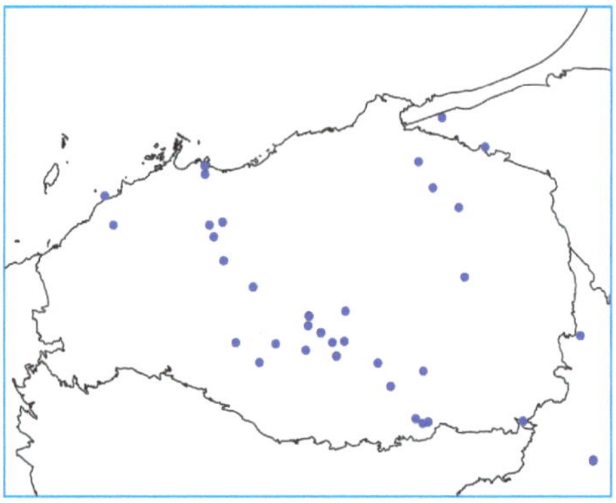

Common names: Wickham's grevillea.

Indigenous names: Ngajarndangu (Yindjibarndi & Ngarluma).

Description: Shrub, 2–5 m tall, lives for 5–10 years, often killed by fire. Leaves 25–70 mm long, 20–35 mm wide, undissected, flat, obovate with several sharply pointed lobes. Flowers red and yellow, with short erect hairs on outside, perianth 5–7 mm long, glandular-hairy, lobes only partly free, pistil 6–8 mm long, stipe 1.5–3 mm long, ovary glabrous, styles hairy, yellow, pollen presenter lateral. Fruit oblong or ellipsoidal, glabrous, to 15 mm long.

Key features for identification: Lobed leaves with sharp points; inflorescence 30–60 mm long; flowers red and yellow; short, erect hairy outside.

Similar species: Similar to *G. wickhamii* subsp. *aprica* and subsp. *macrodonta*. Subsp. *aprica* has appressed biramous hairs on the outer surface of the flowers; inflorescence 15–40 mm long. Subsp. *macrodonta* has glabrous flowers; the margin of the toral rim is dentate with a large triangular tooth at each end; inflorescence 30–60 mm long.

Distribution: Relatively widespread on the ranges of the Pilbara and Little Sandy Desert.

Habitat: Red loams to sandy soils, often near drainage lines.

Flowering period: May to August.

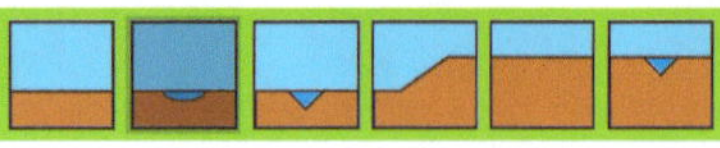

Grevillea wickhamii subsp. *hispidula*

Fruit/Seed collection

Fruit/Seed maturity: September to November.

Appearance of mature fruit/seed: Two seeds are encased within a woody fruit that splits into two segments when mature. The best time to collect the fruit is when the incision line between the two fruit segments begins to split. During maturation, the woody fruit turns from lime green to dark brown. The collection window can occur prior to the fruit splitting provided the fruit has turned 'woody' and is internally dry.

Cleaning requirements

Cleaning techniques: Once the fruits have split open, seeds are easily separated from the woody fruit. Seeds can be separated from the woody segments by passing through a sieve.

Purity and viability: Viability high (>95%).

Seeds per gram: 39 (range 32–51, across all subspecies).

Germination requirements

Dormancy type: Non-dormant.

Embryo type: Investing. **Embryo:seed ratio:** 1.00.

Pre-treatments: Exposure to smoke enhances germination in fresh seeds.

Germination: Maximum seed germination achieved with no pre-treatment is 75% at 25°C. Treatment with smoke increases germination to >95% at 20–25°C, and increases germination to 45–55% at 15°C and 30°C.

Notes: Additional dormancy break pre-treatments such as dry after-ripening or wet/dry cycling may assist in increasing overall germination. Confirming the dormancy status and germination requirements of this species has been difficult. Considerable variation in germination traits may exist amongst populations and between subspecies.

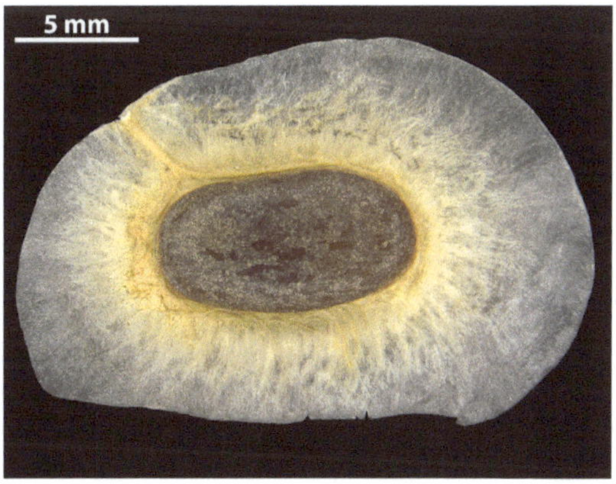

Grevillea wickhamii subsp. *macrodonta*

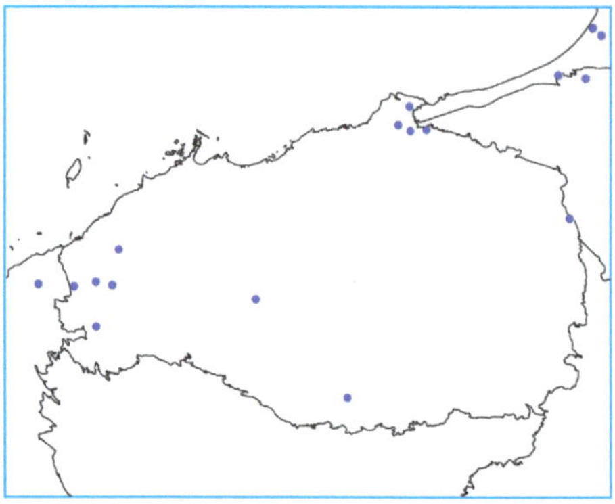

Common names: Wickham's grevillea.

Indigenous names: Ngajarndangu (Yindjibarndi & Ngarluma).

Description: Shrubs, 1.5–2 m high. Branchlets not glaucous. Leaves simple, 30–90 mm long, 25–55 mm wide, undissected, flat, obovate. Margins dentate, flat. Hairs straight. Inflorescence axillary; a raceme. Flowers orange or red (rarely bright yellow), very irregular. Pedicel 3–5 mm long. Perianth 5–7 mm long, glabrous, 4-partite; lobes free and joined. Stamens 4. Pistil 6–8 mm long, stipitate; stipe 1.5–3 mm long. Ovary glabrous. Styles hairy, yellow. Pollen presenter lateral. Fruit oblong or ellipsoidal, glabrous, 10–15 mm long.

Key features for identification: Lobed leaves with sharp points; inflorescence 30–60 mm long; flowers red and yellow, glabrous outside; the margin of the toral rim is dentate with a large triangular tooth at each end.

Similar species: Similar to *G. wickhamii* subsp. *aprica* and subsp. *hispidula*. Subsp. *aprica* has appressed biramous hairs on the outer surface of the flowers; inflorescence 15–40 mm long. Subsp. *hispidula* has short erect hairs on the outer surface of the flowers; inflorescence 30–60 mm long.

Distribution: Scattered across the Pilbara, Great Sandy Desert and south-west Kimberley regions.

Habitat: In woodland in gravelly or loam soil, often on ridges.

Flowering period: Mainly April to August.

Grevillea wickhamii subsp. *macrodonta*

Fruit/Seed collection

Fruit/Seed maturity: September to November.

Appearance of mature fruit/seed: Two seeds are encased within a woody fruit that splits into two segments when mature. The best time to collect the fruit is when the incision line between the two fruit segments begins to split. During maturation, the woody fruit turns from lime green to dark brown. The collection window can occur prior to the fruit splitting provided the fruit has turned 'woody' and is internally dry.

Cleaning requirements

Cleaning techniques: Once the fruits have split open, seeds are easily separated from the woody fruit. Seeds can be separated from the woody segments by passing through a sieve.

Purity and viability: Viability high (>95%).

Seeds per gram: 37 (range 32–51, across all subspecies).

Germination requirements

Dormancy type: Non-dormant.

Embryo type: Investing. **Embryo:seed ratio:** 1.00.

Pre-treatments: Exposure to smoke enhances germination in fresh seeds.

Germination: Maximum seed germination achieved with no pre-treatment is 70–75% at 20–25°C. Treatment with GA_3 and smoke increases germination to >85% at 20–25°C, and increases germination to 35–80% at 15°C and 30°C.

Notes: Additional dormancy break pre-treatments such as dry after-ripening or wet/dry cycling may assist in increasing overall germination. Confirming the dormancy status and germination requirements of this species has been difficult. Considerable variation in germination traits may exist amongst populations and between subspecies.

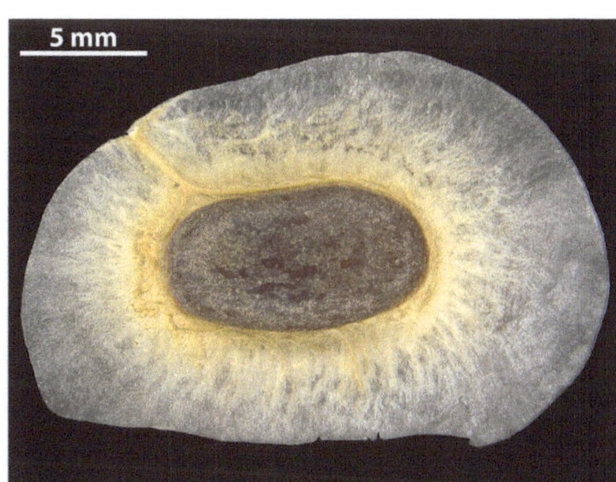

Hakea lorea subsp. *lorea*

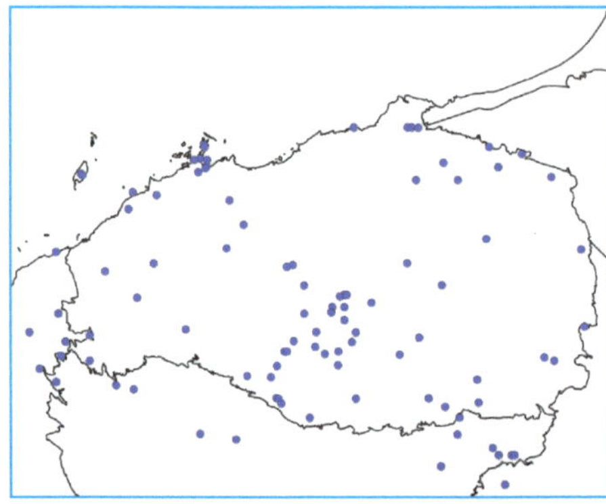

Common names: Corkwood, Long-leaf corkwood, Bootlace tree.

Indigenous names: Garruwa (Yindjibarndi); Garrayin (Ngarluma).

Description: Tall shrub or tree 3–6 m high; bark fissured and corky. New branchlets with appressed or spreading hairs. Leaves terete, 100–330 mm long, 1–1.7 mm diam., hairy to glabrous, entire, pungent to acute. Racemes axillary, 70–130 mm long including peduncle. Pedicels 5–11 mm long, with dense spreading hairs. Receptacle oblique. Calyx cream-yellow, 5–10 mm long, densely hairy outside with spreading hairs and also a few appressed hairs, glabrous inside; limb reflexed. Style 15–20 mm long, glabrous; stigmatic disc oblique, umbonate. Follicle ovoid, 25–40 mm long, 11–15 mm diam., with an excentric apical point, granulate. Seed wing not decurrent.

Key features for identification: New branchlets with appressed or spreading hairs.

Similar species: Similar to *H. chordophylla* which has glabrous new branchlets.

Distribution: From Derby south to Leonora, widespread in the Pilbara. Also occurs in N.T., S.A., Qld and N.S.W.

Habitat: Common on rocky ridges, in open woodland and on hummock-grass sandplains.

Flowering period: May to September.

Hakea lorea subsp. *lorea*

Fruit/Seed collection

Fruit/Seed maturity: October to November.

Appearance of mature fruit/seed: Two seeds are encased within a woody fruit that splits into two segments when mature. The best time to collect the fruit is when the incision line between the two fruit segments begins to split. During maturation, the woody fruit turns from green to dark brown. The collection window of *Hakea* species in the Pilbara appears to be very short. Therefore, collections can occur prior to the fruit splitting provided the fruit has turned 'woody' and is internally dry. Seed retention by capsules appears to be short (<3 months).

Cleaning requirements

Cleaning techniques: Once the fruits have split open, seeds are easily separated from the woody fruit. Seeds can be separated from the woody segments by passing through a sieve.

Purity and viability: Viability generally high (>80%).

Seeds per gram: 18 (range 11–22).

Germination requirements

Dormancy type: Non-dormant.

Embryo type: Investing. **Embryo:seed ratio:** Not determined.

Pre-treatments: No pre-treatment requirements.

Germination: Germination is >90% at 25°C. Germination is reduced to <40% at 30–35°C.

Notes: Collection and germination characteristics for *Hakea chordophylla* are likely to be similar.

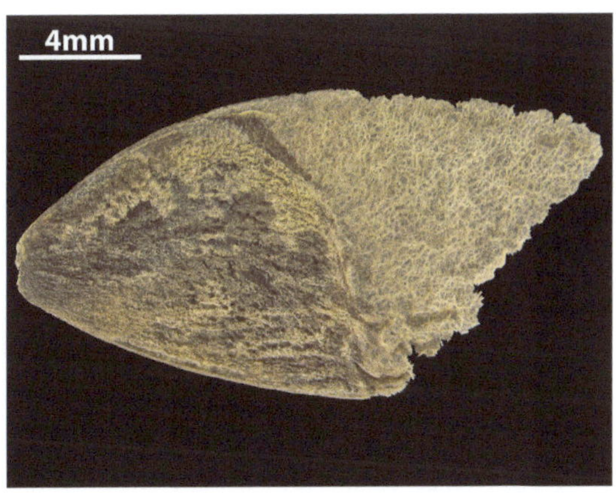

Diplopeltis stuartii var. *stuartii*

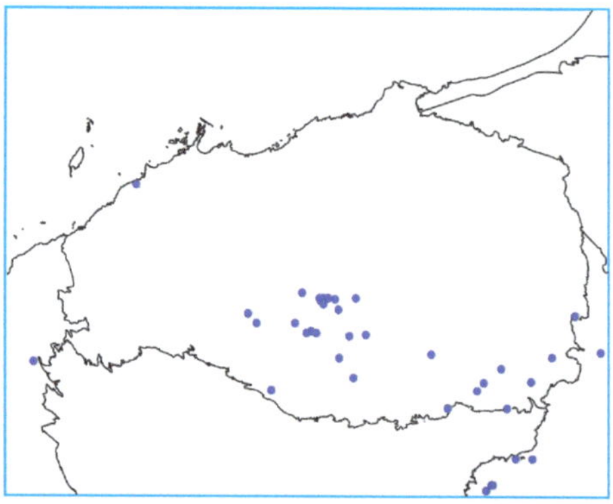

Common names: Desert pepperflower.

Indigenous names: None recorded.

Description: Semi-prostrate shrub or herb to 0.5 m high with indumentum of simple and glandular hairs. Leaves oblong or oblanceolate, lobed or entire, base cuneate, 7–40 mm long, 4–15 mm wide, lower surface densely hairy, upper less so or glabrous. Pedicels 3–10 mm long. Sepals ovate, 2–3 mm long, 1.5–2 mm wide, hirsute outside, sparsely sericeous inside. Petals white or pink, unequal, 5–7 mm long, on slender claws 1 mm long; lamina orbicular, 2–5.5 mm wide, glandular hairs outside, minutely pubescent on margin and inside. Disk 1–1.5 mm long, lobes acute at apex. Filaments 1.5–2.5 mm long. Ovary densely hirsute. Capsule 3-lobed, inflated, 10–15 mm long, 10 mm diam., densely glandular-pubescent.

Key features for identification: Indumentum of simple and glandular hairs; leaves with obtuse or truncate apex with 3–5 acute lobes, 10–25 mm long; petals unequal; fruit a loculicidal capsule.

Similar species: All other species have equal petals and fruit a schizocarp of indehiscent cocci.

Distribution: Widespread in the central arid zone of W.A. Also in N.T.

Habitat: Sandy or lateritic soils, often in *Triodia* grassland associations.

Flowering period: April to September.

Diplopeltis stuartii var. *stuartii*

Fruit/Seed collection

Fruit/Seed maturity: October to November.

Appearance of mature fruit/seed: The multi-seeded capsule has three cylindrically-oblong-shaped segments that are light green when immature. Once mature the capsule splits releasing the seeds.

Cleaning requirements

Cleaning techniques: Seeds should release easily from the capsules once collected and can be separated easily through a sieve. If the capsules require crushing, capsule debris can be separated from the seeds using vacuum separation.

Purity and viability: Viability generally high (>90%). However, collections of this species have had high levels of aborted seeds.

Seeds per gram: 98.

Germination requirements

Dormancy type: Combinational (physical + physiological) (inferred from *Diplopeltis huegelii* in Turner *et al.* 2006).

Embryo type: Linear-coiled. **Embryo:seed ratio:** Not determined.

Pre-treatments: Wet heat for 1–2 mins at 90–95°C.

Germination: Untested. High germination (>80%) expected across a wide range of temperatures once treated to alleviate combinational dormancy.

Notes: Combinational dormancy requires the alleviation of both physical and physiological dormancy. Wet heat treatment has been shown to promote physical dormancy loss in seeds from the Sapindaceae[4,9]. Dry after-ripening can promote the loss of the additional physiological dormancy component of combinational dormancy[9].

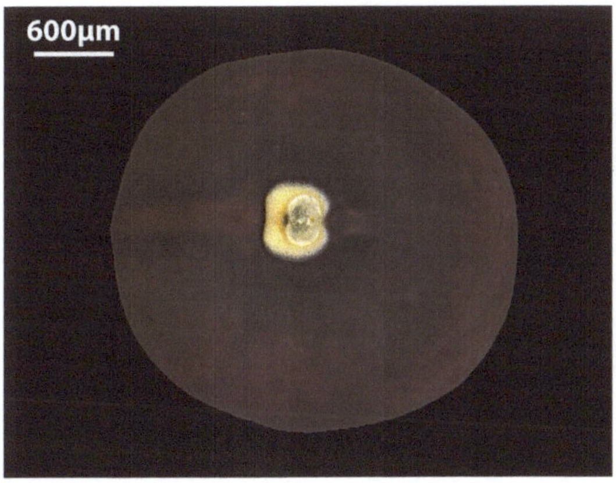

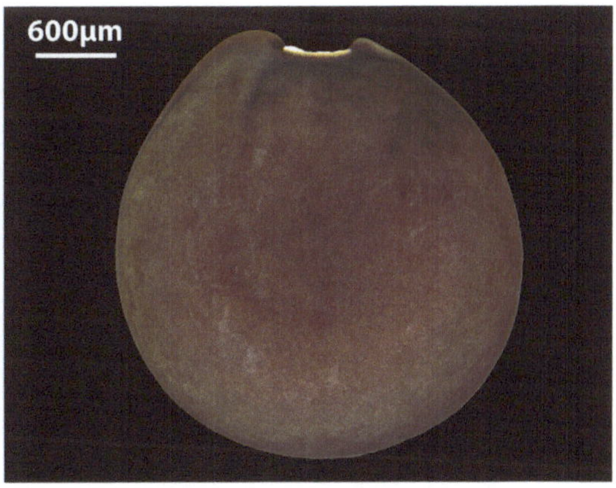

Dodonaea coriacea

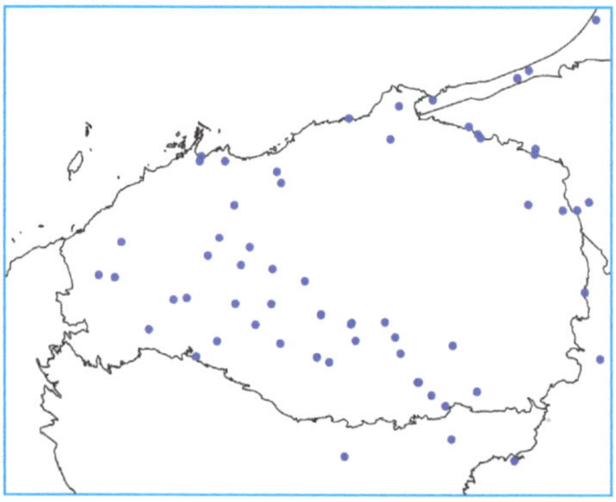

Common names: Hop bush.

Indigenous names: None recorded.

Description: Shrub to 1.5 m high; branchlets, leaves, pedicels, sepals and ovary with small verrucose glands; branchlets viscid and sparsely hairy. Leaves simple, sessile, narrowly obtriangular or narrowly obovate, 10–40 mm long, 4–13 mm wide, coriaceous, midrib prominent, viscid, glabrous or sparsely hairy, base attenuate, margin entire or unevenly serrate or dentate, apex truncate or irregularly dentate. Inflorescence a terminal panicle; pedicels 6–10 mm long, hairy near base; flowers bisexual or rarely unisexual. Sepals 4 or 5, green, narrowly ovate to ovate, 1.5–2.5 mm long, sparsely hairy, acute, persistent. Stamens 4 or 5, usually exserted; filaments 0.5–1 mm long; anthers with small apical appendage, total length 1.3–1.7 mm. Ovary hairy on dorsal side of each carpel. Style sometimes twisted, 1.5–4 mm long. Capsule 3-winged, usually yellow at maturity, sometimes purplish, transversely broadly elliptic in lateral view, 10–16 mm long, 12–17 mm diam., glabrous or sparsely hairy, septifragally dehiscent; wings 3.5–6 mm wide, membranous. Seeds somewhat lenticular, *c.* 3 mm long, 2.5 mm diam., shiny, translucent membrane completely attached.

Key features for identification: Leaves simple, 4–13 mm wide; stamens 4 or 5, sepals persistent; fruit with prominent wings.

Similar species: *D. viscosa* has 6–10 stamens and caducous sepals.

Distribution: From the southern Kimberley to the Pilbara Region. Also occurs in N.T. and Qld.

Habitat: Deep red sand, quartzite or laterite hills.

Flowering period: April to July.

Dodonaea coriacea

Fruit/Seed collection

Fruit/Seed maturity: October to November.

Appearance of mature fruit/seed: The 3- to 4-winged fruit is usually a straw-yellow colour at maturity. Fruits are easily collected by hand when mature. Each fruit contains 1 or 2 black seeds.

Cleaning requirements

Cleaning techniques: The thin-walled fruits can be easily broken down by rubbing by hand. Capsule debris can be separated easily through a sieve, with the debris falling through the sieve. Vacuum separation is an alternative method to separate the fruit debris.

Purity and viability: Viability high (>95%).

Seeds per gram: 189 (range 147–192).

Germination requirements

Dormancy type: Physical.

Embryo type: Linear-coiled. **Embryo:seed ratio:** 2.98.

Pre-treatments: Wet heat for 2 mins at 90–95°C.

Germination: Germination is >95%. Unimpeded across a wide temperature range (10–35°C) once heat treated.

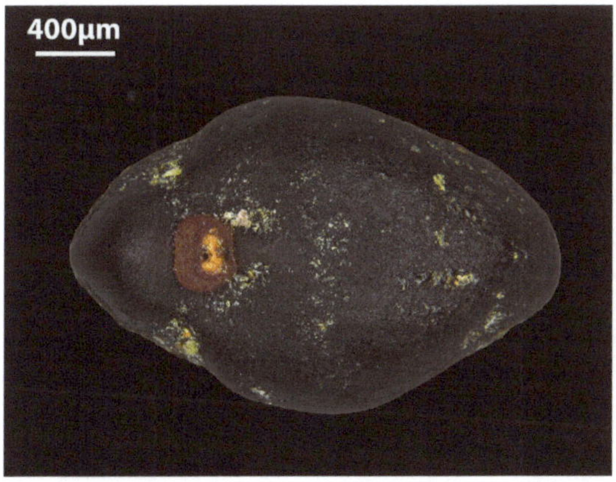

Solanum dioicum

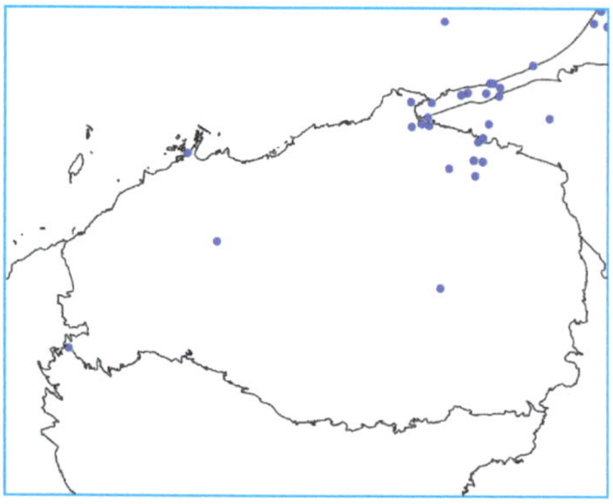

Common names: Gilu.

Indigenous names: None recorded.

Description: Androdioecious, erect or spreading shrub 0.5–1 m high, sparingly clonal, greyish rusty green or silvery green, densely stellate-hairy; prickles common on calyx of female flowers and stems, 5–10 mm long, straight or slightly retrorse. Leaves usually ovate, 35–150 mm long, 12–50 mm wide, concolorous, entire, acute to acuminate. Male inflorescence a cyme with up to 20 flowers. Male flowers: calyx 4–6 mm long, prickles few or absent, the lobes shallowly triangular and 1–2 mm long with small linear apices; corolla purple, shallowly campanulate-rotate, 20–30 mm across; anthers narrowly oblong in outline, 3–5 mm long; ovary vestigial. Bisexual flowers solitary; calyx 10–25 mm long, densely prickly, the lobes linear and 5–15 mm long; corolla purple, broadly star-shaped, 30–50 across; anthers *c.* 5 mm long. Fruiting pedicel 20–30 mm long; fruiting calyx with prickles 5–10 mm long. Berry yellowish green, drying brownish, globular, 20–30 mm wide, surrounded almost completely by the enlarged spiny calyx. Seeds black or dark brown, 2–3 mm long.

Key features for identification: Male flowers with calyx 4–6 mm long, the lobes 1–2 mm long; anthers 3–5 mm long; fruiting pedicel 20–30 mm long; fruiting calyx enclosing berry, with prickles 5–10 mm long.

Similar species: A very variable species belonging to a complex with *S. cunninghamii* and *S. petraeum*. These two species have narrowly elliptic or narrowly ovate leaves, usually 5–25 mm wide and berries 10–20 mm across with prickles 2–5(–7) mm long.

Distribution: Widespread from the Kimberley south to near Port Hedland. Also occurs in N.T.

Habitat: Usually on sand plains, also on ridges.

Flowering period: March to August.

Solanum dioicum

Fruit/Seed collection

Fruit/Seed maturity: October to November.

Appearance of mature fruit/seed: When immature, the globular 'tomato bush' fruits are enclosed in very prickly calyx lobes which are light green in colour. At maturity, the lobes will turn light brown and peel off the fruit exposing the white globular fruit. The fruit wall will also soften when mature. Each fruit contains hundreds of flattened seeds that are black and shiny.

Cleaning requirements

Cleaning techniques: Multiple options exist to clean the seeds out of the fruits. When fresh, the fruits can be cut in half and washed in luke-warm water mixed with a small amount of detergent. The washing process aids in separating the seeds from the slightly tacky fruit pulp. Additionally, washing the fruits in a sieve immersed in the water makes it easy to get the seeds out after cleaning. Immediately after washing the seeds then need to be laid out to dry. An alternative option is to allow the fruits to dry out and then the seeds can be removed from the dried pulp at a later date. However, this can be more difficult than when the fruit is fresh.

Purity and viability: Viability high (>95%).

Seeds per gram: 705 (range 705–750).

Germination requirements

Dormancy type: Physiological.

Embryo type: Linear-curved. **Embryo:seed ratio:** 1.99.

Pre-treatments: Exposure to GA_3 or smoke slightly enhances germination of fresh seeds. Dry after-ripening for 24 months, combined with smoke treatment, yields similar results.

Germination: For fresh seeds tested on water, germination is 0% across a wide temperature range (10–35°C). Once treated with GA_3 or smoke germination increases to 10–15% at 20–25°C. After 24 months of dry after-ripening and exposure to smoke germination increases to 20–30%.

Notes: Previous studies on this species[3] indicate that germination of 60–90% can occur after exposure to GA_3 or smoke. These results were obtained for seeds stored for >10 years indicating that physiological dormancy may have been reduced during the storage period. Therefore, dormancy break pre-treatments such as dry after-ripening (or wet/dry cycling) are recommended.

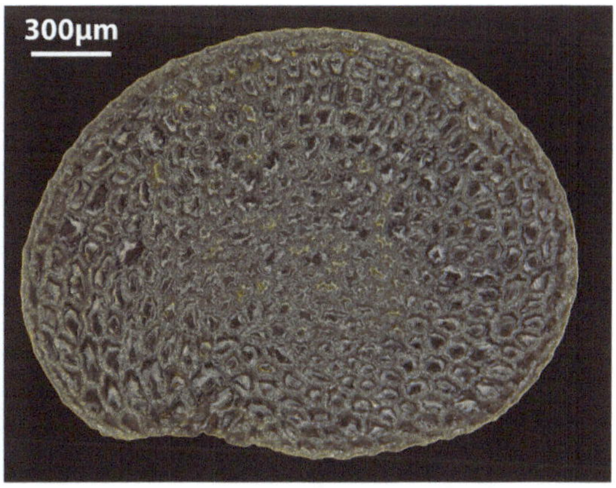

Solanum diversiflorum

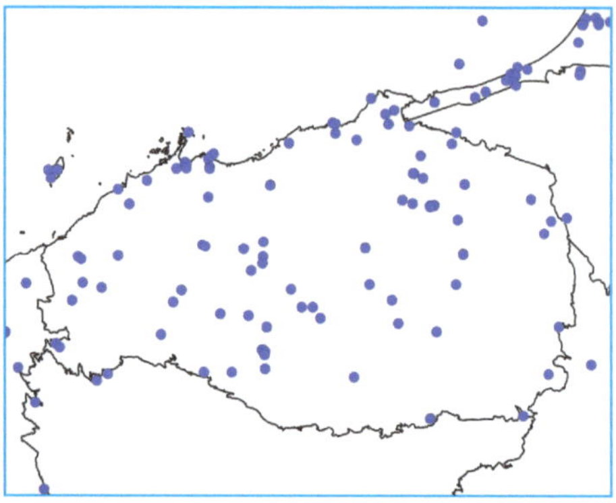

Common names: Bush tomato.

Indigenous names: Garlumbu (Yindjibarndi & Ngarluma).

Description: Sparingly clonal, prickly shrub, 0.2–1 m high, green or grey-green, pubescent with stellate hairs; with many prickles to 10 mm long. Leaves ovate to ovate-oblong; blade 20–40 mm long, 10–20 mm wide, concolorous, deeply lobed; lobes oblong or obovate; petiole to 15 mm long. Inflorescence of one bisexual flower below cyme of up to 20 male flowers, 30–50 mm long. Bisexual flower: pedicels 10–30 mm long, extending in fruit; calyx 5–7 mm long, lobes narrowly triangular, 1.5–3 mm long, extending in fruit; corolla 25–35 mm wide, purple; anthers 5–6 mm long. Male flowers similar but pedicels only 8–15 mm long; calyx-lobes oblong, 2–3 mm long and anthers 4–5 mm long. Berry depressed globular, 20–30 mm wide, pale greenish-yellow; fruiting pedicel 20–40 mm long; fruiting calyx lobes 10–15 mm long. Seeds 3.5–5 mm long, dark brown to black.

Key features for identification: Deeply lobed leaves with prickles; large berry, not surrounded by calyx at maturity.

Similar species: The lobed leaves are very distinctive and it is unlikely to be confused with any native species in the Pilbara region.

Distribution: Widespread in the Pilbara region and the southern Kimberley. Also in the N.T.

Habitat: Often on red sandy plains, sometimes on stony hills, usually with *Triodia* and *Acacia*.

Flowering period: February to November.

Solanum diversiflorum

Fruit/Seed collection

Fruit/Seed maturity: October to January.

Appearance of mature fruit/seed: When immature, the upper quarter of the globular 'tomato bush' fruits are covered in prickly calyx lobes which are light green in colour. At maturity, the lobes will turn from green to light brown-yellow, while the fruit turns from green to white. The fruit wall will also soften when mature. Each fruit contains hundreds of flattened seeds that are black and shiny.

Cleaning requirements

Cleaning techniques: Multiple options exist to clean the seeds out of the fruits. When fresh, the fruits can be cut in half and washed in luke-warm water mixed with a small amount of detergent. The washing process aids in separating the seeds from the slightly tacky fruit pulp. Additionally, washing the fruits in a sieve immersed in the water makes it easy to get the seeds out after cleaning. Immediately after washing the seeds then need to be laid out to dry. An alternative option is to allow the fruits to dry out and then the seeds can be removed from the dried pulp at a later date. However, this can be more difficult than when the fruit is fresh.

Purity and viability: Viability high (>95%).

Seeds per gram: 181 (range 151–190).

Germination requirements

Dormancy type: Physiological.

Embryo type: Linear-curved. **Embryo:seed ratio:** 2.03.

Pre-treatments: Dormancy alleviation treatments are unknown. Germination of fresh seeds after exposure to GA_3 or smoke, or after dry after-ripening for 24 months combined with smoke, is very low.

Germination: For fresh seeds tested on water, GA_3, or smoke, germination is 0% across a wide temperature range (10–35°C). After 24 months of dry after-ripening and exposure to smoke germination is 0–5%.

Notes: Previous studies on this species[3] indicate that germination of *c.* 60% can occur after exposure to GA_3. This indicates that physiological dormancy can be partially reduced by GA_3. Modified dry after-ripening (or wet/dry cycling) treatments need to be re-assessed across multiple seed accessions to resolve dormancy break requirements.

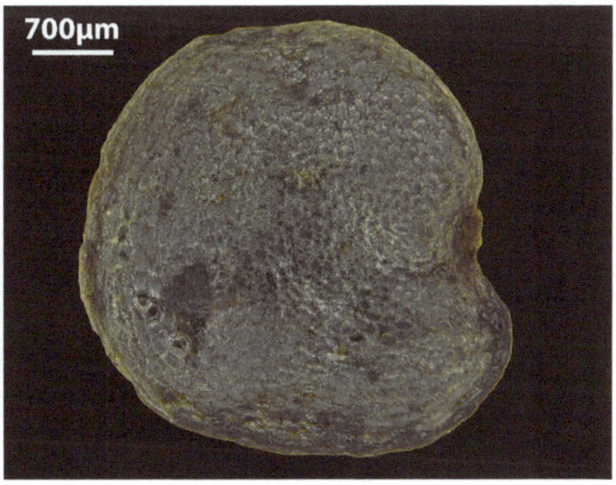

Stylidium desertorum

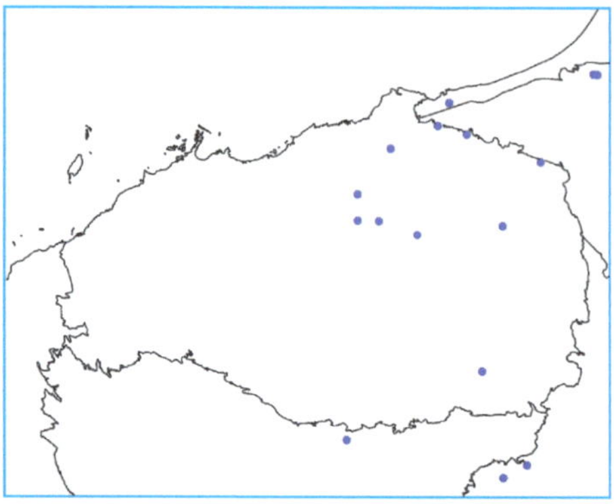

Common names: Desert trigger plant.

Indigenous names: None recorded.

Description: Robust herb 0.07–0.25 m high; stems short, sometimes branched, with the lowermost leaves scattered along the stem and withering early, the upper leaves in a terminal tuft which may appear basal when the stem is very short. Leaves linear, 10–30 mm long, 0.5–1 mm wide, glabrous, margins sometimes minutely white-scabridulous; apex a long white curved hair-like mucro. Inflorescence of 1–several panicles, with slender glandular hairs; peduncle stout, 40–130 mm long; bracts linear, 2–6 mm long; pedicels 1–6 mm long. Ovary with adnate floral tube obovoid, 5–6 mm long, glandular-hairy. Sepals narrowly oblong, 2–3 mm long, free, more or less obtuse. Corolla pink, 7–10 mm long, sparsely glandular-hairy; lobes paired vertically, 3–6 mm long, unequal, entire, obtuse; throat denticulate. Capsule obovoid to ellipsoid but slightly compressed, 4–5 mm long, *c.* 2 mm diam.

Key features for identification: Terete, fleshy leaves in a basal tuft; flowers pink, petals laterally paired.

Similar species: None in the region.

Distribution: From the Edgar Range south to southeast of Carnarvon and the Gibson Desert. Also occurs in N.T.

Habitat: Recorded from seasonal wetlands.

Flowering period: April to September.

Stylidium desertorum

Fruit/Seed collection

Fruit/Seed maturity: August to November.

Appearance of mature fruit/seed: Flowering tends to be fairly staggered throughout the season. Once flowering ceases, immature capsules will develop along the flowering stems. When mature, the small elongated capsule will turn from light green to light brown and split at the capsule tip. Seeds will readily drop from the capsule when upturned. Each capsule contains many seeds. Whole plants can be upturned to collect bulk seeds if the parent plant has senesced.

Cleaning requirements

Cleaning techniques: If collected by upturning stems, the collection should contain near pure seeds. Plant debris and capsules can be separated by vacuum separation if required.

Purity and viability: Viability is very difficult to assess in this species as it has very small seeds. Viability is likely high (>90%) as germination is >95% in multiple germination treatments.

Seeds per gram: 128000.

Germination requirements

Dormancy type: Morphophysiological.

Embryo type: Undifferentiated. **Embryo:seed ratio:** 0.26.

Pre-treatments: Exposure of fresh seeds to GA_3 and smoke significantly increases germination.

Germination: Maximum germination achieved on water is variable with germination ranging between 10–25% at temperatures of 15–35°C. Once treated with GA_3 or smoke germination increases to 55–100%.

Notes: Combinations of either dry after-ripening or wet/dry cycling may assist in alleviating the physiological dormancy component in seeds of this species, removing the need to treat with germination stimulants[5].

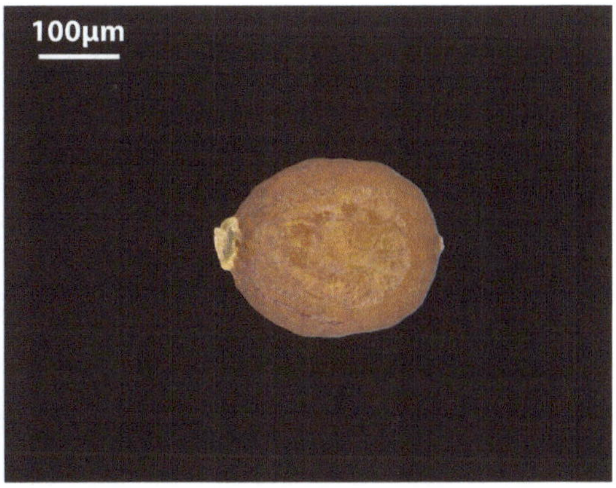

Stylobasium spathulatum

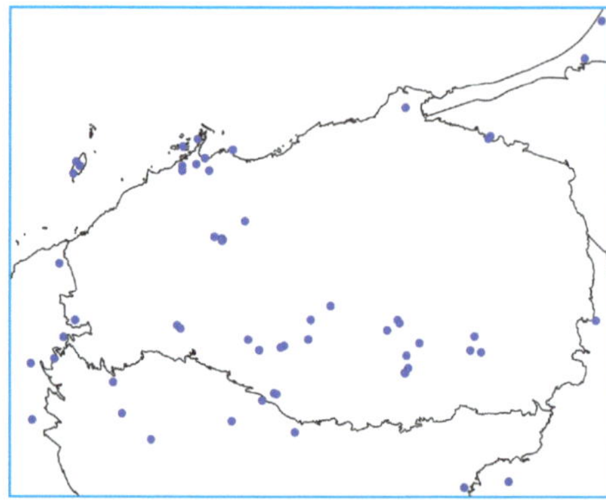

Common names: Pebble bush, Stylobasium.

Indigenous names: None recorded.

Description: Erect bush to 2 m; young shoots hoary-pubescent. Leaves generally pubescent (grey in some areas), cuneate-oblong, lower ones obovate, emarginate, 10–25 mm long, narrowed to a short petiole. Pedicels short, with one or generally two minute bracteoles. Calyx 5 mm long; lobes ciliate, obtuse, 3 mm long, greenish-yellow or brown. Anthers 10, yellow, protruding beyond lobes. Drupe globular, 12 mm diam., surmounted by an expanded calyx.

Key features for identification: Compact shrubs; leaves cuneate-oblong; flowers lacking petals; anthers prominently protruding; fruit a globular drupe.

Similar species: None in the Pilbara.

Distribution: Widespread in the interior of W.A. Also in N.T.

Habitat: Found in a wide range of habitats.

Flowering period: May to October.

Stylobasium spathulatum

Fruit/Seed collection

Fruit/Seed maturity: October to November.

Appearance of mature fruit/seed: When mature, the large spherical fruits, that have a persistent calyx and are covered in a thin outer fruit coat layer (mesocarp), turn brown and fall easily from the plant when touched. Bulk collections can be achieved by knocking branches gently and catching the fruits in large calico bags or buckets.

Cleaning requirements

Cleaning techniques: Collections should contain mainly pure (indehiscent) fruits. However, the thin outer fruit layer can be removed by rubbing the fruits on a sieve that does not allow the fruits to pass through. Considerable pressure can be used as the fruits are very hard and not prone to damage. Progressively, the outer fruit layer fragments will fall through the sieve. The final fruit will have a much smoother, hard seed coat.

Purity and viability: Fruit purity should be high as the majority of collections will consist of the large indehiscent fruits. Fruit viability (i.e. fruit fill via x-ray analysis) is high (>95%).

Seeds per gram: Fruits = 4 (range 4–5).

Germination requirements

Dormancy type: Physical (with some evidence of combinational dormancy – see notes).

Embryo type: Linear-curved. **Embryo:seed ratio:** 1.77.

Pre-treatments: Wet heat for 2–5 mins at 90–95°C or exposure to 98% sulphuric acid for 4 hours results in removal of the impermeable fruit layer.

Germination: Once treated to remove physical dormancy, germination of fresh fruits ranges between 65–80% across a wide temperature range (15–35°C). Increases in germination after exposure to GA_3 or smoke are minimal.

Notes: Caution must be taken when using wet heat as a technique to remove physical dormancy as occasional viability loss has been noted. There is some evidence that dry after-ripening fruits for up to 52 weeks increases the germination rate, with germination occurring 6–7 days earlier than untreated fruits. Therefore, additional pre-treatments may assist in improving the germination potential of the fruits.

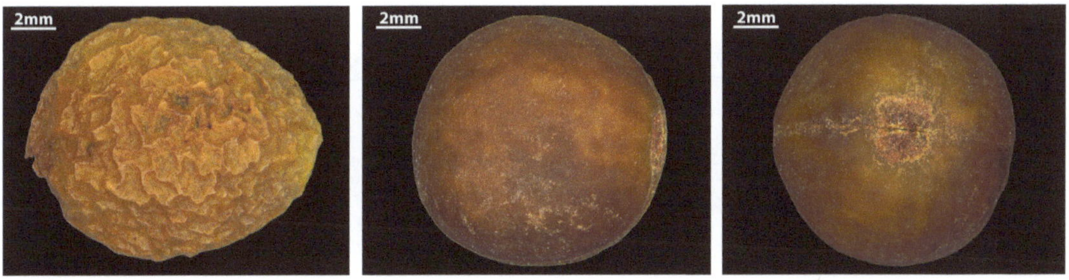

Tribulus suberosus

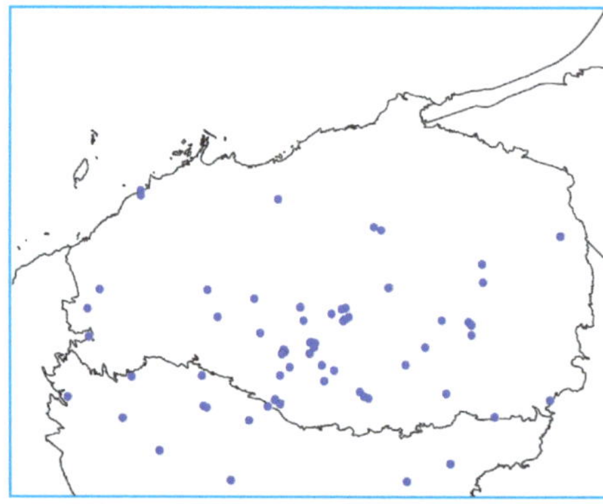

Common names: Cork hopbush, Cork caltrop.

Indigenous names: Gawiwarnda (Yindjibarndi, Ngarluma).

Description: Upright, spreading, subglabrous shrub, 0.3–1 m high, up to 2 m wide; stems with corky bark. Leaves in unequal pairs, with 2–4(–5) pairs of leaflets; petiole 2–5 mm long; leaflets obliquely elliptic, shortly petiolate, glabrous, 3–8 mm long, 2.2–5 mm wide. Flowering pedicel 4–7 mm long, fruiting pedicel 7.5–11 mm long. Sepals 6–7.5 mm long, glabrous, sparsely sericeous or patchily villous. Disc glands 5, between sepals and petals. Petals 5, yellow, elliptic to narrowly elliptic, glabrous, 9–15 mm long, longer than sepals. Stamens usually 10, in 2 slightly unequal whorls, all fertile or some staminodal; filaments 5.5–6 mm long; anthers 0.7–1.1 mm long or sometimes linear and twice as long. Ovary densely white sericeous, 5-celled; ovules 2 per cell; style and stigma 2–4 mm long, style glabrous; stigma 5-ridged, papillose. Fruit a globose, 5-winged schizocarp, 13–20 mm long, 18–25 mm wide, very tardily dissociating into 5 2-winged woody cocci, densely pubescent between wings, lacking any spines.

Key features for identification: Corky bark; 2–4(–5) pairs of leaflets; fruit very tardily dissociating.

Similar species: *T. platypterus* usually lacks corky bark.

Distribution: Widespread in the Pilbara to the edge of the Gibson Desert and Wiluna.

Habitat: On rocky hills and ranges on a variety of soils.

Flowering period: Mainly April to August, but also sporadically in other months.

Tribulus suberosus

Fruit/Seed collection

Fruit/Seed maturity: June to November.

Appearance of mature fruit/seed: Multiple (3–4) papery winged fruits are arranged in a spherical cluster and turn from light green to light brown when mature. Each fruit segment contains two individual wings protruding from a hardened woody indehiscent fruit containing 1 or 2 seeds.

Cleaning requirements

Cleaning techniques: The papery wing of each fruit can easily be removed by crushing the fruits while in calico bags or by rubbing through hands. No damage occurs to the woody fruit segment that contains the seeds. The debris can then be removed via vacuum separation. Seeds have never been successfully removed from the woody fruit.

Purity and viability: The indehiscent fruits of this species can contain up to two individual seeds. The percentage of fruits that contain at least one seed, is 55%. When taking into account the actual number of individual seeds produced, the average seed number per fruit is 1.32.

Seeds per gram: 51 (indehiscent fruits).

Germination requirements

Dormancy type: Physiological.

Embryo type: Linear-straight. **Embryo:seed ratio:** 1.00.

Pre-treatments: Dormancy alleviation treatments unknown.

Germination: Generally 0–2%. Very low germination from fruits across all incubation temperatures tested (25–35°C) and after treatment with GA_3 and smoke.

Notes: Seeds enclosed in indehiscent fruits are commonly difficult to germinate[7,8]. Potential dormancy alleviation treatments include dry after-ripening, warm stratification, or wet/dry cycling.

The botanical descriptions used in Chapter 5 have been adapted from the following sources:

Barker R. M. (2013) Zygophyllaceae. In: Wilson, A.J.G. (ed.) *Flora of Australia Volume 26. Meliaceae, Rutaceae, Zygophyllaceae.* pp. 511–579. (Australian Biological Resources Study: Canberra, and CSIRO Publishing: Melbourne.)

Tribulus suberosus.

Barrett R. L., Henwood M. J. (2015) Iron plant (*Astrotricha hamptonii*, Araliaceae): an enigmatic species with a rich history in bioprospecting. *Telopea* **18**, 297–304.

Astrotricha hamptonii.

Bean A. R. (2011) A taxonomic revision of *Pterocaulon* section *Monenteles* (Labill.) Kuntze (Asteraceae: Inuleae-Plucheinae). *Austrobaileya* **8**, 280–334.

Pterocaulon sphacelatum.

Bentham,G. (1870) *Flora Australiensis: A description of the plants of the Australian Territory. Vol. V. Myoporineae to Proteaceae.* (Lovell Reeve: London.)

Ptilotus auriculifolius.

Carolin R. C. (1992) *Goodenia.* In: George, A.S. (ed.) *Flora of Australia. Volume 35, Brunoniaceae, Goodeniaceae.* pp. 147–327. (Australian Government Publishing Service: Canberra.)

Goodenia stobbsiana; *G. triodiophila*; *G. vilmoriniae.*

Chippendale G. M. (1988) *Eucalyptus.* In: George, A.S. (ed.) *Flora of Australia 19. Myrtaceae-Eucalyptus, Angophora.* pp. 1–448. (Australian Government Publishing Service: Canberra.)

Eucalyptus leucophloia subsp. *leucophloia.*

Fryxell P. A. (1965) A revision of the Australian species of *Gossypium* with observations on the occurrence of *Thespesia* in Australia. *Australian Journal of Botany* **13**, 71–102.

Gossypium robinsonii.

Halford D. A. (2004) Notes on Tiliaceae in Australia, 4. A revision of the stellate-haired species of the genus *Corchorus* L. *Austrobaileya* **6**, 581–629.

Corchorus crozophorifolius.

Harden G. J. (ed.) (1992) *Flora of New South Wales. Volume 3.* (University of New South Wales Press: Kensington.)

Chrysocephalum apiculatum.

Hewson H. J. (1981) The genus *Lepidium* L. (Brassicaceae) in Australia. *Brunonia* **4**, 217–308.

Lepidium catapycnon; *L. echinatum*; *L. pedicellosum.*

Jessop J. P. (ed) [numerous contributors] (1981) *Flora of Central Australia.* (Australian Systematic Botany Society/A.H. & A.W. Reed Pty Ltd: Sydney.)

Eucalyptus gamophylla; *Streptoglossa decurrens.*

Johnson R. W. (2001) A taxonomic revision of *Convolvulus* L. (Convolvulaceae) in Australia. *Austrobaileya* **6**, 1–39.

Convolvulus clementii.

Kakudidi E. K. Z., Lazarides, M., Carnahan, J. A. (1988) A revision of *Enneapogon* (Poaceae. Pappophoreae) in Australia. *Australian Systematic Botany* **1**, 325–353.

Enneapogon caerulescens; *E. polyphyllus*; *E. robustissimus.*

Lazarides M. (1995) The genus *Eriachne* (Eriachneae, Poaceae). *Australian Systematic Botany* **8**, 355–452.

Eriachne pulchella subsp. *dominii.*

Lazarides M. (1997) A revision of *Eragrostis* (Eragrostideae, Eleusininae, Poaceae) in Australia. *Australian Systematic Botany* **10**, 77–187.

Eragrostis dielsii; *E. eriopoda*; *E. tenellula.*

Lazarides M. (1997) A revision of *Triodia* including *Plectrachne* (Poaceae, Eragrostideae, Triodiinae). *Australian Systematic Botany* **10**, 381–489.

Triodia angusta; *T. basedowii*; *T. brizoides*; *T. epactia*; *T. pungens*; *T. wiseana.*

Maslin B.R. (2014) WATTLE2: interactive identification of Australian *Acacia.* Version 2.2. http://www.lucidcentral.org/en-au/keys173;/searchforakey.aspx [accessed June 2014].

Acacia acradenia; *A. adoxa* var. *adoxa*; *A. adsurgens*; *A. ancistrocarpa*; *A. bivenosa*; *A. citrinoviridis*; *A. cowleana*; *A. dictyophleba*; *A. eriopoda*; *A. hamersleyensis*; *A. hilliana*; *A. inaequilatera*; *A. monticola*; *A. ptychophylla*; *A. pyrifolia* var. *pyrifolia*; *A. sclerosperma* subsp. *sclerosperma*; *A. stellaticeps*; *A. tenuissima*; *A. tumida* var. *pilbarensis.*

Mueller F. J. H. von (1883) Definitions of some new Australian plants [Continued.]. *The Southern Science Record* **3**, 67–70.

Hibiscus haynaldii.

Obbens F. J. (2006) A review of the tuberous *Calandrinia* species (section *Tuberosae*), including three new species for Western Australia. *Nuytsia* **16**, 95–115.

Calandrinia schistorhiza.

Palmer J. (1998) A taxonomic revision of *Gomphrena* (Amaranthaceae) in Australia. *Australian Systematic Botany* **11**, 73–161.

Gomphrena cunninghamii.

Randell B. R., Barlow, B. A. (1998) *Senna.* In: Orchard, A.E. (ed.) *Flora of Australia. Volume 12. Mimosaceae (excluding* Acacia*), Caesalpiniaceae.* pp. 89–138. (CSIRO Australia: Canberra.)

Senna artemisioides subsp. *oligophylla*; *S. glutinosa* subsp. *pruinosa*; *S. notabilis*; *S. venusta.*

Ross J. H. (1998) *Petalostylis.* In: Orchard, A.E. (ed.) *Flora of Australia. Volume 12. Mimosaceae (excluding* Acacia*), Caesalpiniaceae.* pp. 157–160. (CSIRO Australia: Canberra.)

Petalostylis labicheoides.

Rye B. L. (1999) A taxonomic revision of the many flowered species of *Trachymene* (Apiaceae) in Western Australia. *Nuytsia* **13**, 193–232.

Trachymene oleracea subsp. *oleracea.*

Simon B. K. (1992) A revision of the genus *Aristida* (Poaceae) in Australia. *Australian Systematic Botany* **5**, 129–226.

Aristida contorta; *A. inaequiglumis*; *A. latifolia.*

Simon B. K., Alfonso Y. (2011–) *AusGrass2*, http://ausgrass2.myspecies.info/ accessed on [23 July 2014].

Cymbopogon ambiguus; *Cymbopogon obtectus.*

Smith P. J. (1992) A revision of the genus *Wahlenbergia* (Campanulaceae) in Australia. *Telopea* **5**, 91–175.

Wahlenbergia tumidifructa.

West J. G. (1985) *Diplopeltis.* In: George, A.S. (ed.) *Flora of Australia. Volume 25.* pp. 101–109. (Australian Government Publishing Service: Canberra.)

Diplopeltis stuartii var. *stuartii.*

Western Australian Herbarium. (1998–) *FloraBase – The Western Australian Flora.* Department of Parks and Wildlife. http://florabase.dpaw.wa.gov.au/ [accessed 6 August 2014].)

Grevillea pyramidalis subsp. *leucadendron*; *G. refracta* subsp. *refracta*; *G. stenobotrya*; *G. wickhamii* subsp. *hispidula*; *G. wickhamii* subsp. *macrodonta*; *Hakea lorea.*

Wheeler J. R. (ed.) [numerous contributors] (1992) *Flora of the Kimberley Region.* (Conservation and Land Management: Perth.)

Abutilon otocarpum; *Boerhavia coccinea*; *Cleome viscosa*; *Cynodon convergens*; *Dodonaea coriacea*; *Goodenia armitiana*; *Gossypium australe*; *Indigofera monophylla*; *Melaleuca glomerata*; *Mirbelia viminalis*; *Neurachne muelleri*; *Newcastelia cladotricha*; *Polycarpaea corymbosa*; *Portulaca oleracea*; *Ptilotus macrocephalus*; *P. nobilis*; *Solanum dioicum*; *S. diversiflorum*; *Stylidium desertorum*; *Stylobasium spathulatum*; *Trichodesma zeylanicum*; *Velleia panduriformis.*

Wilkins C. F., Trudgen, M. E. (2012) A new species of *Gompholobium* (Fabaceae: Mirbelieae) from the Pilbara bioregion of Western Australia. *Nuytsia* **22**, 31–40.

Gompholobium oreophilum.

Wilkins C. F., Whitlock, B. A. (2011) A new Australian genus, *Androcalva*, separated from *Commersonia* (Malvaceae *s.l.* or Byttneriaceae). *Australian Systematic Botany* **24**, 284–349.

Androcalva luteiflora.

Wilson K. L. (1991) Systematic studies in *Cyperus* section *Pinnati* (Cyperaceae). *Telopea* **4**, 361–496.

Cyperus ixiocarpus.

Wilson P. G. (1984) Chenopodiaceae. In: George, A.S. (ed.) *Flora of Australia. Volume 4.* pp. 81–317. (Australian Government Publishing Service: Canberra.)

Dysphania rhadinostachya var. *rhadinostachya.*

Original or unpublished descriptions.

Corchorus aff. *lasiocarpus* (R.L. Barrett); *Hibbertia glaberrima* (R.L. Barrett); *Newcastelia* sp. Hamersley Range (S. van Leeuwen 4264) (R.L. Barrett); *Sida echinocarpa* (R.M. Barker); *Tephrosia* sp. Fortescue (R.L. Barrett); *Triodia* sp. Shovelanna Hill (S. van Leeuwen 3835) (B. Anderson).

Seed biology references in Chapter 5 (fact sheets)

1. Ainsley P., Jones M., Erickson T. (2008) Overcoming physiological dormancy in *Prostanthera eurybioides* (Lamiaceae), a nationally endangered Australian shrub species. *Australian Journal of Botany* **56**, 214–219.
2. Commander L. E., Merritt D. J., Rokich D. P., Dixon K. W. (2009) Seed biology of Australian arid zone species: germination of 18 species used for rehabilitation. *Journal of Arid Environments* **73**, 617–625.
3. Commander L. E., Merritt D. J., Rokich D. P., Flematti G. R., Dixon K. W. (2008) Seed germination of *Solanum* spp. (Solanaceae) for use in rehabilitation and commercial industries. *Australian Journal of Botany* **56**, 333–341.
4. Cook A., Turner S. R., Baskin J. M., Baskin C. C., Steadman K. J., Dixon K. W. (2008) Occurrence of physical dormancy in seeds of Australian Sapindaceae: a survey of 14 species in nine genera. *Annals of Botany* **101**, 1349–1362.
5. Hidayati S. N., Walck J. L., Merritt D. J., Turner S. R., Turner D. W., Dixon K. W. (2012) Sympatric species of *Hibbertia* (Dilleniaceae) vary in dormancy break and germination requirements: implications for classifying morphophysiological dormancy in Mediterranean biomes. *Annals of Botany* **109**, 1111–1123.
6. Hoyle G. L., Daws M. I., Steadman K. J., Adkins S. W. (2008) Mimicking a semi-arid tropical environment achieves dormancy alleviation for seeds of Australian native Goodeniaceae and Asteraceae. *Annals of Botany* **101**, 701–708.
7. Richmond G., Ghisalberti E. (1994) Seed dormancy and germination mechanisms in *Eremophila* (Myoporaceae). *Australian Journal of Botany* **42**, 705–715.
8. Turner S. R., Commander L. E., Baskin J. M., Baskin C. C., Dixon K. W. (2009) Germination behaviour of *Astroloma xerophyllum* (Ericaceae), a species with woody indehiscent endocarps. *Botanical Journal of the Linnean Society* **160**, 299–311.
9. Turner S. R., Merritt D. J., Baskin J. M., Baskin C. C., Dixon K. W. (2006) Combinational dormancy in seeds of the Western Australian endemic species *Diplopeltis huegelii* (Sapindaceae). *Australian Journal of Botany* **54**, 565–570.

References and further reading

Aston H. I. (1973) *Aquatic plants of Australia. A guide to the identification of the aquatic ferns and flowering plants of Australia, both native and naturalized.* (Melbourne University Press: Melbourne.)

Barker R. M. (1986) A taxonomic revision of Australian Acanthaceae. *Journal of the Adelaide Botanic Gardens* **9**, 1–286.

Barker R. M. (2007) Two newly described species and a draft key to the species of *Sida s.lat.* from Western Australia. *Nuytsia* **17**, 13–30.

Barker R. M. (2010) *Australian Solanaceae species: identification and information.* Version 1. An interactive Lucid key and information system at http://www.flora.sa.gov.au/id_tool.html [accessed 8 August 2012].

Barker R. M. (2013) Zygophyllaceae. In: Wilson A.J.G. (ed.) *Flora of Australia Volume 26. Meliaceae, Rutaceae, Zygophyllaceae.* pp. 511–579. (Australian Biological Resources Study: Canberra, and CSIRO Publishing: Melbourne.)

Barker W. R., Greenslade, P.J.M. (eds) (1982) *Evolution of the flora and fauna of arid Australia.* (Peacock Publications: Adelaide.)

Barker W. R., Nesom, G. L., Beardsley, P. M., Fraga, N. S. (2012) A taxonomic conspectus of Phrymaceae: A narrowed circumscription for *Mimulus*, new and resurrected genera, and new names and combinations. *Phytoneuron* **2012-39**, 1–60.

Barrett R. L. (2015). Examining range disjunctions in Australian *Terminalia* (Combretaceae) with taxonomic revision of the *T. canescens* and *T. cunninghamii* species complexes. *Australian Systematic Botany* **28**, 23–45.

Baskin C. C., Baskin, J. M. (2001) *Seeds: ecology, biogeography, and evolution of dormancy and germination.* (Academic Press: San Diego, California.)

Baskin J. M. & Baskin, C. C. (2003) Classification, biogeography, and phylogenetic relationships of seed dormancy. In: Smith R. D., Dickie J. B., Linington S. H., Pritchard H. W., Probert R. J. (eds). *Seed conservation: turning science into practice.* pp. 517–544. (Royal Botanic Gardens, Kew: Kew.)

Baskin J. M., Baskin, C. C. (2004) A classification system for seed dormancy. *Seed Science Research* **14**, 1–16.

Bean A. R. (2013) A taxonomic review of the *Solanum sturtianum* subgroup of subgenus *Leptostemonum* (Solanaceae). *Nuytsia* **23**, 129–161.

Beard J. S. (1990) *Plant life of Western Australia.* (Kangaroo Press: Kenthurst, NSW.)

Beentje H. J. (2010) *The Kew plant glossary: an illustrated dictionary of plant terms.* (Kew Publishing, Royal Botanic Gardens: Kew.)

Bell D. T. (1999) The process of germination in Australian species. *Australian Journal of Botany* **47**, 475–517.

BoM (2013) *Bureau of Meteorology.* Commonwealth of Australia, URL: http://www.bom.gov.au

Bowman D. M. J. S., Latz P. K. Panton, W. J. (1995) Pattern and change in an *Acacia aneura* shrubland and *Triodia* hummock grassland mosaic on rolling hills in central Australia. *Australian Journal of Botany* **43**, 25–37.

Braby M. F. (2000) *Butterflies of Australia: their identification, biology and distribution.* (CSIRO Publishing: Melbourne.)

Brock J. (1993) *Native plants of northern Australia.* (Reed Books: Sydney.)

Brown A. P., Buirchell B. J. (2011) *Field guide to the Eremophilas of Western Australia.* (Simon Nevill Publications: Perth.)

Burbidge N. T. (1945) Ecological notes of the De Grey–Coongan area, with special reference to physiography. *Journal of the Royal Society of Western Australia* **29**, 151–152.

Butcher R., Hurter P. J. H. (2012) *Tephrosia oxalidea* (Fabaceae: Millettieae), a new species from the Pilbara Gascoyne bioregions of Western Australia. *Nuytsia* **22**, 341–349.

Carolin R. C. (1992) *Goodenia. Flora of Australia.* pp. 147–327. (Australian Government Publishing Service: Canberra.)

Chappill J. A., Crisp, M. D., Wilkins, C. F. (2007) Taxonomic revision of *Jacksonia* (Leguminosae: Mirbeliae). *Australian Systematic Botany* **20**, 473–626.

Chinnock R. J. (2007) Eremophila *and allied genera: a monograph of the Myoporaceae.* (Rosenberg Publishing: Kenthurst, NSW.)

Cowie I. D., Short, P. S., Osterkamp Madsen, M. (2000) *Floodplain flora. A flora of the coastal floodplains of the Northern Territory, Australia.* (Australian Biological Resources Study: Canberra, and Parks & Wildlife Commission of the Northern Territory: Darwin.)

Craven L.A. (1987) A taxonomic revision of *Calytrix* Labill. (Myrtaceae). *Brunonia* **10**, 1–138.

Craven L.A. (1996) A taxonomic revision of *Heliotropium* (Boraginaceae) in Australia. *Australian Systematic Botany* **9**, 521–657.

Crawford A. D., Steadman K. J., Plummer J. A., Cochrane A., Probert R. J. (2007) Analysis of seed-bank data confirms suitability of international seed-storage standards for the Australian flora. *Australian Journal of Botany* **55**, 18–29.

Datson B. (2002) *Samphires in Western Australia. A field guide to Chenopodiaceae tribe Salicornieae.* (Department of Conservation and Land Management: Perth.)

Dillon S. J. (2014) *Grevillea saxicola* (Proteaceae), a new species from the Pilbara of Western Australia. *Nuytsia* **24**, 103–108.

Dixon D. J. (2003) A taxonomic revision of the Australian *Ficus* species in the section *Malvanthera* (*Ficus* subg. *Urostigma*: Moraceae). *Telopea* **10**, 125–153.

DPAW (2007–) *NatureMap: Mapping Western Australia's Biodiversity.* Department of Environment and Conservation, URL: http://naturemap.dpaw.wa.gov.au/default.aspx

Dunlop C. R., Leach G. J., Cowie I. D. (1995) *Flora of the Darwin Region. Vol. 2. Northern Territory Botanical Bulletin No. 20.* (Conservation Commission of the Northern Territory: Darwin.)

Halford D. A. (1992) Review of the genus *Oldenlandia* L. (Rubiaceae) and related genera in Australia. *Austrobaileya* **3**, 683–722.

Halford D. A. (1995) Notes on Tiliaceae in Australia, 2: A revision of the simple-haired species of the genus *Corchorus* L. *Austrobaileya* **4**, 297–320.

Halford D. A. (1997) Notes on Tiliaceae in Australia, 3: A revision of the genus *Triumfetta* L. *Austrobaileya* **4**, 495–587.

Halford D. A. (2004) Notes on Tiliaceae in Australia, 4. A revision of the stellate-haired species of the genus *Corchorus* L. *Austrobaileya* **6**, 581–629.

Halford D. A., Harris, W. K. (2012) A taxonomic revision of *Euphorbia* section *Anisophyllum* Roeper (Euphorbiaceae) in Australia. *Austrobaileya* **8**, 441–600.

Harris J. G., Harris M. W. (2000) *Plant identification terminology.* 2nd edition. (Spring Lake Publishing: Spring Lake, Utah.)

Hayes A., Hayes, S. (2007) *Ngambunyjarri: Thalanyjibarndi Yininjyarri Thalanyji plant names and uses.* (Wangka Maya Pilbara Aboriginal Language Centre: Onslow, W.A.)

Hewson H. J. (1982) Capparaceae. In: George A.S. (ed.) *Flora of Australia. Volume 8.* pp. 207–230. (Australian Government Publishing Service: Canberra.)

Hewson H. J. (1982) Brassicaceae. In: George A.S. (ed.) *Flora of Australia. Volume 8.* pp. 231–357. (Australian Government Publishing Service: Canberra.)

Hewson H. J., Meikle R. D. (1984) Nyctaginaceae. In: George, A.S. (ed.) *Flora of Australia. Volume 4.* pp. 5–19. (Australian Government Publishing Service: Canberra.)

Hislop M. C., Thiele, K. R., Brassington, D. (2013) *Cochlospermum macnamarae* (Bixaceae), a rare, new endemic from the Pilbara bioregion of Western Australia. *Nuytsia* **23**, 89–94.

Holland A. E. (2002) A review of *Crotalaria* L. (Fabaceae) in Australia. *Austrobaileya* **6**, 293–324.

James J. J., Sheley R. L. Erickson T., Rollins K. S., Taylor M. H., Dixon K. W. (2013) A systems approach to restoring degraded drylands. *Journal of Applied Ecology* **50**, 730–739.

James J. J., Svejcar T. J., Rinella M. J. (2011) Demographic processes limiting seedling recruitment in arid grassland restoration. *Journal of Applied Ecology* **48**, 961–969.

Jessop J. P. (ed) (1981) *Flora of Central Australia.* (Australian Systematic Botany Society/A.H. & A.W. Reed Pty Ltd: Sydney.)

Jessop J. P., Dashorst G. R. M., James F. M. (2006) *Grasses of South Australia. An illustrated guide to the native and naturalised species.* (Wakefield Press: Kent Town, SA.)

Johnson R. W. (2011) Convolvulaceae (version 1). In: Kellermann J. (ed.) *Flora of South Australia* (ed. 5). 20 pp. (State Herbarium of South Australia: Adelaide.)

Juluwarlu Aboriginal Corporation. (2003) *Wanggalili: Yindjibarndi and Ngarluma plants.* (Juluwarlu Aboriginal Corporation: Roebourne, W.A.)

Lazarides M. (1995) The genus *Eriachne* (Eriachneae, Poaceae). *Australian Systematic Botany* **8**, 355–452.

Lazarides M. (1997) A revision of *Eragrostis* (Eragrostideae, Eleusininae, Poaceae) in Australia. *Australian Systematic Botany* **10**, 77–187.

Lazarides M. (1997) A revision of *Triodia* including *Plectrachne* (Poaceae, Eragrostideae, Triodiinae). *Australian Systematic Botany* **10**, 381–489.

Leach G. J. (1989) Taxonomic revision of *Bergia* (Elatinaceae) in Australia. *Journal of the Adelaide Botanic Gardens* **11**, 75–100.

Marchant N. G., Wheeler J. R., Rye B. L., Bennett E. M., Lander N. S., Macfarlane T. D. (1987) *Flora of the Perth Region Part 1 and 2.* (Western Australian Herbarium: Perth.)

McDonald M. W., Brooker M. I. H., Butcher P. A. (2009). A taxonomic revision of *Eucalyptus camaldulensis* (Myrtaceae). *Australian Systematic Botany* **22**, 257–285.

McKenzie N. L., van Leeuwen S. J., Pinder A. M. (2009) Introduction to the Pilbara Biodiversity Survey, 2002–2007. In: George, A. S., McKenzie, N. L. & Doughty, P. (eds.). *A Biodiversity Survey of the Pilbara Region of Western Australia, 2002–2007. Records of the Western Australian Museum Supplement No. 78.* pp. 3–89.

Merritt D. J., Dixon K. W. (2011) Restoration seed banks – a matter of scale. *Science* **332**, 424–425.

Merritt D. J., Martyn A. J., Ainsley P., Young R. E., Seed L. U., Thorpe M., Hay F. R., Commander L. E., Shackelford N., Offord C. A., Dixon K. W., Probert R. J. (2014) A continental-scale study of seed lifespan in experimental storage examining seed, plant, and environmental traits associated with longevity. *Biodiversity and Conservation* **23**, 1081–1104.

Merritt D. J., Turner S. R., Clarke S., Dixon K. W. (2007) Seed dormancy and germination stimulation syndromes for Australian temperate species. *Australian Journal of Botany* **55**, 336–344.

Mitchell A. A., Wilcox D. G. (1994) *Arid shrubland plants of Western Australia.* 2nd edition. (University of Western Australia Press: Nedlands, WA.)

Orchard A. E. (ed.) (1999) *Flora of Australia. Volume 1, Introduction.* (Australian Biological Resources Study: Canberra, and CSIRO Publishing: Melbourne.)

Paczkowska G. & Chapman A. R. (2000) *The Western Australian flora. A descriptive catalogue.* (Wildflower Society of Western Australia (Inc.), the Western Australian Herbarium, CALM and the Botanic Gardens and Parks Authority: Perth.)

Pedley L. (1999) *Desmodium* Desv. (Fabaceae) and related genera in Australia: a taxonomic revision. *Austrobaileya* **5**, 209–261.

Prober R., Adams J., Coneybeer J., Crawford A., Hay F. (2007) Seed quality for conservation is critically affected by pre-storage factors. *Australian Journal of Botany* **55**, 326–335.

Rye B. L. (1996). A taxonomic review of the genera *Lachnostachys*, *Newcastelia* and *Physopsis* (Chloanthaceae) in Western Australia. *Nuytsia* **11**, 79–107.

Rye B. L. (1999) A taxonomic revision of the many flowered species of *Trachymene* (Apiaceae) in Western Australia. *Nuytsia* **13**, 193–232.

Shepherd K .A., Trudgen M. E. (2011) *Eragrostis surreyana* (Poaceae) an uncommon, habitat restricted new species from the Pilbara Bioregion of Western Australia. *Telopea* **13**, 143–148.

Short P. S., Cowie I. D. (eds) (2011) *Flora of the Darwin region. Vol. 1.* (Northern Territory Herbarium, Department of Natural Resources, Environment, the Arts and Sport: Darwin.)

Simon B. K. (1992) A revision of the genus *Aristida* (Poaceae) in Australia. *Australian Systematic Botany* **5**, 129–226.

Telford I. R. H., Naaykens J. (2015) *Synostemon hamersleyensis* (Phyllanthaceae), a new species endemic to the Pilbara, Western Australia. *Nuytsia* **25**, 31–37.

Turner S. R., Steadman K. J., Vlahos S., Koch J. M., Dixon K. W. (2013) Seed treatment optimizes benefits of seed bank storage for restoration-ready seeds: the feasibility of prestorage dormancy alleviation for mine-site revegetation. *Restoration Ecology* **21**, 186–192.

West J. G. (1984) A revision of *Dodonaea* Miller (Sapindaceae) in Australia. *Brunonia* **7**, 1–194.

Westbrooke M. E., Florentine S. K., Milberg P. (2005) Arid land vegetation dynamics after a rare flooding event: influence of fire and grazing. *Journal of Arid Environments* **61**, 249–260.

Western Australian Herbarium. (1998–) *FloraBase – The Western Australian Flora.* Department of Parks and Wildlife. http://florabase.dpaw.wa.gov.au/ [accessed 6 August 2013].)

Wheeler J. R. (ed.) (1992) *Flora of the Kimberley Region.* (Conservation and Land Management: Perth.)

Wright B. R., Clarke P. J. (2009) Fire, aridity and seed banks. What does seed bank composition reveal about community processes in fire-prone desert? *Journal of Vegetation Science* **20**, 663–674.

Appendix A. Summary of seed dormancy classes and suggested dormancy alleviating pre-treatments for a selection of framework species or species of conservation value from the Pilbara arid zone

Family	Species	Dormancy Class	Suggested pre-treatments
Malvaceae	*Abutilon otocarpum*	PY+PD	DH10 C100, HW5 C95, SA4, AR*, WD*
Fabaceae	*Acacia acradenia*	PY	HW2 C80-95
Fabaceae	*Acacia adoxa* var. *adoxa*	PY	HW1-2 C95
Fabaceae	*Acacia adsurgens*	PY	HW2 C95
Fabaceae	*Acacia ancistrocarpa*	PY	HW2 C95
Fabaceae	*Acacia bivenosa*	PY	HW2 C95
Fabaceae	*Acacia citrinoviridis*	PY	HW2 C95
Fabaceae	*Acacia cowleana*	PY	HW2 C80-95
Fabaceae	*Acacia dictyophleba*	PY	HW1 C95
Fabaceae	*Acacia eriopoda*	PY	HW1 C95
Fabaceae	*Acacia hamersleyensis*	PY	HW1 C95
Fabaceae	*Acacia hilliana*	PY	HW2 C95
Fabaceae	*Acacia inaequilatera*	PY	HW2 C70-95
Fabaceae	*Acacia monticola*	PY	HW1 C95
Fabaceae	*Acacia ptychophylla*	PY	HW1-2 C95
Fabaceae	*Acacia pyrifolia*	PY	HW2 C80-95
Fabaceae	*Acacia sclerosperma*	PY	HW2 C95, SA*
Fabaceae	*Acacia stellaticeps*	PY	HW1 C95
Fabaceae	*Acacia tenuissima*	PY	HW1 C95
Fabaceae	*Acacia tumida* var. *pilbarensis*	PY	HW2 C95
Malvaceae	*Androcalva luteiflora*	PY	HW2 C95
Poaceae	*Aristida contorta*	PD	AR*, WD*, G+
Poaceae	*Aristida inaequiglumis*	ND	No pre-treatment required
Poaceae	*Aristida latifolia*	PD	AR*, WD*, G+, K+
Araliaceae	*Astrotricha hamptonii*	MPD	AR*, WD*, ST*
Nyctaginaceae	*Boerhavia coccinea*	PD	AR*, N*, K+
Portulacaceae	*Calandrinia schistorhiza*	ND	AR*, G+
Asteraceae	*Chrysocephalum apiculatum*	ND	AR*, WD*, G+
Cleomaceae	*Cleome viscosa*	PD	AR, WD*, G+, K+
Convolvulaceae	*Convolvulus clementii*	PY	HW1 C95, SA1
Malvaceae	*Corchorus* aff. *lasiocarpus*	PY	HW1-2 C70-90
Malvaceae	*Corchorus crozophorifolius*	PY	HW<1 C70-90
Poaceae	*Cymbopogon ambiguus*	ND (for seeds)	CTS, AR* or WD* for florets
Poaceae	*Cymbopogon obtectus*	ND (for seeds)	CTS, G+, K+, AR* or WD* for florets
Poaceae	*Cynodon convergens*	PD (for seeds)	CTS, AR*, WD* G+
Cyperaceae	*Cyperus ixiocarpus*	ND	AR*, G+, K+
Sapindaceae	*Diplopeltis stuartii*	PY+PD	HW1-2 C95, AR*
Sapindaceae	*Dodonaea coriacea*	PY	HW2 C95
Chenopodiaceae	*Dysphania rhadinostachya*	PD	WD*, G+
Poaceae	*Enneapogon caerulescens*	PD (for seeds)	CTS, AR*, WD*, G+
Poaceae	*Enneapogon polyphyllus*	PD (for seeds)	CTS, AR*, WD*, G+
Poaceae	*Enneapogon robustissimus*	ND (for seeds)	CTS, AR*, WD*, G+, K+

Family	Species	Dormancy Class	Suggested pre-treatments
Poaceae	*Eragrostis dielsii*	PD (for seeds)	CTS, AR*, WD*
Poaceae	*Eragrostis eriopoda*	PD (for seeds)	CTS, AR*, WD*
Poaceae	*Eragrostis tenellula*	PD (for seeds)	CTS, AR*, WD*, G+
Poaceae	*Eriachne pulchella* subsp. *dominii*	ND (for seeds)	CTS, AR*, WD*, G+, K+
Myrtaceae	*Eucalyptus gamophylla*	ND	No pre-treatment required
Myrtaceae	*Eucalyptus leucophloia*	ND	No pre-treatment required
Fabaceae	*Gompholobium oreophilum*	PY+PD	HW1-2 C95, AR*
Amaranthaceae	*Gomphrena cunninghamii*	PD	CTS, AR*, WD*, G+, K+
Goodeniaceae	*Goodenia armitiana*	ND	AR*, WD*, G+, K+
Goodeniaceae	*Goodenia stobbsiana*	PD	AR*, WD*, G+
Goodeniaceae	*Goodenia triodiophila*	ND	AR*, WD*, G+, K+
Goodeniaceae	*Goodenia vilmoriniae*	PD	AR*, WD*, G+
Malvaceae	*Gossypium australe*	PY (or PY+PD)	HW1-2 C90 (?), SA*, N, AR*
Malvaceae	*Gossypium robinsonii*	PY (or PY+PD)	HW1-2 C80-95, AR*
Proteaceae	*Grevillea pyramidalis*	ND	No pre-treatment required
Proteaceae	*Grevillea refracta* ssp. *refracta*	ND	AR*, WD*
Proteaceae	*Grevillea stenobotrya*	ND	AR*, WD*, G+, K+
Proteaceae	*Grevillea wickhamii* ssp. *hispidula*	ND	AR*, WD*, K+
Proteaceae	*Grevillea wickhamii* ssp. *macrodonta*	ND	AR*, WD*, G+, K+
Proteaceae	*Hakea lorea*	ND	No pre-treatment required
Dilleniaceae	*Hibbertia glaberrima*	MPD	AR*, WD, ST, K+
Malvaceae	*Hibiscus haynaldii*	PY+PD	HW2 C70-90, SA8, AR*, WD*, G+
Fabaceae	*Indigofera monophylla*	PY	HW1-2 C70-90
Brassicaceae	*Lepidium catapycnon*	PD	AR*, WD*, G+
Brassicaceae	*Lepidium echinatum*	ND	G+, K+
Brassicaceae	*Lepidium pedicellosum*	PD	AR*, WD*, G+, K+
Myrtaceae	*Melaleuca glomerata*	ND	No pre-treatment required
Fabaceae	*Mirbelia viminalis*	PY	HW1-2 C70-90
Poaceae	*Neurachne muelleri*	PD (for seeds)	CTS, AR*, WD*, G+
Lamiaceae	*Newcastelia cladotricha*	PD	AR*, WD*, ST*, N*
Lamiaceae	*Newcastelia* sp. Hamersley Range	PD	AR*, WD*, ST*, N*
Fabaceae	*Petalostylis labicheoides*	PY	HW1 C80-90
Caryophyllaceae	*Polycarpaea corymbosa*	ND	No pre-treatment required
Portulacaceae	*Portulaca oleracea*	PD	AR, G+
Asteraceae	*Pterocaulon sphacelatum*	PD	AR*, WD*, G+, K+
Amaranthaceae	*Ptilotus auriculifolius*	ND (for seeds)	CTS, AR*, WD*, G+, K+
Amaranthaceae	*Ptilotus macrocephalus*	ND (for seeds)	CTS, AR*, WD*, G+, K+
Amaranthaceae	*Ptilotus nobilis*	ND (for seeds)	CTS, AR*, WD*
Fabaceae	*Senna artemisioides* ssp. *oligophylla*	PY	HW1 C90
Fabaceae	*Senna glutinosa* ssp. *pruinosa*	PY	HW1-2 C70-90
Fabaceae	*Senna notabilis*	PY	HW1 C70-90
Fabaceae	*Senna venusta*	PY	HW1 C70-90

Family	Species	Dormancy Class	Suggested pre-treatments
Malvaceae	*Sida echinocarpa*	PY+PD	HW2 C80-90, DH5-30 C80-90, SA1, AR*
Solanaceae	*Solanum dioicum*	PD	AR, WD*, G+, K+
Solanaceae	*Solanum diversiflorum*	PD	AR*, WD*
Asteraceae	*Streptoglossa decurrens*	ND	AR*, WD*, G+, K+
Stylidiaceae	*Stylidium desertorum*	MPD	AR*, WD*, G+, K+
Surianaceae	*Stylobasium spathulatum*	PY (or PY+PD)	HW2-5 C95, SA4, AR
Fabaceae	*Tephrosia* sp. Fortescue	PY+PD	HW1-2 C80-90, AR*
Araliaceae	*Trachymene oleracea*	MPD	AR*, WD*, ST*
Zygophyllaceae	*Tribulus suberosus*	PD	AR*, WD* ST*, N*
Boraginaceae	*Trichodesma zeylanicum*	PD	AR*, WD*
Poaceae	*Triodia angusta*	PD	CTS, AR, WD*, K+
Poaceae	*Triodia basedowii*	PD	CTS, AR, WD*, K+
Poaceae	*Triodia brizoides*	PD	CTS, AR, WD*, K+
Poaceae	*Triodia epactia*	PD	CTS, AR, WD*, K+
Poaceae	*Triodia pungens*	PD	CTS, AR, WD*, K+
Poaceae	*Triodia* sp. Shovelanna Hill	PD	CTS, AR, WD*, K+
Poaceae	*Triodia wiseana*	PD	CTS, AR, WD*, K+
Goodeniaceae	*Velleia panduriformis*	PD	AR*, WD*, G+, K+
Campanulaceae	*Wahlenbergia tumidifructa*	MD	No pre-treatment required

Dormancy class: ND = non-dormant (germination in water >75% for at least one incubation temperature);
PY = physical dormancy;
PD = physiological dormancy;
MD = morphological dormancy;
MPD = morphophysiological dormancy;
PY + PD = combinational dormancy.

Suggested pre-treatments: $HW^{minutes}$ $C^{degrees\ celcius}$ = hot water treatment for a defined time (e.g. 1–5 mins) at a certain temperature (e.g. 90°C);
AR = dry after-ripening treatment (e.g. store seeds at 30°C and 50% RH for 3–12 months, or longer);
N = nicking with scalpel blade or mechanical scarifier; SA^{hours} = immersion in 98% sulphuric acid;
$DH^{minutes}$ $C^{degrees\ celcius}$ = dry convection oven heat treatment for a defined time (e.g. 1–5 mins) at a certain temperature (e.g. 90°C);
ST = warm stratification (e.g. incubate fully hydrated seeds at 30–35°C for 6–8 weeks then transfer to a cooler temperature for germination);
WD = wet/dry cycling (e.g. store seeds at 30°C and 50% RH for 12 days, hydrate seeds for 2 days, then return to 30°C and 50% RH; repeat cycle 3–4 times);
CTS = clean to seeds by removing covering fruit/floret structures;
* = suggested pre-treatment, but no data is available to verify the response;
G+ = positive germination response to GA_3;
K+ = positive germination response to KAR_1.

Note that positive responses to GA_3 and/or KAR_1 have only been reported for species for which data is available; seeds of other species may respond, particularly seeds with PD. Note also that where species have been labelled as ND, some pre-treatments are still suggested as these may increase the range of temperatures suitable for germination, or increase germination speed.

Smoke, or products derived from smoke, has long been known to promote seed germination of a diverse range of plant species. Throughout the germination testing conducted to gather the data presented in this book we used the pure, synthetic form of the primary chemical in smoke responsible for promoting germination – karrikinolide, or KAR_1. However, for the restoration practitioner there are well-established techniques for treating seeds with smoke, and the use of either smoke water, or aerosol smoke, is recommended (Figure B.1). Now widely used in nursery production, landscape gardening, and smaller-scale mine-site restoration programs, the following information describes how to produce and use smoke derived products.

Smoke water is produced by burning plant material in a steel drum (the 'smoke generator') and drawing the smoke through a 20–30 L plastic drum filled with water for 60 minutes (Figure B.1 A). Most dried plant and woody material is suitable, however, the use of hay or straw which is cheap and readily available, is recommended. The smoke water produced is quite concentrated and requires dilution prior to using as a seed pre-treatment (Figure B.1 B). Dilutions of 1:10, 1:100, or even 1:1000 can be effective, but a 1:10 dilution (i.e. a 10% solution) is most commonly used. Seeds can be soaked directly in the diluted smoke water for up to 24 hours and then re-dried prior to sowing in the nursery or field. Alternatively, dry seeds can be aerosol smoked for up to 60 minutes in a plastic tent, or small shed (Figure B.1 B).

Equipment required

- 200 L steel drum with a removable and sealable lid.
- Fan and voltage regulator to control fan speed (for inlet of drum).
- Cooling pipe(s) to carry smoke to water filled drum (or to a plastic tent or shed for the application of aerosol smoke – Figure B.1 – B).
- A 20–30 L plastic drum through which smoke is drawn (or a plastic tent or small shed).
- A vacuum capable of drawing sufficient air and smoke through plastic drums.
- Hay or straw bales, or dry leaves and other plant material, for burning and producing smoke.

How to make smoke water

1. Position a 200 L steel drum in a well ventilated outdoor area.
2. Attach an inlet pipe to the side of the drum wall at the lowest point possible to allow air to be passed into the drum. This air inlet promotes the production of smoke and the exit of the smoke through the drum/pipe outlet.
3. Attach an outlet (cooling) pipe to the side of the drum wall at the highest point possible that allows the smoke produced from burning material to exit the drum (i.e. via the pipe when the drum is sealed with the lid).
4. Attach a cooling pipe of at least 1 m in length to a smaller 20–30 L plastic drum.
5. Attach a flexible hose or conduit to the end of the cooling pipe to direct smoke into the drum (inlet) that is positioned in the drum below the water line.
6. Attach a second hose to the same smoke water drum that is above the water line (outlet) to allow excess air/smoke to exit the drum. A second empty drum can be connected to the first drum in series to prevent water entering the vacuum.
7. Attach a vacuum to provide a source of suction to draw the smoke from the smoke generator through the cooling pipe and the plastic drum full of water.
8. Once all the components above have been connected, place sufficient plant material in the drum to burn for approximately 60 minutes (extra material can be added if required at 15–20 minute intervals). Ignite the plant material and seal the drum with the drum lid. Turn on both the inlet fan and vacuum.
9. The inlet fan and vacuum might need adjusting to optimise smoke movement and smoke production from the burning plant material. Allow up to 60 minutes to make the smoke water.
10. Smoke water remains vital for up to 3 years if stored in a cool, dark location, or indefinitely if frozen.

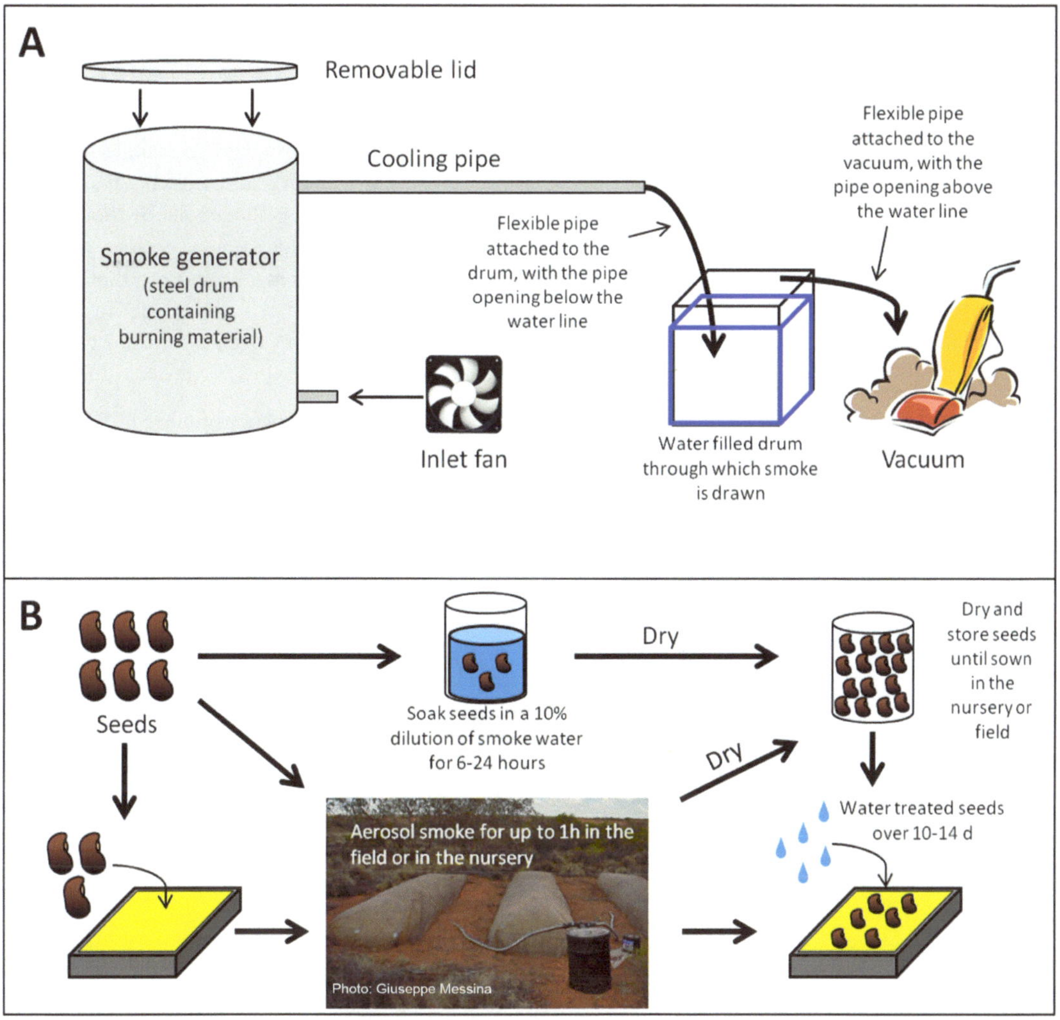

Figure B.1 Smoke water preparation involves burning plant material (straw and leafy materials but preferably not woody material) and drawing smoke through a water-filled drum using an optional inlet fan and a vacuum to generate airflow and suction (A). If necessary, a second, empty drum can be connected in series to the water-filled drum, with the vacuum attached to this empty drum, to prevent water entering the vacuum. Once smoke water is produced (60 minutes), seeds can be soaked in a 10% dilution of the smoke water for up to 24 hours and re-dried for storage until required (B). An alternative option is to apply aerosol smoke directly to dry seeds or seeds that are sown in seed trays, by forcing the smoke into an enclosed space, such as a plastic tent or small shed, for 60 minutes. Aerosol smoke can also be applied directly to the soil surface in the field to stimulate the soil seed-bank.

Appendix C. Checklist of vascular plants in the Pilbara Bioregion

Prepared by Steve Dillon, 1 October 2015.

Notes regarding the Pilbara vascular plant species list

Every three months all the Western Australain Herbarium records from within the rectangle shown in Figure C.1 are extracted from the main database. All the non-vascular records and any taxa that are not on the current WA Census are removed. Records of taxa not on the previous Pilbara list or those with a dubious distribution are double-checked to ensure that the identification and location are correct. A species list is then produced and any changes since the previous list are documented. Please note that this is a snapshot of the WA Herbarium records and therefore will quickly be out of date – please refer to *FloraBase* (www.florabase.dpaw.wa.gov.au) for the most up to date information. Species occuring within the query box, but outside the biogeographic region, have been excluded from this list.

Species in **bold** are included in this book and the name is followed by the voucher number (**TEE** = T.E. Erickson), held at the Western Australian Herbarium. (**NS** = Nindethana Seeds). (**BNSS** = Bushland Native Seed Supply). An '*' denotes introduced species that are considered naturalised in the Pilbara.

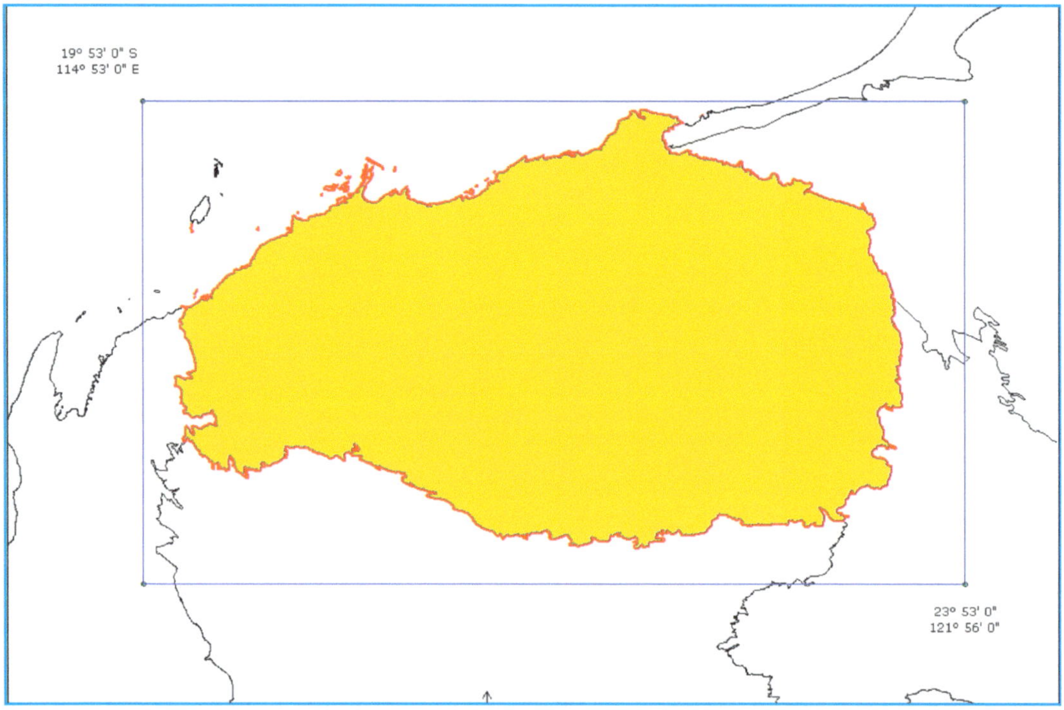

Figure C.1 Map showing area from which records are extracted to create the Pilbara plant species list.

Acanthaceae Juss.
Avicennia marina (Forssk.) Vierh. subsp. *marina*
Dicladanthera forrestii F.Muell.
Dicladanthera glabra R.M.Barker
Dicliptera armata F.Muell.
Dipteracanthus australasicus F.Muell. subsp. *australasicus*
Dipteracanthus chichesterensis Trudgen & de Koch
Harnieria kempeana subsp. *muelleri* (R.M.Barker) R.M.Barker
Rostellularia adscendens var. *clementii* (Domin) R.M.Barker
Rostellularia adscendens var. *latifolia* (R.Br.) R.M.Barker
Rostellularia adscendens var. *pogonanthera* (F.Muell.) R.M.Barker

Aizoaceae Martinov
Gunniopsis propinqua Chinnock
Sesuvium portulacastrum (L.) L.
Trianthema cussackianum F.Muell.
Trianthema glossostigmum F.Muell.
Trianthema oxycalyptrum F.Muell. var. *oxycalyptrum*
Trianthema pilosum F.Muell.
**Trianthema portulacastrum* L.
Trianthema sp. Python Pool (G.R. Guerin & M.E. Trudgen GG 1023)
Trianthema triquetrum Willd.
Trianthema turgidifolium F.Muell.
Zaleya galericulata (Melville) H.Eichler subsp. *galericulata*

Amaranthaceae Juss.
Achyranthes aspera L.
**Aerva javanica* (Burm.f.) Schult.
Alternanthera angustifolia R.Br.
Alternanthera denticulata R.Br.
Alternanthera nana R.Br.
Alternanthera nodiflora R.Br.
**Alternanthera pungens* Kunth
Amaranthus centralis J.Palmer & Mowatt
Amaranthus cochleitepalus Domin
Amaranthus cuspidifolius Domin
Amaranthus induratus J.Palmer & Mowatt
Amaranthus interruptus R.Br.
Amaranthus mitchellii Benth.
Amaranthus undulatus R.Br.
**Amaranthus viridis* L.
Gomphrena affinis subsp. *pilbarensis* J.Palmer
Gomphrena canescens R.Br. subsp. *canescens*
**Gomphrena celosioides* Mart.
Gomphrena cucullata J.Palmer
Gomphrena cunninghamii (Moq.) Druce **TEE 406**
Gomphrena kanisii J.Palmer
Gomphrena lanata R.Br.
Gomphrena leptoclada Benth. subsp. *leptoclada*
Gomphrena leptophylla (Benth.) J.Palmer
Gomphrena pusilla Benth.
Gomphrena sordida Farmar
Gomphrena sp. Martins Well (K.F. Kenneally 6116)
Ptilotus aervoides (F.Muell.) F.Muell.
Ptilotus aphyllus Benl
Ptilotus appendiculatus Benl
Ptilotus arthrolasius F.Muell.
Ptilotus astrolasius F.Muell.
Ptilotus auriculifolius (Moq.) F.Muell. **TEE 715**
Ptilotus axillaris (Benth.) F.Muell.
Ptilotus calostachyus F.Muell.
Ptilotus carinatus Benl
Ptilotus chamaecladus Diels
Ptilotus clementii (Farmar) Benl
Ptilotus divaricatus (Gaudich.) F.Muell.
Ptilotus drummondii (Moq.) F.Muell.
Ptilotus fusiformis (R.Br.) Poir., Lam.
Ptilotus gaudichaudii (Steud.) J.M.Black subsp. *eremita* (S.Moore) Lally
Ptilotus gaudichaudii (Steud.) J.M.Black subsp. *gaudichaudii*
Ptilotus gomphrenoides var. *conglomeratus* (Farmar) Benl
Ptilotus gomphrenoides Benth. var. *gomphrenoides*
Ptilotus gomphrenoides var. *roseo-albus* (Farmar) Benl
Ptilotus helipteroides (F.Muell.) F.Muell.
Ptilotus incanus (R.Br.) Poir.
Ptilotus latifolius R.Br.
Ptilotus macrocephalus (R.Br.) Poir. **TEE 437**
Ptilotus mollis Benl
Ptilotus murrayi F.Muell.
Ptilotus nobilis (Lindl.) F.Muell. subsp. ***nobilis*** **TEE 492**
Ptilotus obovatus (Gaudich.) F.Muell. var. *obvatus*
Ptilotus petiolatus Farmar
Ptilotus polystachyus (Gaudich.) F.Muell.
Ptilotus roei (Benth.) F.Muell.
Ptilotus rotundifolius (F.Muell.) F.Muell.
Ptilotus schwartzii var. *georgei* (Diels) Benl
Ptilotus schwartzii Tate var. *schwartzii*
Ptilotus subspinescens R.W.Davis
Ptilotus trichocephalus Benl
Ptilotus villosiflorus F.Muell.
Ptilotus wilsonii Benl
**Pupalia lappacea* (L.) Juss.
Surreya diandra (R.Br.) R.Masson & G.Kadereit

Apiaceae Lindl.
Centella asiatica (L.) Urb.
**Cyclospermum leptophyllum* (Pers.) Britton & P.Wilson
Daucus glochidiatus (Labill.) Fisch., C.A.Mey. & Ave-Lall.

Apocynaceae Juss.
**Asclepias curassavica* L.
**Calotropis procera* (Aiton) W.T.Aiton
Carissa lanceolata R.Br.
**Catharanthus roseus* (L.) G.Don
Cynanchum carnosum (R.Br.) Schltr.
Cynanchum floribundum R.Br.
Cynanchum pedunculatum R.Br.
Gymnanthera cunninghamii (Benth.) P.I.Forst.
Marsdenia angustata P.I.Forst.
Marsdenia australis (R.Br.) Druce
Rhyncharrhena linearis (Decne.) K.L.Wilson
Sarcostemma viminale subsp. *australe* (R.Br.) P.I.Forst.
Tylophora cinerascens (R.Br.) P.I.Forst.
Tylophora flexuosa R.Br.

Araliaceae Juss.
Astrotricha hamptonii F.Muell. **TEE 797**
Trachymene bialata (Domin) B.L.Burtt
Trachymene didiscoides (F.Muell.) B.L.Burtt
Trachymene oleracea (Domin) B.L.Burtt subsp. ***oleracea*** **TEE 724**
Trachymene pilbarensis Rye

Arecaceae Bercht. & J.Presl
Livistona alfredii F.Muell.
**Phoenix dactylifera* L.
**Washingtonia filifera* H.Wendl.

Asparagaceae Juss.
Thysanotus exiliflorus F.Muell.
Thysanotus manglesianus Kunth
Thysanotus sp. Eremaean (S. van Leeuwen 1067)

Asphodelaceae Juss.
**Aloe vera* var. *officinalis* (Forssk.) Baker
**Asphodelus fistulosus* L.
Bulbine pendula Keighery

Aspleniaceae Newman
Pleurosorus rutifolius (R.Br.) Fee

Asteraceae Bercht. & J. Presl
**Acanthospermum hispidum* DC.
Actinobole uliginosum (A.Gray) H.Eichler
Angianthus acrohyalinus Morrison
Angianthus cunninghamii (DC.) Benth.
Angianthus cyathifer P.S.Short
Angianthus milnei Benth.
Angianthus tomentosus J.C.Wendl.
Apowollastonia hamersleyensis Orchard
**Arctotheca calendula* (L.) Levyns
**Bidens pilosa* L. var. *pilosa*
**Bidens subalternans* DC. var. *subalternans*
Bidens subalternans var. *areneosa* Orchard
**Bidens subalternans* var. *simulans* Scherff
Blumea tenella DC.
Brachyscome blackii G.L.Davis
Brachyscome ciliaris (Labill.) Less.
Brachyscome iberidifolia Benth.
Brachyscome rudallensis P.S.Short
Calocephalus beardii P.S.Short ms
Calocephalus francisii (F.Muell.) Benth.
Calocephalus knappii (F.Muell.) Ewart & Jean White
Calocephalus multiflorus (Turcz.) Benth.
Calocephalus sp. Wittenoom (A.S. George 1082)
Calotis hispidula (F.Muell.) F.Muell.
Calotis latiuscula F.Muell. & Tate
Calotis multicaulis (Turcz.) Druce
Calotis plumulifera F.Muell.
Calotis porphyroglossa Benth.
Calotis squamigera C.T.White
Centipeda crateriformis N.G.Walsh subsp. *crateriformis*
Centipeda minima subsp. *macrocephala* N.G.Walsh
Centipeda minima (L.) A.Braun & Asch. subsp. *minima*
Centipeda thespidioides F.Muell.
Chrysocephalum apiculatum (Labill.) Steetz **TEE 486**
Chrysocephalum eremaeum (Haegi) Anderb.
Chrysocephalum gilesii (F.Muell.) Paul G.Wilson
Chrysocephalum pterochaetum F.Muell.
Chrysocephalum sp. Pilbara (H. Demarz 2852)
**Conyza bonariensis* (L.) Cronquist
**Cyanthillium cinereum* (L.) H.Rob.
Decazesia hecatocephala F.Muell.
Dichromochlamys dentatifolia (F.Muell.) Dunlop
Flaveria sp. Tom Price (M.E. Trudgen 11246)
**Flaveria trinervia* (Spreng.) C.Mohr
Gnephosis arachnoidea Turcz.
Gnephosis brevifolia (A.Gray) Benth.
Helichrysum luteoalbum (L.) Rchb.
Helichrysum oligochaetum F.Muell.
Iotasperma sessilifolium (F.Muell.) G.L.Nesom
Ixiochlamys cuneifolia (R.Br.) Sond.
**Lactuca saligna* L.
**Lactuca serriola* L. forma *serriola*
Leiocarpa semicalva (F.Muell.) Paul G.Wilson subsp. *semicalva*
Leiocarpa tomentosa (Sond.) Paul G.Wilson
Minuria integerrima (DC.) Benth.
Minuria leptophylla DC.
Myriocephalus oldfieldii (F.Muell.) Paul G.Wilson
Myriocephalus rudallii Benth.
Myriocephalus scalpellus Paul G.Wilson
Olearia fluvialis Lander
Olearia mucronata Lander
Olearia plucheacea Lander
Olearia sp. Kennedy Range (G. Byrne 66)
Olearia stuartii (F.Muell.) Benth.
Olearia xerophila (F.Muell.) Benth.
**Parthenium hysterophorus* L.
Pentalepis trichodesmoides subsp. *hispida* Orchard
Pentalepis trichodesmoides F.Muell. subsp. *trichodesmoides*
Peripleura (Vittadinia) arida (N.T.Burb.) G.L.Nesom
Peripleura (Vittadinia) obovata (N.T.Burb.) G.L.Nesom
Peripleura (Vittadinia) virgata (N.T.Burb.) G.L.Nesom
Pilbara trudgenii Lander
Pleurocarpaea gracilis Lander & P.J.H.Hurter
Pluchea dentex Benth.
Pluchea dunlopii Hunger
Pluchea ferdinandi-muelleri Domin
Pluchea longiseta A.R.Bean
Pluchea rubelliflora (F.Muell.) B.L.Rob.
Pluchea tetranthera F.Muell.
Podolepis capillaris (Steetz) Diels
Podolepis eremaea Jeanes
Podolepis remota Jeanes
Pterocaulon intermedium (DC.) A.R.Bean
Pterocaulon serrulatum var. *velutinum* (Ewart & O.B.Davies) Guillaumin
Pterocaulon sphacelatum (Labill.) F.Muell. **TEE 658**
Pterocaulon sphaeranthoides (DC.) F.Muell.
Pterocaulon xenicum A.R.Bean
Rhodanthe ascendens Paul G.Wilson
Rhodanthe charsleyae (F.Muell.) Paul G.Wilson
Rhodanthe citrina (Benth.) Paul G.Wilson
Rhodanthe floribunda (DC.) Paul G.Wilson
Rhodanthe humboldtiana (Gaudich.) Paul G.Wilson
Rhodanthe margarethae (F.Muell.) Paul G.Wilson
Rhodanthe polakii (F.Muell.) Paul G.Wilson
Rhodanthe propinqua (W.Fitzg.) Paul G.Wilson
Rhodanthe psammophila Paul G.Wilson
Rhodanthe sterilescens (F.Muell.) Paul G.Wilson
Rhodanthe stricta (Lindl.) Paul G.Wilson
Rhodanthe tietkensii (F.Muell.) Paul G.Wilson
Roebuckiella cheilocarpa (F.Muell.) P.S.Short var. *cheilocarpa*
Roebuckiella ciliocarpa (W.Fitzg.) P.S.Short
Roebuckiella cuneata (P.S.Short) P.S.Short
Roebuckiella oncocarpa (Diels) P.S.Short
Roebuckiella similis (P.S.Short) P.S.Short
Rutidosis helichrysoides DC. subsp. *helichrysoides*
Schoenia ayersii (F.Muell.) J.M.Black
Schoenia cassiniana (Gaudich.) Steetz
Senecio gregorii F.Muell.
Senecio hamersleyensis I.Thomps.
Senecio magnificus F.Muell.
Senecio pinnatifolius A.Rich.
**Sigesbeckia orientalis* L.
**Solidago canadensis* L.
Sonchus hydrophilus Boulos
**Sonchus oleraceus* L.

Streptoglossa adscendens (Benth.) Dunlop
Streptoglossa bubakii (Domin) Dunlop
Streptoglossa cylindriceps (J.M.Black) Dunlop
Streptoglossa decurrens (DC.) Dunlop **TEE 524**
Streptoglossa liatroides (Turcz.) Dunlop
Streptoglossa macrocephala (F.Muell.) Dunlop
Streptoglossa odora (F.Muell.) Dunlop
Streptoglossa tenuiflora Dunlop
**Symphyotrichum squamatum* (Spreng.) G.L.Nesom
Taplinia saxatilis Lander
**Taraxacum officinale* F.H.Wigg.
**Tridax procumbens* L.
Vittadinia cervicularis N.T.Burb.
Vittadinia dissecta var. *hirta* N.T.Burb.
Vittadinia eremaea N.T.Burb.
Vittadinia sp. Coondewanna Flats (S. van Leeuwen 4684)
Vittadinia sulcata N.T.Burb.
Xerochrysum bracteatum (Vent.) Tzvelev

Bignoniaceae Juss.
Dolichandrone heterophylla (R.Br.) F.Muell.
Pandorea pandorana (Andrews) Steenis

Bixaceae Kunth
Cochlospermum macnamarae Hislop, K.R.Thiele & Brassington

Boraginaceae Juss.
Cordia subcordata Lam.
Ehretia saligna R.Br. var. *saligna*
Halgania cyanea var. Allambi Stn (B.W. Strong 676)
Halgania erecta Ewart & B.Rees
Halgania gustafsenii F.Muell. var. *gustafsenii*
Halgania gustafsenii var. Mid West (G. Perry 370)
Halgania solanacea var. Mt Doreen (G.M. Chippendale 4206)
Heliotropium ammophilum Craven
Heliotropium chrysocarpum Craven
Heliotropium conocarpum Benth.
Heliotropium crispatum Benth.
Heliotropium cunninghamii Benth.
Heliotropium curassavicum L.
Heliotropium diversifolium Benth.
**Heliotropium europaeum* L.
Heliotropium foliatum R.Br.
Heliotropium glanduliferum Craven
Heliotropium heteranthum (F.Muell.) Ewart & O.B.Davies
Heliotropium inexplicitum Craven
Heliotropium murinum Craven
Heliotropium muticum Domin
Heliotropium ovalifolium Forssk.
Heliotropium pachyphyllum Craven
Heliotropium parviantrum Craven
Heliotropium skeleton Craven
Heliotropium sp. Ord River (W. Fitzgerald 1611)
Heliotropium tanythrix Craven
Heliotropium tenuifolium R.Br.
Heliotropium transforme Craven
Heliotropium ventricosum R.Br.
Heliotropium vestitum Benth.
Trichodesma zeylanicum var. *grandiflorum* Randell
Trichodesma zeylanicum var. *latisepalum* F.Muell.
Trichodesma zeylanicum (Burm.f.) R.Br. var. ***zeylanicum*** **TEE 566**

Brassicaceae Burnett
**Brassica × napus* L.
**Brassica tournefortii* Gouan
Cuphonotus andraeanus (F.Muell.) E.A.Shaw
Lepidium amelum Lepschi
Lepidium catapycnon Hewson **TEE 645**
**Lepidium didymum* L.
Lepidium echinatum Hewson **TEE 780**
Lepidium foliosum Desv.
Lepidium muelleri-ferdinandii Thell.
Lepidium oxytrichum Sprague
Lepidium pedicellosum F.Muell. **TEE 799**
Lepidium phlebopetalum (F.Muell.) F.Muell.
Lepidium pholidogynum F.Muell.
Lepidium platypetalum Hewson
Menkea villosula (F.Muell. & Tate) J.M.Black
**Sisymbrium orientale* L.
Stenopetalum anfractum E.A.Shaw
Stenopetalum decipiens E.A.Shaw
Stenopetalum nutans F.Muell.
Stenopetalum pedicellare Benth.
Stenopetalum velutinum F.Muell.

Byblidaceae Domin
Byblis pilbarana Lowrie

Cactaceae Juss.
**Cylindropuntia imbricata* (Haw.) F.M.Knuth
**Opuntia elata* Salm-Dyck
**Opuntia stricta* (Haw.) Haw.

Campanulaceae Juss.
Isotoma petraea F.Muell.
Lobelia arnhemiaca E.Wimm.
Lobelia heterophylla Labill. subsp. *heterophylla*
Lobelia heterophylla subsp. *pilbarensis* N.G.Walsh
Wahlenbergia caryophylloides P.J.Sm.
Wahlenbergia gracilenta Lothian
Wahlenbergia queenslandica P.J.Sm.
Wahlenbergia tumidifructa P.J.Sm. **TEE 717**

Capparaceae Juss.
Capparis lasiantha DC.
Capparis mitchellii Lindl.
Capparis spinosa var. *nummularia* (DC.) Bailey
Capparis umbonata Lindl., T.Mitch.

Caryophyllaceae Juss.
Polycarpaea corymbosa (L.) Lam. var. ***corymbosa*** **TEE 719**
Polycarpaea holtzei Maiden & Betche
Polycarpaea involucrata F.Muell.
Polycarpaea longiflora F.Muell.

Casuarinaceae R. Br.
Allocasuarina decaisneana (F.Muell.) L.A.S.Johnson

Celastraceae R. Br.
Denhamia cunninghamii (Hook.) M.P.Simmons
Macgregoria racemigera F.Muell.
Maytenus sp. Mt Windell (S. van Leeuwen 846)
Stackhousia clementii Domin
Stackhousia intermedia F.M.Bailey
Stackhousia muricata Lindl. subsp. Annual (W.R. Barker 2172)
Stackhousia sp. swollen gynophore (W.R. Barker 2041)

Chenopodiaceae Vent.
Atriplex amnicola Paul G.Wilson
Atriplex bunburyana F.Muell.
Atriplex codonocarpa Paul G.Wilson
Atriplex eremitis Cranfield
Atriplex flabelliformis Paul G.Wilson
Atriplex holocarpa F.Muell.
Atriplex isatidea Moq.
Atriplex lindleyi subsp. *conduplicata* (F.Muell.) Paul G.Wilson
Atriplex quadrivalvata Diels
Atriplex semilunaris Aellen
Atriplex spinulosa Paul G.Wilson
**Bassia scoparia* (L.) A.J.Scott
Chenopodium auricomum Lindl., T.Mitch.
Chenopodium gaudichaudianum (Moq.) Paul G.Wilson
**Chenopodium murale* L.
Dissocarpus paradoxus (R.Br.) Ulbr., Engl. & Prantl
Dysphania cristata (F.Muell.) Mosyakin & Clemants
Dysphania glomulifera subsp. *eremaea* Paul G.Wilson
Dysphania kalpari Paul G.Wilson
Dysphania melanocarpa forma *leucocarpa* (Aellen) Paul G.Wilson & K.A.Sheph.
Dysphania melanocarpa (J.M.Black) Mosyakin & Clemants forma *melanocarpa*
Dysphania plantaginella F.Muell.
Dysphania pumilio (R.Br.) Mosyakin & Clemants
Dysphania rhadinostachya subsp. *inflata* (Aellen) Paul G.Wilson
Dysphania rhadinostachya (F.Muell.) A.J.Scott subsp. ***rhadinostachya*** **TEE 713**
Dysphania saxatilis (Paul G.Wilson) Mosyakin & Clemants
Dysphania sp. Fortescue Marsh (A. Markey & S. Dillon FM 9709)
Dysphania sphaerosperma Paul G.Wilson
Enchylaena tomentosa R.Br. var. *tomentosa*
Eremophea spinosa (Ewart & O.B.Davies) Paul G.Wilson
Maireana amoena (Diels) Paul G.Wilson
Maireana aphylla (R.Br.) Paul G.Wilson
Maireana carnosa (Moq.) Paul G.Wilson
Maireana eriosphaera Paul G.Wilson
Maireana georgei (Diels) Paul G.Wilson
Maireana integra (Paul G.Wilson) Paul G.Wilson
Maireana luehmannii (F.Muell.) Paul G.Wilson
Maireana melanocoma (F.Muell.) Paul G.Wilson
Maireana planifolia (F.Muell.) Paul G.Wilson
Maireana platycarpa Paul G.Wilson
Maireana pyramidata (Benth.) Paul G.Wilson
Maireana stipitata Paul G.Wilson
Maireana suaedifolia (Paul G.Wilson) Paul G.Wilson
Maireana thesioides (C.A.Gardner) Paul G.Wilson
Maireana tomentosa Moq. subsp. *tomentosa*
Maireana triptera (Benth.) Paul G.Wilson
Maireana villosa (Lindl.) Paul G.Wilson
Neobassia astrocarpa (F.Muell.) A.J.Scott
Rhagodia eremaea Paul G.Wilson
Rhagodia preissii subsp. *obovata* (Moq.) Paul G.Wilson
Rhagodia sp. Hamersley (M. Trudgen 17794)
Salsola australis R.Br.
Sclerolaena beaugleholei (Ising) A.J.Scott
Sclerolaena bicornis Lindl. var. *bicornis*
Sclerolaena burbidgeae (Ising) A.J.Scott
Sclerolaena convexula (R.H.Anderson) A.J.Scott
Sclerolaena cornishiana (F.Muell.) A.J.Scott
Sclerolaena costata (R.H.Anderson) A.J.Scott
Sclerolaena cuneata Paul G.Wilson
Sclerolaena densiflora (W.Fitzg.) A.J.Scott
Sclerolaena deserticola Paul G.Wilson
Sclerolaena diacantha (Nees) Benth.
Sclerolaena eriacantha (F.Muell.) Ulbr.
Sclerolaena gardneri (Ising) A.J.Scott
Sclerolaena glabra (F.Muell.) Domin
Sclerolaena hostilis (Diels) Domin
Sclerolaena lanicuspis (F.Muell.) Benth.
Sclerolaena minuta (Ising) A.J.Scott
Sclerolaena recurvicuspis (W.Fitzg.) Domin
Sclerolaena tetragona Paul G.Wilson
Sclerolaena uniflora R.Br.
Suaeda arbusculoides L.S.Sm.
Tecticornia auriculata (Paul G.Wilson) K.A.Sheph. & Paul G.Wilson
Tecticornia disarticulata (Paul G.Wilson) K.A.Sheph. & Paul G.Wilson
Tecticornia globulifera K.A.Sheph.
Tecticornia halocnemoides subsp. *longispicata* (Paul G.Wilson) K.A.Sheph. & Paul G.Wilson
Tecticornia halocnemoides subsp. *tenuis* (Paul G.Wilson) K.A.Sheph. & Paul G.Wilson
Tecticornia indica subsp. *bidens* (Nees) K.A.Sheph. & Paul G.Wilson
Tecticornia indica (Willd.) K.A.Sheph. & Paul G.Wilson subsp. *indica*
Tecticornia indica subsp. *juliacea* (Paul G.Wilson) K.A.Sheph. & Paul G.Wilson
Tecticornia indica subsp. *leiostachya* (Benth.) K.A.Sheph. & Paul G.Wilson
Tecticornia medusa K.A.Sheph. & S.J.van Leeuwen
Tecticornia pergranulata subsp. *elongata* (Paul G.Wilson) K.A.Sheph. & Paul G.Wilson
Tecticornia pruinosa (Paulsen) K.A.Sheph. & Paul G.Wilson
Tecticornia pterygosperma subsp. *denticulata* (Paul G.Wilson) K.A.Sheph. & Paul G.Wilson

Tecticornia sp. Christmas Creek (K.A. Shepherd & T. Colmer *et al.* KS 1063)
Tecticornia sp. Dennys Crossing (K.A. Shepherd & J. English KS 552)
Tecticornia sp. Yoothapina Station (A.A. Mitchell 883)
Tecticornia verrucosa Paul G.Wilson
Threlkeldia diffusa R.Br.

Cleomaceae Bercht. & J.Presl
Cleome oxalidea F.Muell.
Cleome uncifera Kers subsp. *uncifera*
Cleome viscosa L. **TEE 410**

Combretaceae R. Br.
Terminalia circumalata F.Muell.
Terminalia platyphylla F.Muell.
Terminalia supranitifolia Byrnes

Commelinaceae Mirb.
Commelina ensifolia R.Br.
Murdannia graminea (R.Br.) G.Brueckn.

Convolvulaceae Juss.
Bonamia alatisemina R.W.Johnson
Bonamia erecta R.W.Johnson
Bonamia linearis (R.Br.) Hallier f.
Bonamia media (R.Br.) Hallier f.
Bonamia pannosa (R.Br.) Hallier f.
Bonamia pilbarensis R.W.Johnson
Bonamia rosea (F.Muell.) Hallier f.
Convolvulus clementii Domin **TEE 439**
Convolvulus remotus R.Br.
Cressa australis R.Br.
Cuscuta victoriana Yunck.
Davenportia davenportii (F.Muell.) R.W.Johnson
Duperreya commixta (Staples) Staples
Evolvulus alsinoides var. *decumbens* (R.Br.) Ooststr.
Evolvulus alsinoides var. *villosicalyx* Ooststr.
Ipomoea calobra W.Hill & F.Muell.
Ipomoea coptica (L.) Roth
Ipomoea costata Benth.
Ipomoea diamantinensis Eardley
Ipomoea lonchophylla J.M.Black
Ipomoea muelleri Benth.
Ipomoea pes-caprae subsp. *brasiliensis* (L.) Ooststr.
Ipomoea plebeia R.Br.
Ipomoea polymorpha Roem. & Schult.
Ipomoea racemigera F.Muell. & Tate
Jacquemontia pannosa (R.Br.) Mabb.
**Merremia dissecta* (Jacq.) Hallier f. var. dissecta
Operculina aequisepala (Domin) R.W.Johnson
Operculina brownii Ooststr.
Polymeria ambigua R.Br.
Polymeria calycina R.Br.
Polymeria distigma Benth.
Polymeria lanata R.Br.
Polymeria longifolia Lindl.

Crassulaceae J. St.-Hil.
Crassula peduncularis (Sm.) Meigen

Cucurbitaceae Juss.
Austrobryonia pilbarensis I.Telford
**Citrullus colocynthis* (L.) Schrad.
**Citrullus lanatus* (Thunb.) Matsum. & Nakai
**Coccinia grandis* (L.) Voigt
Cucumis argenteus (Domin) P.Sebastian & I.Telford
Cucumis melo L.
**Cucumis myriocarpus* Naudin
Cucumis picrocarpus F.Muell.
Cucumis variabilis P.Sebastian & I.Telford
Cucumis sp. Barrow Island (D.W. Goodall 1264)
**Cucurbita pepo* L.
Trichosanthes cucumerina L. var. *cucumerina*

Cupressaceae Gray
Callitris columellaris F.Muell.

Cymodoceaceae Vines
Halodule uninervis (Forssk.) Asch.
Syringodium isoetifolium (Asch.) Dandy

Cyperaceae Juss.
Baumea rubiginosa (Spreng.) Boeckeler
Bulbostylis barbata (Rottb.) C.B.Clarke
Bulbostylis burbidgeae K.L.Wilson
Bulbostylis turbinata S.T.Blake
Cladium procerum S.T.Blake
Cyperus betchei subsp. *commiscens* K.L.Wilson
Cyperus bifax C.B.Clarke
Cyperus blakeanus K.L.Wilson
Cyperus bulbosus Vahl
Cyperus castaneus var. *brevimucronatus* Kuek.
Cyperus concinnus R.Br.
Cyperus conicus (R.Br.) Boeckeler
Cyperus cunninghamii (C.B.Clarke) C.A.Gardner subsp. *cunninghamii*
Cyperus dactylotes Benth.
Cyperus difformis L.
Cyperus gilesii Benth.
Cyperus hesperius K.L.Wilson
**Cyperus involucratus* Poir.
Cyperus iria L.
Cyperus ixiocarpus F.Muell. **TEE 412**
Cyperus macrostachyos Lam.
Cyperus microcephalus R.Br. subsp. *microcephalus*
Cyperus microcephalus subsp. saxicola K.L.Wilson
Cyperus nervulosus (Kuek.) S.T.Blake
**Cyperus polystachyos* Rottb.
Cyperus pulchellus R.Br.
Cyperus pygmaeus Rottb.
Cyperus rigidellus (Benth.) J.M.Black
**Cyperus rotundus* L.
Cyperus squarrosus L.
Cyperus vaginatus R.Br.
Eleocharis atropurpurea (Retz.) Kunth
Eleocharis dulcis (Burm.f.) Henschel
Eleocharis geniculata (L.) Roem. & Schult.

Eleocharis pallens S.T.Blake
Eleocharis papillosa Latz
Eleocharis sphacelata R.Br.
Eleocharis spiralis (Rottb.) Schult.
Fimbristylis cephalophora F.Muell.
Fimbristylis depauperata R.Br.
Fimbristylis dichotoma Vahl
Fimbristylis elegans S.T.Blake
Fimbristylis ferruginea Vahl
Fimbristylis littoralis Gaudich.
Fimbristylis microcarya F.Muell.
Fimbristylis neilsonii F.Muell.
Fimbristylis oxystachya F.Muell.
Fimbristylis phaeoleuca S.T.Blake
Fimbristylis rara R.Br.
Fimbristylis sieberiana Kunth
Fimbristylis simulans Latz
Fimbristylis sp. Shay Gap (K.R. Newbey 10293)
Fuirena ciliaris (L.) Roxb.
Fuirena incrassata S.T.Blake
Lipocarpha microcephala (R.Br.) Kunth
Schoenoplectus dissachanthus (S.T.Blake) J.Raynal
Schoenoplectus laevis (S.T.Blake) J.Raynal
Schoenoplectus lateriflorus (J.F.Gmel.) Lye
Schoenoplectus subulatus (Vahl) Lye
Schoenus falcatus R.Br.
Schoenus punctatus R.Br.

Dilleniaceae Salisb.
Hibbertia glaberrima F.Muell. **TEE 708**

Droseraceae Salisb.
Drosera burmanni Vahl
Drosera hartmeyerorum Schlauer
Drosera finlaysoniana Wall.

Elaeocarpaceae Juss.
Tetratheca fordiana R.Butcher

Elatinaceae Dumort.
Bergia ammannioides Roth
Bergia henshallii G.J.Leach
Bergia pedicellaris (F.Muell.) Benth.
Bergia perennis subsp. *exigua* G.J.Leach
Bergia perennis subsp. *obtusifolia* G.J.Leach
Bergia trimera Fisch. & C.A.Mey.

Eriocaulaceae Martinov
Eriocaulon cinereum R.Br.

Euphorbiaceae Juss.
Adriana tomentosa var. *hookeri* (F.Muell.) C.L.Gross & M.A.Whalen
Adriana tomentosa Gaudich. var. *tomentosa*
Euphorbia australis Boiss. var. *australis*
Euphorbia australis var. *glabra* Halford & W.K.Harris
Euphorbia australis var. *hispidula* Halford & W.K.Harris
Euphorbia australis var. *subtomentosa* Domin
Euphorbia biconvexa Domin
Euphorbia boophthona C.A.Gardner
Euphorbia careyi F.Muell.
Euphorbia clementii Domin
Euphorbia coghlanii F.M.Bailey
Euphorbia drummondii Boiss.
Euphorbia ferdinandi Baill. var. *ferdinandi*
Euphorbia fitzroyensis Halford & W.K.Harris
**Euphorbia hirta* L.
Euphorbia inappendiculata Domin subsp. *inappendiculata*
Euphorbia inappendiculata Domin var. *queenslandica* Domin
Euphorbia mitchelliana Boiss., A.DC.
Euphorbia multifaria Halford & W.K.Harris
Euphorbia myrtoides Boiss., A.DC.
Euphorbia parvicaruncula D.C.Hassall
Euphorbia porcata Halford & W.K.Harris
Euphorbia psilosperma Halford & W.K.Harris
**Euphorbia prostrata* Aiton
Euphorbia sharkoensis Baill.
Euphorbia schultzii Benth.
Euphorbia stevenii Bailey
Euphorbia tannensis subsp. *eremophila* (A.Cunn.) Hassall
**Euphorbia tirucalli* L.
Euphorbia trigonosperma Halford & W.K.Harris
Euphorbia vaccaria var. *erucoides* Halford & W.K.Harris
Euphorbia vaccaria Baill. var. *vaccaria*
Euphorbia wheeleri Baill.
Mallotus nesophilus Muell.Arg.
Microstachys chamaelea (L.) Hook.f.
**Ricinus communis* L.

Fabaceae Lindl.
Acacia acradenia F.Muell. **TEE 574**
Acacia adoxa Pedley var. ***adoxa*** **NS38117**
Acacia adoxa var. *subglabra* Pedley
Acacia adsurgens Maiden & Blakely **TEE 597**
Acacia adsurgens × *rhodophloia*
Acacia ampliceps Maslin
Acacia ampliceps × *bivenosa*
Acacia ampliceps × *sclerosperma* subsp. *sclerosperma*
Acacia ancistrocarpa Maiden & Blakely **TEE 544**
Acacia ancistrocarpa × *citrinoviridis*
Acacia aneura Benth.
Acacia aphanoclada Maslin
Acacia aphanoclada × *pyrifolia*
Acacia aptaneura Maslin & J.E.Reid
Acacia arida Benth.
Acacia arrecta Maslin
Acacia atkinsiana Maslin
Acacia ayersiana Maconochie
Acacia balsamea R.S.Cowan & Maslin
Acacia bivenosa DC. **TEE 601**
Acacia bivenosa × *sclerosperma* subsp. *sclerosperma*
Acacia bromilowiana Maslin
Acacia camptoclada C.R.P.Andrews
Acacia catenulata subsp. *occidentalis* Maslin
Acacia citrinoviridis Tindale & Maslin **TEE 666**
Acacia colei Maslin & L.Thomson var. *colei*
Acacia colei var. *ileocarpa* M.W.McDonald & Maslin
Acacia coriacea DC. subsp. *coriacea*

Acacia coriacea subsp. *pendens* R.S.Cowan & Maslin
Acacia cowleana Tate **TEE 568**
Acacia cuspidifolia Maslin
Acacia cuthbertsonii Luehm. subsp. *cuthbertsonii*
Acacia cyperophylla var. *omearana* Maslin
Acacia daweana Maslin
Acacia dictyophleba F.Muell. **TEE 656**
Acacia distans Maslin
Acacia drepanocarpa subsp. *latifolia* Pedley
Acacia effusa Maslin
Acacia elachantha Maslin & M.W.McDonald
Acacia eriopoda Maiden & Blakely **TEE 582**
Acacia eriopoda × trachycarpa
Acacia eriopoda × tumida var. *pilbarensis*
Acacia exigua I.M.Turner
Acacia fecunda Maslin
Acacia fuscaneura Maslin & J.E.Reid
Acacia glaucocaesia Domin
Acacia gregorii F.Muell.
Acacia hamersleyensis Maslin **TEE 660**
Acacia heteroneura Benth.
Acacia hilliana Maiden, Ewart & O.B.Davies **TEE 595**
Acacia holosericea G.Don
Acacia inaequilatera Domin **TEE 607**
Acacia incurvaneura Maslin & J.E.Reid
Acacia intorta Maslin
Acacia kempeana F.Muell.
Acacia leeuweniana Maslin
Acacia levata R.S.Cowan & Maslin
Acacia ligulata Benth.
Acacia macraneura Maslin & J.E.Reid
Acacia maitlandii F.Muell.
Acacia marramamba Maslin
Acacia melleodora Pedley
Acacia minutissima Maslin
Acacia minyura Randell
Acacia monticola J.M.Black **TEE 613**
Acacia monticola × trachycarpa
Acacia monticola × tumida var. *pilbarensis*
Acacia mulganeura Maslin & J.E.Reid
Acacia murrayana Benth.
Acacia nyssophylla F.Muell.
Acacia orthocarpa F.Muell.
Acacia oswaldii F.Muell.
Acacia pachyacra Maiden & Blakely
Acacia pachycarpa Benth.
Acacia paraneura Randell
Acacia pruinocarpa Tindale
Acacia pteraneura Maslin & J.E.Reid
Acacia ptychophylla F.Muell. **BNSS465**
Acacia pyrifolia var. *morrisonii* Maiden
Acacia pyrifolia DC. var. ***pyrifolia*** **TEE 603**
Acacia retivenea subsp. *clandestina* R.S.Cowan & Maslin
Acacia rhodophloia Maslin
Acacia rhodophloia × sibirica
Acacia robeorum Maslin
Acacia sabulosa Maslin
Acacia sclerosperma F.Muell. subsp. ***sclerosperma*** **TEE 599**
Acacia sericophylla F.Muell.
Acacia sibilans Maslin
Acacia sibirica S.Moore
Acacia sp. Airlie Island (V. Long VL 163)
Acacia sp. Barklys (J.L. Egan 124)
Acacia sp. East Fortescue (J. Bull & D. Roberts ONS A 27.01)
Acacia sp. Jimblebar (S. van Leeuwen 1342)
Acacia sp. Marble Bar (J.G. & M.H. Simmons 3499)
Acacia sp. Mulga Paraburdoo (B.R. Maslin *et al.* BRM 9201)
Acacia sp. Nullagine (B.R. Maslin 4955)
Acacia sp. Ripon Hills (B.R. Maslin 8460)
Acacia sphaerostachya E.Pritz.
Acacia spondylophylla F.Muell.
Acacia steedmanii subsp. *borealis* Maslin
Acacia stellaticeps Kodela, Tindale & D.A.Keith **TEE 593**
Acacia stenophylla Benth.
Acacia subcontorta Maslin
Acacia subtiliformis Maslin
Acacia synchronicia Maslin
Acacia tenuissima F.Muell. **TEE 605**
Acacia tetragonophylla F.Muell.
Acacia thoma Maslin
Acacia trachycarpa E.Pritz.
Acacia trachycarpa × tumida var. *pilbarensis*
Acacia trudgeniana Maslin
Acacia tumida var. ***pilbarensis*** M.W.McDonald **TEE 591**
Acacia tumida Benth. var. *tumida*
Acacia victoriae Benth. subsp. *victoriae*
Acacia wanyu Tindale
Acacia xiphophylla E.Pritz.
Aenictophyton reconditum subsp. *macrophyllum* I.Thomps.
Aeschynomene indica L.
Albizia lebbeck (L.) Benth.
Alysicarpus muelleri Schindl.
Bauhinia (Lysiphyllum) cunninghamii (Benth.) Benth.
Cajanus acutifolius (Benth.) Maesen
Cajanus cinereus (Benth.) F.Muell.
Cajanus marmoratus (Benth.) F.Muell.
Cajanus pubescens (Ewart & Morrison) Maesen
Canavalia rosea (Sw.) DC.
Chamaecrista symonii Pedley
**Clitoria ternatea* L.
Crotalaria cunninghamii subsp. *sturtii* (R.Br.) A.E.Holland
Crotalaria dissitiflora subsp. *benthamiana* (Pritzel) A.E.Holland
**Crotalaria juncea* L.
Crotalaria medicaginea var. *neglecta* (Wight & Arn.) Baker
Crotalaria novae-hollandiae DC. subsp. *novae-hollandiae*
Crotalaria ramosissima Roxb.
Crotalaria smithiana A.T.Lee
Cullen badocanum (Blanco) Verdc.
Cullen cinereum (Lindl.) J.W.Grimes
Cullen discolor (Domin) J.W.Grimes
Cullen graveolens (Domin) J.W.Grimes
Cullen lachnostachys (F.Muell.) J.W.Grimes
Cullen leucanthum (F.Muell.) J.W.Grimes

Cullen leucochaites J.W.Grimes
Cullen martinii (F.Muell.) J.W.Grimes
Cullen pallidum (N.T.Burb.) J.W.Grimes
Cullen patens (Lindl.) J.W.Grimes
Cullen pogonocarpum J.W.Grimes
Cullen stipulaceum (Decne.) J.W.Grimes
Daviesia eremaea Crisp
Desmodium campylocaulon Benth.
Desmodium filiforme Zoll. & Moritzi
Desmodium muelleri Benth.
**Desmodium scorpiurus* (Sw.) Desv.
Dichrostachys spicata (F.Muell.) Domin
Erythrina vespertilio Benth., T.Mitch.
Erythrophleum chlorostachys (F.Muell.) Baill.
Gastrolobium grandiflorum F.Muell.
Glycine canescens F.J.Herm.
Glycine falcata Benth.
Glycine tabacina (Labill.) Benth.
Glycine tomentella Hayata
Gompholobium karijini Chappill
Gompholobium oreophilum C.F.Wilkins & Trudgen **BNSS 2009**
Gompholobium polyzygum F.Muell.
Gompholobium simplicifolium (F.Muell. & Tate) Crisp
Indigastrum parviflorum (Wight & Arn.) Schrire
Indigofera ammobia Maconochie
Indigofera boviperda Morrison subsp. *boviperda*
Indigofera boviperda Morrison subsp. *eremaea* Peter G.Wilson & Rowe
Indigofera brevidens Benth. in T.Mitch.
Indigofera chamaeclada Peter G.Wilson & Rowe subsp. *chamaeclada*
Indigofera colutea (Burm.f.) Merr.
Indigofera cuspidata Peter G.Wilson & Rowe
Indigofera decipiens Peter G.Wilson & Rowe
Indigofera fractiflexa Peter G.Wilson & Rowe subsp. *fractiflexa*
Indigofera georgei E.Pritz.
Indigofera gilesii Peter G.Wilson & Rowe
Indigofera hirsuta L.
**Indigofera hochstetteri* Baker
Indigofera ixocarpa Peter G.Wilson & Rowe
Indigofera linifolia (L.f.) Retz.
Indigofera linnaei Ali
Indigofera monophylla DC. **TEE 563**
**Indigofera oblongifolia* Forssk.
Indigofera petraea Peter G.Wilson & Rowe
Indigofera roseola Peter G.Wilson & Rowe
Indigofera rugosa Benth.
**Indigofera sessiliflora* DC.
Indigofera sp. Bungaroo Creek (S. van Leeuwen 4301)
Indigofera trita L.f. subsp. *trita*
Isotropis atropurpurea F.Muell.
Isotropis forrestii F.Muell.
Isotropis parviflora Benth.
Jacksonia aculeata W.Fitzg.
Kennedia prorepens (F.Muell.) F.Muell.
Leptosema anomalum (Ewart & Morrison) Crisp
Leptosema chambersii F.Muell.
**Leucaena leucocephala* (Lam.) de Wit
Lotus australis Andrews
Lotus cruentus Court
**Macroptilium atropurpureum* (Moc. & Sesse) Urb.
**Medicago polymorpha* L.
Mirbelia viminalis (Benth.) C.A.Gardner **TEE 654**
Neptunia dimorphantha Domin
Neptunia gracilis Benth. forma *gracilis*
Neptunia monosperma Benth.
**Parkinsonia aculeata* L.
Petalostylis cassioides (F.Muell.) Symon
Petalostylis labicheoides R.Br. in C.Sturt **TEE 668**
**Prosopis glandulosa* × *velutina*
**Prosopis pallida* (Willd.) Kunth
Rhynchosia australis Benth.
Rhynchosia bungarensis Maesen
Rhynchosia minima (L.) DC.
Rothia indica subsp. *australis* A.E.Holland
Senna artemisioides subsp. *filifolia* Randell
Senna artemisioides subsp. *helmsii* (Symon) Randell
Senna artemisioides subsp. ***oligophylla*** (F.Muell.) Randell **TEE 664**
Senna artemisioides subsp. *petiolaris* Randell
Senna artemisioides subsp. × *artemisioides*
Senna artemisioides subsp. × *sturtii* (R.Br.) Randell
Senna charlesiana (Symon) Randell
Senna costata (Bailey & C.T.White) Randell
Senna curvistyla (J.M.Black) Randell
Senna ferraria (Symon) Randell
Senna glaucifolia (Randell) Randell
Senna glutinosa subsp. *chatelainiana* (Gaudich.) Randell
Senna glutinosa (DC.) Randell subsp. *glutinosa*
Senna glutinosa subsp. ***pruinosa*** (F.Muell.) Randell **TEE 615**
Senna glutinosa subsp. × *luerssenii* (Domin) Randell
Senna hamersleyensis (Symon) Randell
Senna notabilis (F.Muell.) Randell **TEE 561**
**Senna occidentalis* (L.) Link
Senna pleurocarpa var. *angustifolia* (Symon) Randell
Senna pleurocarpa (F.Muell.) Randell var. *pleurocarpa*
Senna sericea (Symon) Albr. & Symon
Senna sp. Karijini (M.E. Trudgen 10392)
Senna sp. Meekatharra (E. Bailey 1-26)
Senna sp. Millstream (E. Leyland s.n. 30/8/1990)
Senna stricta (Randell) Randell
Senna symonii (Randell) Randell
Senna venusta (F.Muell.) Randell **TEE 512**
Sesbania cannabina (Retz.) Pers.
Sesbania formosa (F.Muell.) N.T.Burb.
**Stylosanthes guianensis* (Aubl.) Sw.
**Stylosanthes hamata* (L.) Taub.
Swainsona campylantha F.Muell.
Swainsona canescens (Benth.) F.Muell.
Swainsona complanata Joy Thomps.
Swainsona decurrens A.T.Lee
Swainsona formosa (G.Don) Joy Thomps.
Swainsona forrestii A.T.Lee
Swainsona incei W.R.Price
Swainsona kingii F.Muell.

Swainsona leeana J.Z.Weber
Swainsona maccullochiana F.Muell.
Swainsona microphylla A.Gray
Swainsona oliveri F.Muell.
Swainsona oroboides Benth.
Swainsona paucifoliolata Joy Thomps.
Swainsona pterostylis (DC.) Bakh.f.
Swainsona stenodonta F.Muell.
Swainsona tanamiensis Joy Thomps.
Swainsona tenuis E.Pritz.
Swainsona thompsoniana R.W.Davis & P.J.H.Hurter
**Tamarindus indica* L.
Templetonia egena (F.Muell.) Benth.
Templetonia hookeri (F.Muell.) Benth.
Tephrosia arenicola Maconochie
Tephrosia brachyodon var. *longifolia* (Benth.) Domin
Tephrosia clementii Skan
Tephrosia leptoclada Benth.
Tephrosia oxalidea R.Butcher & P.J.H.Hurter
Tephrosia remotiflora Benth.
Tephrosia rosea var. *clementii* Domin
Tephrosia rosea var. Fortescue creeks (M.I.H. Brooker 2186)
Tephrosia rosea var. Port Hedland (A.S. George 1114)
Tephrosia rosea Benth. var. *rosea*
Tephrosia simplicifolia Benth.
Tephrosia sp. B Kimberley Flora (C.A. Gardner 7300)
Tephrosia sp. Bungaroo Creek (M.E. Trudgen 11601)
Tephrosia sp. Clay soils (S. van Leeuwen *et al.* PBS 0273)
Tephrosia sp. D Kimberley Flora (R.D. Royce 1848)
Tephrosia sp. Dunes (J.R. Maconochie 938)
***Tephrosia* sp. Fortescue** (A.A. Mitchell 606) **TEE 611**
Tephrosia sp. Meentheena (S. van Leeuwen 4479)
Tephrosia sp. Newman (A.A. Mitchell PRP 29)
Tephrosia sp. Northern (K.F. Kenneally 11950)
Tephrosia sp. NW Eremaean (S. van Leeuwen *et al.* PBS 0356)
Tephrosia sp. O.T. Station (S.T. Blake 17659)
Tephrosia sp. Onslow (K.R. Newbey 10571)
Tephrosia spechtii Pedley
Tephrosia stipuligera W.Fitzg.
Tephrosia supina Domin
Tephrosia uniovulata F.Muell.
Tephrosia virens Pedley
Thinicola incana (J.H.Ross) J.H.Ross
Trigonella suavissima Lindl.
**Vachellia farnesiana* (L.) Wight & Arn.
Vigna lanceolata Benth. var. *lanceolata*
Vigna sp. Central (M.E. Trudgen 1626)
Vigna sp. Hamersley Clay (A.A. Mitchell PRP 113)
Vigna sp. Rockpiles (R. Butcher *et al.* RB 1400)
Zornia albiflora Mohlenbr.
Zornia muelleriana subsp. *congesta* S.T.Reynolds & A.E.Holland

Frankeniaceae Desv.
Frankenia ambita Ostenf.
Frankenia cinerea A.DC.
Frankenia hispidula Summerh.
Frankenia pauciflora DC. var. *pauciflora*
Frankenia setosa W.Fitzg.
Gentianaceae Juss.
Schenkia australis (R.Br.) G.Mans.
Schenkia clementii (Domin) G.Mans.

Geraniaceae Juss.
Erodium crinitum Carolin
Erodium cygnorum Nees

Goodeniaceae R. Br.
Brunonia australis R.Br.
Brunonia sp. Long hairs (D.E. Symon 2440)
Brunonia australis var. A Kimberley Flora (K.F. Kenneally 5452)
Dampiera anonyma Lepschi & Trudgen
Dampiera atriplicina Rajput & Carolin
Dampiera candicans F.Muell.
Dampiera cinerea Ewart & O.B.Davies
Dampiera dentata Rajput
Dampiera metallorum Lepschi & Trudgen
Goodenia armitiana F.Muell. **TEE 795**
Goodenia azurea subsp. *hesperia* L.W.Sage & Albr.
Goodenia berringbinensis Carolin
Goodenia cusackiana (F.Muell.) Carolin
Goodenia forrestii F.Muell.
Goodenia hartiana L.W.Sage
Goodenia havilandii Maiden & Betche
Goodenia heterochila F.Muell.
Goodenia lamprosperma F.Muell.
Goodenia lyrata Carolin
Goodenia microptera F.Muell.
Goodenia mueckeana F.Muell.
Goodenia muelleriana Carolin
Goodenia nuda E.Pritz.
Goodenia pallida Carolin
Goodenia pascua Carolin
Goodenia pedicellata L.W.Sage & K.W.Dixon
Goodenia prostrata Carolin
Goodenia ramelii F.Muell.
Goodenia scaevolina F.Muell.
Goodenia sp. East Pilbara (A.A. Mitchell PRP 727)
Goodenia sp. Sandy Creek (R.D. Royce 1653)
Goodenia stellata Carolin
Goodenia stobbsiana F.Muell. **TEE 609**
Goodenia tenuiloba F.Muell.
Goodenia triodiophila Carolin **TEE 479**
Goodenia vilmoriniae F.Muell. **TEE 482**
Goodenia wilunensis Carolin
Scaevola acacioides Carolin
Scaevola amblyanthera F.Muell. var. *amblyanthera*
Scaevola amblyanthera var. *centralis* Carolin
Scaevola browniana Carolin subsp. *browniana*
Scaevola collaris F.Muell.
Scaevola crassifolia Labill.
Scaevola cunninghamii DC.
Scaevola parvifolia Benth. subsp. *parvifolia*
Scaevola parvifolia subsp. *pilbarae* Carolin
Scaevola pulchella Carolin

Scaevola revoluta subsp. *stenostachya* (W.Fitzg.) Carolin
Scaevola sericophylla Benth.
Scaevola sp. Hamersley Range basalts (S. van Leeuwen 3675)
Scaevola sp. Mt Bruce (M.E. Trudgen 1333)
Scaevola sp. Mt Nameless (P.A.S. Wurm 1443)
Scaevola spinescens R.Br.
Velleia connata F.Muell.
Velleia discophora F.Muell.
Velleia glabrata Carolin
Velleia panduriformis Benth. **TEE 587**

Gyrostemonaceae A.Juss.
Codonocarpus cotinifolius (Desf.) F.Muell.
Gyrostemon ramulosus Desf.
Gyrostemon tepperi (H.Walter) A.S.George

Haloragaceae R. Br.
Gonocarpus ephemerus Orchard
Haloragis gossei F.Muell. var. *gossei*
Haloragis gossei var. *inflata* Orchard
Haloragis maierae Orchard
Haloragis odontocarpa forma *pterocarpa* Orchard
Haloragis trigonocarpa F.Muell.
Myriophyllum verrucosum Lindl.

Hemerocallidaceae R.Br.
Corynotheca asperata R.J.F.Hend.
Corynotheca flexuosissima R.J.F.Hend.
Corynotheca pungens R.J.F.Hend.
Tricoryne sp. Hamersley Range (S. van Leeuwen 915)

Hernandiaceae Blume
Gyrocarpus americanus subsp. *pachyphyllus* Kubitzki

Hydrocharitaceae Juss.
Enhalus acoroides (L.f.) Royle
Halophila decipiens Ostenf.
Halophila minor (Zoll.) Hartog
Halophila ovalis (R.Br.) Hook.f.
Najas marina L.
Najas tenuifolia R.Br.
Thalassia hemprichii (Ehrenb.) Asch.
Vallisneria annua S.W.L.Jacobs & K.A.Frank
Vallisneria nana R.Br.

Juncaceae Juss.
Juncus kraussii subsp. *australiensis* (Buchenau) Snogerup

Lamiaceae Martinov
Basilicum polystachyon (L.) Moench
Clerodendrum floribundum var. *angustifolium* Moldenke
Clerodendrum floribundum R.Br. var. *floribundum*
Clerodendrum floribundum var. *ovatum* (R.Br.) Domin
Clerodendrum tomentosum var. *lanceolatum* (F.Muell.) Munir
Clerodendrum tomentosum (Vent.) R.Br. var. *tomentosum*
Cyanostegia cyanocalyx (F.Muell.) C.A.Gardner
Dicrastylis cordifolia Munir
Dicrastylis doranii F.Muell.
Dicrastylis mitchellii Rye
Newcastelia cephalantha F.Muell.
Newcastelia cladotricha F.Muell. **TEE 585**
Newcastelia hexarrhena F.Muell.
Newcastelia roseoazurea Rye
Newcastelia **sp. Hamersley Range** (S. van Leeuwen 4264) **TEE 776**
Ocimum basilicum L.
Pityrodia sp. Marble Bar (G. Woodman & D. Coultas GWDC Opp 4)
Plectranthus intraterraneus S.T.Blake
Prostanthera albiflora B.J.Conn
Prostanthera campbellii F.Muell.
Quoya loxocarpa (F.Muell.) B.J.Conn & Henwood
Quoya paniculata F.Muell.
Spartothamnella puberula (F.Muell.) Maiden & Betche
Spartothamnella teucriiflora (F.Muell.) Moldenke
Teucrium pilbaranum B.J.Conn
Teucrium racemosum R.Br.
**Vitex trifolia* var. *subtrisecta* (Kuntze) Moldenke

Lauraceae Juss.
Cassytha capillaris Meisn.
Cassytha filiformis L.
Cassytha racemosa forma *pilosa* (Benth.) J.Z.Weber

Lentibulariaceae Rich.
Utricularia australis R.Br.

Loganiaceae Mart.
Mitrasacme connata R.Br.
Mitrasacme exserta F.Muell.

Loranthaceae Juss.
Amyema benthamii (Blakely) Danser
Amyema bifurcata (Benth.) Tiegh.
Amyema fitzgeraldii (Blakely) Danser
Amyema gibberula (Tate) Danser var. *gibberula*
Amyema hilliana (Blakely) Danser
Amyema miquelii (Miq.) Tiegh.
Amyema preissii (Miq.) Tiegh.
Amyema sanguinea var. *pulchra* (Ewart) Barlow
Amyema sanguinea (F.Muell.) Danser var. *sanguinea*
Amyema sp. Fortescue (M.E. Trudgen 5358)
Diplatia grandibractea (F.Muell.) Tiegh.
Lysiana casuarinae (Miq.) Tiegh.
Lysiana murrayi (F.Muell. & Tate) Tiegh.
Lysiana subfalcata (Hook.) Barlow

Lythraceae J.St.-Hil.
Ammannia auriculata Willd.
Ammannia baccifera L.
Ammannia multiflora Roxb.
Rotala diandra (F.Muell.) Koehne
Rotala mexicana Cham. & Schltdl.

Malvaceae Juss.
Abelmoschus ficulneus (L.) Wight
Abutilon amplum Benth.

Abutilon cryptopetalum (F.Muell.) Benth.
Abutilon cunninghamii Benth.
Abutilon fraseri (Hook.) Walp. subsp. fraseri
Abutilon indicum var. *australiense* Britten
Abutilon lepidum (F.Muell.) Baker f.
Abutilon leucopetalum (F.Muell.) Benth.
Abutilon macrum F.Muell.
Abutilon malvifolium (Benth.) J.M.Black
Abutilon otocarpum F.Muell. **TEE 408**
Abutilon oxycarpum subsp. Prostrate (A.A. Mitchell PRP 1266)
Abutilon sp. Dioicum (A.A. Mitchell PRP 1618)
Abutilon sp. Onslow (F. Smith s.n. 10/9/61)
Abutilon sp. Pilbara (W.R. Barker 2025)
Abutilon sp. Pritzelianum (S. van Leeuwen 5095)
Alyogyne pinoniana (Gaudich.) Fryxell var. *pinoniana*
Androcalva loxophylla (F.Muell.) C.F.Wilkins & Whitlock
Androcalva luteiflora (E.Pritz.) C.F.Wilkins & Whitlock **TEE 617**
Brachychiton acuminatus Guymer
Brachychiton gregorii F.Muell.
Corchorus aestuans L.
Corchorus carnarvonensis Halford
Corchorus congener Halford
Corchorus crozophorifolius (Baill.) Burret **TEE 722**
Corchorus elachocarpus F.Muell.
Corchorus incanus Halford subsp. *incanus*
Corchorus incanus subsp. *lithophilus* Halford
Corchorus laniflorus Rye
Corchorus lasiocarpus Halford subsp. *lasiocarpus*
Corchorus lasiocarpus subsp. *parvus* Halford
Corchorus aff. ***lasiocarpus*** **TEE 514**
Corchorus parviflorus (Benth.) Domin
Corchorus sidoides F.Muell. subsp. *sidoides*
Corchorus sidoides subsp. *vermicularis* (F.Muell.) Halford
Corchorus sp. Hamersley Range (S. van Leeuwen 3586)
Corchorus sp. Hamersley Range hilltops (S. van Leeuwen 3826)
Corchorus sp. Meentheena (S. van Leeuwen 4537)
Corchorus sp. Yarrie (J. Bull & D. Roberts CAL 01.05)
Corchorus tectus Halford
Corchorus tridens L.
Corchorus trilocularis L.
Corchorus walcottii F.Muell.
Gossypium australe F.Muell. **NS 36715**
**Gossypium hirsutum* L.
Gossypium robinsonii F.Muell. **BNSS 11.047**
Gossypium sturtianum J.H.Willis var. *sturtianum*
Hannafordia bissillii F.Muell. subsp. *bissillii*
Hannafordia quadrivalvis subsp. *recurva* C.F.Wilkins
Hibiscus austrinus Juswara & Craven var. *austrinus*
Hibiscus brachychlaenus F.Muell.
Hibiscus brachysiphonius F.Muell.
Hibiscus burtonii F.M.Bailey
Hibiscus coatesii F.Muell.
Hibiscus goldsworthii F.Muell.
Hibiscus haynaldii F.Muell. **TEE 510**
Hibiscus leptocladus Benth.
Hibiscus sp. Canga (P.J.H. Hurter & J. Naaykens 11013)
Hibiscus sp. Gardneri (A.L. Payne PRP 1435)
Hibiscus sp. Gurinbiddy Range (M.E. Trudgen MET 15708)
Hibiscus sp. Mt Brockman (E. Thoma ET 1354)
Hibiscus sp. Mt Robinson (G. Byrne 3537)
Hibiscus sturtii var. *campylochlamys* Benth.
Hibiscus sturtii var. *grandiflorus* Benth.
Hibiscus sturtii var. Meentheena (S. van Leeuwen 4589)
Hibiscus sturtii var. *platychlamys* Benth.
Hibiscus sturtii var. *truncatus* Fryxell,
Hibiscus verdicourtii Craven
Keraudrenia nephrosperma (F.Muell.) F.Muell.
Keraudrenia velutina subsp. *elliptica* C.F.Wilkins ms
Lawrencia cinerea Lander
Lawrencia densiflora (Baker f.) Melville
Lawrencia glomerata Hook.
Lawrencia helmsii (F.Muell. & Tate) Lander
Lawrencia viridigrisea Lander
**Malvastrum americanum* (L.) Torr.
**Malvastrum coromandelianum* (L.) Garcke
Melhania oblongifolia F.Muell.
**Melochia pyramidata* L.
Sida arenicola S.T.Reynolds & A.E.Holland
Sida arsiniata R.M.Barker
Sida brownii Clement
Sida calyxhymenia DC.
Sida cardiophylla F.Muell.
Sida clementii Domin
Sida echinocarpa F.Muell. **TEE 539**
Sida ectogama W.R.Barker & R.M.Barker
Sida fibulifera Lindl. in T.L.Mitch.
Sida intricata F.Muell.
Sida laevis Domin
Sida macropoda Benth.
Sida platycalyx Benth.
Sida rohlenae Domin subsp. *rohlenae*
Sida sp. A Kimberley Flora (P.A. Fryxell & L.A. Craven 3900)
Sida sp. Articulation below (A.A. Mitchell PRP 1605)
Sida sp. Barlee Range (S. van Leeuwen 1642)
Sida sp. Dark green fruits (S. van Leeuwen 2260)
Sida sp. Excedentifolia (J.L. Egan 1925)
Sida sp. Golden calyces glabrous (H.N. Foote 32)
Sida sp. Hamersley Range (K. Newbey 10692)
Sida sp. Pilbara (A.A. Mitchell PRP 1543)
Sida sp. Pindan (B.G. Thomson 3398)
Sida sp. Rabbit Flat (B.J. Carter 626)
Sida sp. Shovelanna Hill (S. van Leeuwen 3842)
Sida sp. Spiciform panicles (E. Leyland s.n. 14/8/90)
Sida sp. Supplejack Station (T.S. Henshall 2345)
Sida sp. Tiny glabrous fruit (A.A. Mitchell PRP1152)
Sida sp. Verrucose glands (F.H. Mollemans 2423)
Sida spinosa L.
Sida trichopoda F.Muell.
Triumfetta appendiculata F.Muell.
Triumfetta chaetocarpa F.Muell.
Triumfetta clementii (Domin) Rye
Triumfetta deserticola Halford
Triumfetta echinata Halford

Triumfetta incana Halford
Triumfetta johnstonii F.Muell.
Triumfetta leptacantha F.Muell.
Triumfetta maconochieana Halford
Triumfetta plumigera F.Muell.
Triumfetta propinqua Halford
Triumfetta ramosa Sprague & Hutch
Waltheria indica L.
Waltheria virgata Ewart & Cookson

Marsileaceae Mirb.
Marsilea angustifolia R.Br.
Marsilea drummondii A.Braun
Marsilea exarata A.Braun
Marsilea hirsuta R.Br.
Marsilea mutica Mett.

Meliaceae Juss.
Owenia acidula F.Muell.
Owenia reticulata F.Muell.

Menispermaceae Juss.
Tinospora smilacina Benth.

Molluginaceae Bartl.
Glinus lotoides L.
Glinus oppositifolius (L.) A.DC.
Mollugo cerviana (L.) Ser.
Mollugo molluginea (F.Muell.) Druce

Moraceae Gaudich.
Ficus aculeata var. *indecora* (Miq.) D.J.Dixon
Ficus brachypoda (Miq.) Miq.
Ficus geniculata var. *insignis* (Kurz) C.C.Berg
Ficus platypoda (Miq.) Miq.
Ficus virens Aiton var. *virens*

Moringaceae Martinov
**Moringa oleifera* Lam.

Myrtaceae Juss.
Aluta maisonneuvei (F.Muell.) Rye & Trudgen subsp. *maisonneuvei*
Aluta quadrata Rye & Trudgen
Calytrix carinata Craven
Corymbia aspera (F.Muell.) K.D.Hill & L.A.S.Johnson
Corymbia candida K.D.Hill & L.A.S.Johnson subsp. *candida*
Corymbia candida subsp. *dipsodes* K.D.Hill & L.A.S.Johnson
Corymbia candida subsp. *lautifolia* K.D.Hill & L.A.S.Johnson
Corymbia deserticola (D.J.Carr & S.G.M.Carr) K.D.Hill & L.A.S.Johnson subsp. *deserticola*
Corymbia dichromophloia (F.Muell.) K.D.Hill & L.A.S.Johnson
Corymbia ferriticola (Brooker & Edgecombe) K.D.Hill & L.A.S.Johnson
Corymbia flavescens K.D.Hill & L.A.S.Johnson
Corymbia hamersleyana (D.J.Carr & S.G.M.Carr) K.D.Hill & L.A.S.Johnson
Corymbia opaca (D.J.Carr & S.G.M.Carr) K.D.Hill & L.A.S.Johnson
Corymbia zygophylla (Blakely) K.D.Hill & L.A.S.Johnson
Eucalyptus aridimontana D.Nicolle & M.E.French
Eucalyptus camaldulensis subsp. *obtusa* (Blakely) Brooker & M.W.McDonald
Eucalyptus camaldulensis subsp. *refulgens* Brooker & M.W.McDonald
Eucalyptus ewartiana Maiden
Eucalyptus gamophylla F.Muell. **TEE 506**
Eucalyptus kingsmillii Maiden & Blakely subsp. *kingsmillii*
Eucalyptus leucophloia Brooker subsp. ***leucophloia*** **TEE 726**
Eucalyptus lucasii Blakely
Eucalyptus lucens Brooker & Dunlop
Eucalyptus odontocarpa F.Muell.
Eucalyptus pilbarensis Brooker & Edgecombe
Eucalyptus repullulans D.Nicolle
Eucalyptus rowleyi D.Nicolle & M.E.French
Eucalyptus socialis subsp. *eucentrica* (L.A.S.Johnson & K.D.Hill) D.Nicolle
Eucalyptus tephrodes L.A.S.Johnson & K.D.Hill
Eucalyptus trivalva Blakely
Eucalyptus victrix L.A.S.Johnson & K.D.Hill
Eucalyptus xerothermica L.A.S.Johnson & K.D.Hill
Lamarchea sulcata A.S.George
Melaleuca argentea W.Fitzg.
Melaleuca bracteata F.Muell.
Melaleuca eleuterostachya F.Muell.
Melaleuca glomerata F.Muell. **TEE 517**
Melaleuca lasiandra F.Muell.
Melaleuca leiocarpa F.Muell.
Melaleuca leucadendra (L.) L.
Melaleuca linophylla F.Muell.
Melaleuca xerophila Barlow
Osbornia octodonta F.Muell.
Thryptomene wittweri J.W.Green
Verticordia forrestii F.Muell.

Nyctaginaceae Juss.
Boerhavia burbidgeana Hewson
Boerhavia coccinea Mill. **TEE 418**
Boerhavia gardneri Hewson
Boerhavia paludosa (Domin) Meikle
Boerhavia repleta Hewson
Boerhavia schomburgkiana Oliv.
Commicarpus australis Meikle

Nymphaeaceae Salisb.
Nymphaea macrosperma Merr. & L.M.Perry

Oleaceae Hoffmanns. & Link
Jasminum didymum subsp. *lineare* (R.Br.) P.S.Green

Onagraceae Juss.
Ludwigia perennis L.

Ophioglossaceae Martinov
Ophioglossum lusitanicum L.

Orobanchaceae Vent.
Buchnera linearis R.Br.
Striga curviflora (R.Br.) Benth.
Striga squamigera W.R.Barker

Oxalidaceae R. Br.
**Oxalis corniculata* L.
Oxalis perennans Haw.
Oxalis sp. Pilbara (M.E. Trudgen 12725)

Papaveraceae Juss.
**Argemone ochroleuca* Sweet subsp. *ochroleuca*

Passifloraceae Roussel
**Passiflora foetida* var. *hispida* (Triana & Planch.) Killip

Pedaliaceae R. Br.
Josephinia eugeniae F.Muell.
Josephinia sp. Marandoo (M.E. Trudgen 1554)
Josephinia sp. Mt Edgar Station (N.T. Burbidge 1194)
Josephinia sp. Style-topped fruit (M. Maier BES MM 1017)
Josephinia sp. Woodstock (A.A. Mitchell PRP 989)

Phrymaceae Schauer
Elacholoma hornii F.Muell.
Glossostigma diandrum (L.) Kuntze
Mimulus gracilis R.Br.
Peplidium aithocheilum W.R.Barker
Peplidium maritimum (L.f.) Wettst.
Peplidium muelleri Benth.
Peplidium sp. C Evol. Fl. Fauna Arid Aust. (N.T. Burbidge & A. Kanis 8158)
Peplidium sp. E Evol. Fl. Fauna Arid Aust. (A.S. Weston 12768)

Phyllanthaceae Martinov
Bridelia tomentosa Blume
Flueggea virosa subsp. *melanthesoides* (F.Muell.) G.L.Webster
Notoleptopus decaisnei (Benth.) Voronts. & Petra Hoffm. var. *decaisnei*
Notoleptopus decaisnei var. Orbicularis (A.B. Craig 428)
Phyllanthus eremicus R.L.Barrett & I.Telford
Phyllanthus erwinii J.T.Hunter & J.J.Bruhl
Phyllanthus exilis S.Moore
Phyllanthus hamelinii I.Telford & R.L.Barrett
Phyllanthus hebecarpus Benth.
Phyllanthus maderaspatensis L.
Phyllanthus reticulatus var. *glaber* Muell.Arg.
**Phyllanthus tenellus* Roxb.
Sauropus crassifolius (Muell.Arg.) Airy Shaw
Sauropus trachyspermus (F.Muell.) Airy Shaw
Synostemon hamersleyensis I.Telford & Naaykens

Pittosporaceae R. Br.
Pittosporum angustifolium Lodd.
Pittosporum phillyreoides DC.

Plantaginaceae Juss.
Plantago cunninghamii Decne
Stemodia glabella W.R.Barker
Stemodia grossa Benth.
Stemodia kingii F.Muell.
Stemodia linophylla F.Muell.
Stemodia sp. Battle Hill (A.L. Payne 1006)
Stemodia sp. Onslow (A.A. Mitchell 76/148)
Stemodia viscosa Roxb.

Plumbaginaceae Juss.
Aegialitis annulata R.Br.
Muellerolimon salicorniaceum (F.Muell.) Lincz.
Plumbago zeylanica L.

Poaceae Barnhart & Barnh.
Acrachne racemosa (B.Heyne) Ohwi
Alloteropsis cimicina (L.) Stapf
Amphipogon caricinus F.Muell. var. *caricinus*
Amphipogon sericeus (Vickery) T.Macfarlane
**Andropogon gayanus* Kunth.
Aristida anthoxanthoides (Domin) Henrard
Aristida burbidgeae B.K.Simon
***Aristida contorta* F.Muell. TEE 639**
Aristida holathera Domin var. *holathera*
Aristida hygrometrica R.Br.
***Aristida inaequiglumis* Domin TEE 515**
Aristida ingrata Domin
Aristida jerichoensis var. *subspinulifera* Henrard
***Aristida latifolia* Domin TEE 441**
Aristida lazaridis B.K.Simon
Aristida nitidula (Henrard) J.M.Black
Aristida obscura Henrard
Aristida personata Henrard
Aristida pruinosa Domin
**Arundo donax* L.
Astrebla elymoides F.M.Bailey
Astrebla lappacea (Lindl.) Domin
Astrebla pectinata (Lindl.) Benth.
Austrostipa nitida (Summerh. & C.E.Hubb.) S.W.L.Jacobs & J.Everett
Bothriochloa bladhii (Retz.) S.T.Blake subsp. *bladhii*
Bothriochloa decipiens var. *cloncurrensis* (Domin) C.E.Hubb.
Bothriochloa ewartiana (Domin) C.E.Hubb.
**Cenchrus ciliaris* L.
**Cenchrus echinatus* L.
**Cenchrus setaceus* (Forssk.) Morrone
**Cenchrus setiger* Vahl
**Chloris barbata* Sw.
Chloris pectinata Benth.
Chloris pumilio R.Br.
**Chloris virgata* Sw.
Chrysopogon fallax S.T.Blake
***Cymbopogon ambiguus* A.Camus TEE 420**
Cymbopogon bombycinus (R.Br.) Domin
***Cymbopogon obtectus* S.T.Blake TEE 484**
***Cynodon convergens* F.Muell. TEE 635**

**Cynodon dactylon* (L.) Pers.
Cynodon prostratus (C.A.Gardner & C.E.Hubb.) P.M.Peterson
Cynodon tenellus R.Br.
Dactyloctenium radulans (R.Br.) P.Beauv.
Dichanthium fecundum S.T.Blake
Dichanthium sericeum subsp. *humilius* (J.M.Black) B.K.Simon
Dichanthium sericeum subsp. *polystachyum* (Benth.) B.K.Simon
Dichanthium sericeum (R.Br.) A.Camus subsp. *sericeum*
Digitaria ammophila (F.Muell.) Hughes
Digitaria brownii (Roem. & Schult.) Hughes
**Digitaria ciliaris* (Retz.) Koeler
Digitaria coenicola (F.Muell.) Hughes
Digitaria ctenantha (F.Muell.) Hughes
Digitaria longiflora (Retz.) Pers.
**Echinochloa colona* (L.) Link
Elytrophorus spicatus (Willd.) A.Camus
Enneapogon avenaceus (Lindl.) C.E.Hubb.
Enneapogon caerulescens (Gaudich.) N.T.Burb. **TEE 630**
Enneapogon cylindricus N.T.Burb.
Enneapogon lindleyanus (Domin) C.E.Hubb.
Enneapogon pallidus (R.Br.) P.Beauv.
Enneapogon polyphyllus (Domin) N.T.Burb. **TEE 447**
Enneapogon purpurascens (R.Br.) P.Beauv.
Enneapogon robustissimus (Domin) N.T.Burb. **TEE 424**
Enteropogon ramosus B.K.Simon
Eragrostis crateriformis Lazarides
Eragrostis cumingii Steud.
Eragrostis desertorum Domin
Eragrostis dielsii Pilg. **TEE 502**
Eragrostis elongata (Willd.) J.Jacq.
Eragrostis eriopoda Benth. **TEE 476**
Eragrostis exigua Lazarides
Eragrostis falcata (Gaudich.) Steud.
Eragrostis fallax Lazarides
Eragrostis lanicaulis Lazarides
Eragrostis laniflora Benth.
Eragrostis lanipes C.E.Hubb.
Eragrostis leptocarpa Benth.
**Eragrostis minor* Host
Eragrostis olida Lazarides
Eragrostis parviflora (R.Br.) Trin.
Eragrostis pergracilis S.T.Blake
**Eragrostis pilosa* (L.) P.Beauv.
Eragrostis setifolia Nees
Eragrostis sp. Erect spikelets (P.K. Latz 2122)
Eragrostis sp. Mt Robinson (S. van Leeuwen 4109)
Eragrostis speciosa (Roem. & Schult.) Steud.
Eragrostis surreyana K.A.Sheph. & Trudgen
Eragrostis tenellula (Kunth) Steud. **TEE 504**
Eragrostis xerophila Domin
Eriachne aristidea F.Muell.
Eriachne benthamii Hartley
Eriachne ciliata R.Br.
Eriachne festucacea F.Muell.
Eriachne flaccida Hartley
Eriachne gardneri Hartley
Eriachne glauca R.Br. var. *glauca*
Eriachne helmsii Hartley
Eriachne lanata Lazarides
Eriachne melicacea F.Muell.
Eriachne mucronata R.Br.
Eriachne obtusa R.Br.
Eriachne ovata Nees
Eriachne pulchella subsp. ***dominii*** (Hartley) Lazarides **TEE 422**
Eriachne pulchella Domin subsp. *pulchella*
Eriachne tenuiculmis Hartley
Eriochloa fatmensis (Hochst. & Steud.) Clayton
Eriochloa procera (Retz.) C.E.Hubb.
Eriochloa pseudoacrotricha (Thell.) S.T.Blake
Eulalia aurea (Bory) Kunth
Heteropogon contortus (L.) Roem. & Schult.
Imperata cylindrica (L.) P.Beauv.
Ischaemum albovillosum B.K.Simon
Iseilema dolichotrichum C.E.Hubb.
Iseilema eremaeum S.T.Blake
Iseilema macratherum Domin
Iseilema membranaceum (Lindl.) Domin
Iseilema vaginiflorum Domin
**Lamarckia aurea* (L.) Moench
Leptochloa digitata (R.Br.) Domin
Leptochloa fusca (L.) Kunth. subsp. *fusca*
Leptochloa fusca subsp. *muelleri* (Benth.) N.Snow
**Leptochloa fusca* subsp. *uninervia* (J.Presl) N.Snow
**Melinis repens* (Willd.) Zizka
Mnesithea formosa (R.Br.) de Koning & Sosef
Monachather paradoxus Steud.
Neurachne minor S.T.Blake
Neurachne muelleri (Hack.) S.T.Blake **TEE 429**
Oxychloris scariosa (F.Muell.) Lazarides
Panicum decompositum R.Br.
Panicum effusum R.Br.
Panicum laevinode Lindl.
Panicum mindanaense Merr.
Paractaenum refractum (F.Muell.) R.D.Webster
Paspalidium basicladum Hughes
Paspalidium clementii (Domin) C.E.Hubb.
Paspalidium constrictum (Domin) C.E.Hubb.
Paspalidium gracile (R.Br.) Hughes
Paspalidium jubiflorum (Trin.) Hughes
Paspalidium rarum (R.Br.) Hughes
Paspalidium reflexum R.D.Webster
Paspalidium retiglume (Domin) Hughes
Paspalidium tabulatum (Hack.) C.E.Hubb.
**Paspalum dilatatum* Poir.
**Paspalum fasciculatum* Fluegge
Perotis rara R.Br.
Phragmites karka (Retz.) Steud.
**Polypogon monspeliensis* (L.) Desf.
Schizachyrium fragile (R.Br.) A.Camus
Setaria dielsii R.A.W.Herrm.
**Setaria italica* (L.) P.Beauv.
**Setaria sphacelata* (Schumach.) Stapf & C.E.Hubb.
Setaria surgens Stapf
Setaria verticillata (L.) P.Beauv.

Sorghum plumosum (R.Br.) P.Beauv. var. *plumosum*
Sorghum stipoideum (Ewart & Jean White) C.A.Gardner & C.E.Hubb.
Sorghum timorense (Kunth) De Vriese
Spathia neurosa Ewart & M.E.L.Archer
Spinifex longifolius R.Br.
Sporobolus actinocladus (F.Muell.) F.Muell.
Sporobolus australasicus Domin
Sporobolus mitchellii (Trin.) S.T.Blake
Sporobolus virginicus (L.) Kunth
Themeda avenacea (F.Muell.) Maiden & Betche
Themeda sp. Hamersley Station (M.E. Trudgen 11431)
Themeda sp. Mt Barricade (M.E. Trudgen 2471)
Themeda triandra Forssk.
Thyridolepis mitchelliana (Nees) S.T.Blake
Thyridolepis xerophila (Domin) S.T.Blake
Tragus australianus S.T.Blake
Triodia angusta N.T.Burb. **TEE 773**
Triodia basedowii E.Pritz. **TEE 771**
Triodia basitricha M.D.Barrett
Triodia biflora Lazarides
Triodia bitextura Lazarides
Triodia brizoides N.T.Burb. **TEE 769**
Triodia epactia S.W.L.Jacobs **TEE 637**
Triodia lanigera Domin
Triodia longiceps J.M.Black
Triodia melvillei (C.E.Hubb.) Lazarides
Triodia plurinervata N.T.Burb.
Triodia pungens R.Br. **TEE 627**
Triodia schinzii (Henrard) Lazarides
Triodia secunda N.T.Burb.
Triodia sp. Karijini (S. van Leeuwen 4111)
Triodia sp. Mt Ella (M.E. Trudgen 12739)
Triodia sp. Peedamulla (A.A. Mitchell PRP 1636)
Triodia sp. Robe River (M.E. Trudgen *et al.* MET 12367)
Triodia sp. **Shovelanna Hill** (S. van Leeuwen 3835) **TEE 623**
Triodia sp. Warrawagine (A.L. Payne PRP 1859)
Triodia wiseana C.A.Gardner **TEE 632**
Tripogon loliiformis (F.Muell.) C.E.Hubb.
Triraphis mollis R.Br.
Urochloa holosericea subsp. *velutina* R.D.Webster
Urochloa occidentalis var. *ciliata* (C.A.Gardner & C.E.Hubb.) B.K.Simon
Urochloa occidentalis (C.A.Gardner & C.E.Hubb.) B.K.Simon var. *occidentalis*
Urochloa piligera (Benth.) R.D.Webster
Urochloa pubigera (Roem. & Schult.) R.D.Webster
Urochloa subquadripara (Trin.) R.D.Webster
Whiteochloa airoides (R.Br.) Lazarides
Whiteochloa capillipes (Benth.) Lazarides
Whiteochloa cymbiformis (Hughes) B.K.Simon
Xerochloa barbata R.Br.
Xerochloa imberbis R.Br.
Xerochloa laniflora Benth.
Yakirra australiensis (Domin) Lazarides & R.D.Webster var. *australiensis*
Yakirra majuscula (Benth.) Lazarides & R.D.Webster

Polygalaceae Hoffmanns. & Link
Polygala galeocephala R.A.Kerrigan
Polygala glaucifolia R.A.Kerrigan
Polygala isingii Pedley
Polygala longifolia Poir.

Polygonaceae Juss.
**Acetosa vesicaria* (L.) A.Love
Duma florulenta (Meisn.) T.M.Schust.

Portulacaceae Juss.
Calandrinia balonensis Lindl.
Calandrinia eremaea Ewart
Calandrinia pentavalvis Obbens
Calandrinia ptychosperma F.Muell.
Calandrinia pumila (Benth.) F.Muell.
Calandrinia quadrivalvis F.Muell.
Calandrinia reticulata Syeda
Calandrinia schistorhiza Morrison **TEE 778**
Calandrinia sp. Black angular seeds (A.A. Mitchell PRP 1661)
Calandrinia sp. The Pink Hills (F. Obbens FO 19/06)
Calandrinia stagnensis J.M.Black
Portulaca australis Endl.
Portulaca bicolor F.Muell.
Portulaca conspicua Domin
Portulaca cyclophylla F.Muell.
Portulaca decipiens Poelln.
Portulaca filifolia F.Muell.
Portulaca intraterranea J.M.Black
Portulaca oleracea L. **TEE 416B**
**Portulaca pilosa* L.

Potamogetonaceae Bercht. & J. Presl
Potamogeton drummondii Benth.
Potamogeton tepperi A.Benn.
Potamogeton tricarinatus A.Benn.
Stuckenia pectinata (L.) Börner

Primulaceae Borkh.
Aegiceras corniculatum (L.) Blanco
**Lysimachia arvensis* (L.) U.Manns & Anderb.
Samolus junceus R.Br.
Samolus repens var. *floribundus* Benth.
Samolus sp. Fortescue Marsh (A. Markey & R. Coppen FM 9702)
Samolus sp. Millstream (M.I.H. Brooker 2076)

Proteaceae Juss.
Grevillea berryana Ewart & Jean White
Grevillea eriostachya Lindl.
Grevillea juncifolia Hook. subsp. *juncifolia*
Grevillea pyramidalis subsp. *leucadendron* (R.Br.) Makinson
Grevillea pyramidalis R.Br. subsp. ***pyramidalis*** **TEE 621**
Grevillea refracta R.Br. subsp. ***refracta*** **TEE 589**
Grevillea saxicola S.J.Dillon
Grevillea stenobotrya F.Muell. **TEE 784**
Grevillea striata R.Br.

Grevillea wickhamii subsp. *aprica* McGill.
Grevillea wickhamii subsp. ***hispidula*** Makinson **TEE 662**
Grevillea wickhamii subsp. ***macrodonta*** Makinson **TEE 664**
Hakea chordophylla F.Muell.
Hakea leucoptera subsp. *sericipes* W.R.Barker
Hakea lorea (R.Br.) R.Br. subsp. ***lorea*** **NS 39846**
Hakea macrocarpa R.Br.
Hakea preissii Meisn.
Hakea rhombales F.Muell.
Hakea stenophylla R.Br. subsp. *stenophylla*

Psilotaceae J.W.Griff. & Henfr.
Psilotum nudum (L.) P.Beauv.

Pteridaceae E.D.M.Kirchn.
Adiantum capillus-veneris L.
Ceratopteris thalictroides (L.) Brongn.
Cheilanthes austrotenuifolia H.M.Quirk & T.C.Chambers
Cheilanthes brownii (Kuhn) Domin
Cheilanthes contigua Baker
Cheilanthes lasiophylla Pic.Serm.
Cheilanthes sieberi subsp. *pseudovellea* H.M.Quirk & T.C.Chambers
Cheilanthes sieberi Kunze subsp. *sieberi*
Paraceterach reynoldsii (F.Muell.) Tindale
Pteris vittata L.

Rhamnaceae Juss.
Cryptandra monticola Rye & Trudgen
Stenanthemum petraeum Rye
Ventilago viminalis Hook.

Rhizophoraceae Pers.
Bruguiera exaristata Ding Hou
Ceriops australis (C.T.White) Ballment, T.J.Sm. & J.A.Stoddart
Rhizophora stylosa Griff.

Rubiaceae Juss.
Dentella asperata Airy Shaw
Dentella minutissima C.T.White & W.D.Francis
Oldenlandia argillacea (Halford) Halford
Oldenlandia crouchiana (F.Muell.) F.Muell.
Oldenlandia galioides (F.Muell.) F.Muell.
Oldenlandia pterospora (F.Muell.) F.Muell.
Oldenlandia sp. Hamersley Station (A.A. Mitchell PRP 1479)
Pomax rupestris F.Muell., Linnaea 25:395 (1853)
Psydrax latifolia (Benth.) S.T.Reynolds & R.J.F.Hend.
Psydrax rigidula S.T.Reynolds & R.J.F.Hend.
Psydrax suaveolens (S.Moore) S.T.Reynolds & R.J.F.Hend.
Spermacoce brachystema Benth.
Spermacoce hillii (Chippend.) Govaerts
Spermacoce occidentalis Harwood
Synaptantha tillaeacea (F.Muell.) Hook.f. var. *tillaeacea*

Ruppiaceae Horan.
Ruppia maritima L.
Ruppia polycarpa R.Mason

Rutaceae Juss.
Geijera salicifolia Schott

Salviniaceae Martinov
**Salvinia molesta* D.S.Mitch.

Santalaceae R. Br.
Anthobolus leptomerioides F.Muell.
Exocarpos sparteus R.Br.
Santalum acuminatum (R.Br.) A.DC.
Santalum lanceolatum R.Br.
Santalum spicatum (R.Br.) A.DC.

Sapindaceae Juss.
Alectryon oleifolius (Desf.) S.T.Reynolds subsp. *oleifolius*
Atalaya hemiglauca (F.Muell.) Benth.
Diplopeltis eriocarpa (Benth.) Hemsl.
Diplopeltis stuartii F.Muell. var. ***stuartii*** **TEE 782**
Dodonaea coriacea (Ewart & O.B.Davies) McGill. **TEE 619**
Dodonaea lanceolata F.Muell. var. *lanceolata*
Dodonaea pachyneura F.Muell.
Dodonaea petiolaris F.Muell.
Dodonaea viscosa subsp. *mucronata* J.G.West
Dodonaea viscosa subsp. *spatulata* (Sm.) J.G.West

Scrophulariaceae Juss.
Eremophila accrescens Chinnock
Eremophila appressa Chinnock
Eremophila caespitosa Chinnock
Eremophila canaliculata Chinnock
Eremophila clarkei A.F.Oldfield & F.Muell.
Eremophila coacta Chinnock
Eremophila compacta S.Moore subsp. *compacta*
Eremophila cryptothrix Chinnock
Eremophila cuneifolia Kraenzl.
Eremophila decipiens Ostenf. subsp. *decipiens*
Eremophila enata Chinnock
Eremophila exilifolia F.Muell.
Eremophila flaccida Chinnock subsp. *flaccida*
Eremophila forrestii F.Muell. subsp. *forrestii*
Eremophila forrestii subsp. *hastieana* (W.Fitzg.) Chinnock
Eremophila forrestii subsp. Pingandy (M.E. Trudgen 2662)
Eremophila forrestii subsp. *viridis* Chinnock
Eremophila fraseri F.Muell. subsp. *fraseri*
Eremophila fraseri subsp. *parva* Chinnock
Eremophila galeata Chinnock
Eremophila incisa Chinnock
Eremophila jucunda Chinnock subsp. *jucunda*
Eremophila jucunda subsp. *pulcherrima* Chinnock
Eremophila lachnocalyx C.A.Gardner
Eremophila lanceolata Chinnock
Eremophila latrobei subsp. *filiformis* Chinnock
Eremophila latrobei subsp. *glabra* (L.S.Sm.) Chinnock
Eremophila latrobei F.Muell. subsp. *latrobei*
Eremophila longifolia (R.Br.) F.Muell.
Eremophila maculata subsp. *brevifolia* (Benth.) Chinnock
Eremophila maculata subsp. *filifolia* Chinnock

Eremophila maculata (Ker Gawl.) F.Muell. subsp. *maculata*
Eremophila magnifica Chinnock subsp. *magnifica*
Eremophila magnifica subsp. *velutina* Chinnock
Eremophila margarethae S.Moore
Eremophila oppositifolia subsp. *angustifolia* (S.Moore) Chinnock
Eremophila petrophila Chinnock subsp. *petrophila*
Eremophila phyllopoda subsp. *obliqua* Chinnock
Eremophila phyllopoda Chinnock subsp. *phyllopoda*
Eremophila pilosa Chinnock
Eremophila platycalyx subsp. *pardalota* Chinnock
Eremophila platycalyx F.Muell. subsp. *platycalyx*
Eremophila reticulata Chinnock
Eremophila rigida Chinnock
Eremophila sp. Hamersley Range (K. Walker KW 136)
Eremophila sp. Little Sandy Desert (A.A. Mitchell PRP 1263)
Eremophila sp. Rudall River (P.G. Wilson 10512)
Eremophila sp. Snowy Mountain (S. van Leeuwen 3737)
Eremophila sp. West Angelas (S. van Leeuwen 4068)
Eremophila spongiocarpa Chinnock
Eremophila strongylophylla F.Muell.
Eremophila tietkensii F.Muell. & Tate
Eremophila youngii subsp. *lepidota* Chinnock
Myoporum montanum R.Br.

Solanaceae Juss.
**Datura leichhardtii* Benth.
**Datura metel* L.
Duboisia hopwoodii (F.Muell.) F.Muell.
Nicotiana benthamiana Domin
Nicotiana cavicola N.T.Burb.
Nicotiana heterantha Symon & Kenneally
Nicotiana occidentalis subsp. *obliqua* N.T.Burb.
Nicotiana occidentalis H.-M.Wheeler subsp. *occidentalis*
Nicotiana rosulata subsp. *ingulba* (J.M.Black) P.Horton
Nicotiana rosulata (S.Moore) Domin subsp. *rosulata*
Nicotiana simulans N.T.Burb.
Nicotiana umbratica N.T.Burb.
**Physalis angulata* L.
Solanum albostellatum R.W.Davis & P.J.H.Hurter
Solanum ashbyae Symon
Solanum austropiceum A.R.Bean
Solanum centrale J.M.Black
Solanum chippendalei Symon
Solanum cleistogamum Symon
Solanum coactiliferum J.M.Black
Solanum dioicum W.Fitzg. **TEE 578**
Solanum diversiflorum F.Muell. **TEE 580**
Solanum elatius A.R.Bean
Solanum esuriale Lindl.
Solanum ferocissimum Lindl.
Solanum gabrielae Domin
Solanum gilesii Symon
Solanum horridum Poir.
Solanum kentrocaule A.R.Bean
Solanum lasiophyllum Poir.
Solanum lucani F.Muell.
Solanum morrisonii Domin
**Solanum nigrum* L.
Solanum octonum A.R.Bean
Solanum orbiculatum Poir. subsp. *orbiculatum*
Solanum phlomoides Benth.
Solanum piceum A.R.Bean
Solanum pycnotrichum A.R.Bean
Solanum sturtianum F.Muell.

Stylidiaceae R.Br.
Stylidium desertorum Carlquist **TEE 790**
Stylidium fluminense F.L.Erickson & J.H.Willis
Stylidium weeliwolli Lowrie & Kenneally

Surianaceae Arn.
Stylobasium spathulatum Desf. **TEE 792**

Tamaricaceae Link
**Tamarix aphylla* (L.) H.Karst.

Thelypteridaceae Pic.Serm.
Ampelopteris prolifera (Retz.) Copel.

Thymelaeaceae Juss.
Pimelea ammocharis F.Muell.
Pimelea forrestiana F.Muell.
Pimelea holroydii F.Muell.
Pimelea microcephala R.Br. subsp. *microcephala*

Typhaceae Juss.
Typha domingensis Pers.

Urticaceae Juss.
Parietaria cardiostegia Greuter

Verbenaceae J. St.-Hil.
**Phyla nodiflora* (L.) Greene var. *nodiflora*

Violaceae Batsch
Hybanthus aurantiacus (Benth.) F.Muell.

Zygophyllaceae R. Br.
Tribulopis angustifolia R.Br.
Tribulus astrocarpus F.Muell.
Tribulus cistoides L.
Tribulus forrestii F.Muell.
Tribulus hirsutus Benth.
Tribulus macrocarpus Benth.
Tribulus minutus Benth.
Tribulus occidentalis R.Br.
Tribulus platypterus Benth.
Tribulus sp. long-styled eichlerianus (A.S. George 10666)
Tribulus suberosus R.M.Barker **TEE 445**
**Tribulus terrestris* L.
Zygophyllum eichleri R.M.Barker
Zygophyllum iodocarpum F.Muell.
Zygophyllum kochii Tate
Zygophyllum retivalve Domin
Zygophyllum simile H.Eichler

Glossary

This glossary primarily has been adapted from three Western Australian Herbarium publications, *Flora of the Perth Region* (Marchant *et al.* 1987), *Flora of the Kimberley* (Wheeler 1992) and *The Western Australian Flora. A Descriptive Catalogue* (Paczkowska & Chapman 2000), with additional terms from *Plant identification terminology* (Harris & Harris 2000) and *The Kew plant glossary: an illustrated dictionary of plant terms* (Beentje 2010).

A

abaxial: away from the axis, referring to the surface of an organ that is furthest from the axis in bud. *cf.* **adaxial**

achene: a dry, 1-seeded, indehiscent fruit formed from a superior ovary of one carpel. Used to describe the fruit of the Asteraceae formed from an inferior ovary; equivalent to the term cypsela

acicular: needle-shaped, usually stiff, often applied to leaf-tips

aciculate: a surface with fine lines, as if scratched

actinomorphic: having radially arranged floral segments which are more or less equal in size and shape, referring to a flower, calyx or corolla. *cf.* **zygomorphic**

aculeate: covered in prickles

acuminate: tapering gradually to a protracted point. *cf.* **attenuate**

acute: terminating in a distinct but not protracted point, the converging edges forming an angle of less than 90 degrees. *cf.* **obtuse**

adaxial: towards the axis, referring to the surface of an organ that is closest to the axis in bud. *cf.* **abaxial**

adnate: fused to an organ of a different kind, e.g. a stamen fused to a petal or an anther fused for its whole length to the filament. Also used to describe the type of anther in some eucalypts which is **basifixed**, immobile and about as wide as long. *cf.* **adherent**, **coherent**, **connate**

after-ripening (dry): the process of dormancy loss in mature, dry seeds. A type of pre-treatment used to alleviate physiological dormancy (and morphophysiological dormancy). Commonly, dry after-ripening conditions involve storing seeds 'dry' in constant temperatures of 25–40°C and 30–50% relative humidity

alluvium: deposits of earth, sand, etc., left by water flowing over land that is not permanently submerged adj. **alluvial**

alternate: borne singly and spaced around and along the axis, applied to leaves or other organs on an axis. Also used to describe the position of floral parts of different whorls on different radii, e.g. stamens with respect to petals. *cf.* **opposite**

amplexicaul: clasping the stem

anastomosing: connecting with one another, particularly applied to veins

androdioecious: having male flowers and bisexual flowers on separate plants. *cf.* **gynodioecious**

anfractuose: wavy hairs that are tightly twisted together

annual: completing the full cycle of germination to fruiting within a single year and then dying. *cf.* **biennial**, **perennial**

anther: that part of the stamen in which the pollen is produced

antrorse: directed upwards. *cf.* **retrorse**

apex: the tip of an organ. pl. **apices**

apiculum: a short, sharp, flexible point. adj. **apiculate**

appressed: pressed closely to another organ but not united

areole: more or less circular areas on a surface divided from similar areas by a vein or rib. adj. **areolate**

aril: a fleshy to hard structure which develops from the funicle or ovule after fertilisation and encloses all or part of the seed. adj. **arillate**

aristate: having a stiff, bristle-like tip

aristulate: having a small, stiff, bristle-like tip; a diminutive of **aristate**

aromatic: producing volatile oils with a distinct odour

articulate: jointed; usually fracturing easily at the nodes or point of articulation into segments or articles

ascending: arched upwards in the lower part and becoming erect in the upper part. *cf.* **decumbent**, **procumbent**

asperulate: slightly rough with small hairs

asymmetric: two sides or parts unuequal, uneven halves produced when cut down the middle

attenuate: tapering gradually. *cf.* **acuminate**

auricle: an ear-shaped appendage at the base of a leaf, leaflet or corolla lobe; or an ear-like outgrowth at the base of the sheath of some grasses and other monocots. adj. **auriculate**, also used to describe a leaf base which has lobes on both sides of the petiole

awn: a slender, bristle-like projection, e.g. from the back or tip of the glumes and lemmas in some grasses and on the fruit of some Geraniaceae. adj. **awned**

axil: the angle between one part of a plant and another part, e.g. a branch and a leaf. adj. **axillary**

axis: main stem of the plant or the flower-bearing portion of an inflorescence or the ovule-bearing structure in ovaries with axile placentation

B

bark: outermost layer on stems and roots of woody plants, down to the cambium layer

barren: sterile, not producing seed

basal: arising from or positioned at the base

basalt: dark, fine-grained volcanic rock, typically composed largely of plagioclase with pyroxene and olivine, weathering to dark brown soils

basifixed: anthers attached at the base to the filament

beak: a prominent terminal projection, especially of a carpel or fruit. adj. **beaked**

beard: a tuft of long hairs

berry: a fleshy or pulpy indehiscent fruit with the seed(s) embedded in the fleshy tissue of the pericarp

biramous: two-armed, may be unequally so

biconvex: convex on both sides

biennial: completing the full cycle of germination to fruiting in more than one, but not more than two years, and then dying

bipinnate: 2-pinnate; twice pinnately divided

bisexual: having both sexes, as in a flower bearing both fertile anthers and a fertile ovary. *cf.* **unisexual**

bitextured: a surface with two distinct textures, often sharply divided (e.g. the lemma surface in some *Triodia*)

blade: lamina; part of the leaf above the sheath or petiole

blistered: a surface with the appearance of raised blisters *cf.* **bullate**

bract: a leaf-like structure, usually different in form from the foliage leaves, associated with an inflorescence or flower. adj. **bracteate**

bracteole: a small bract-like structure borne singly or in a pair on the pedicel or calyx of a flower. adj. **bracteolate**

branchlets: a small branch, the final (or youngest) division of the branching system

broadly: having a length:breadth ratio between 6:5 and 1:1; if the ratio is less, then the shape is described as very broadly. *cf.* **narrowly**

bulb: a modified underground axis that is short and crowned by a mass of usually fleshy, imbricate scales. adj. **bulbous**

bullate: blistered in appearance, often between veins

C

caducous: falling early before associated organs are mature. *cf.* **deciduous**, **persistent**

callus, **callosity**: a thickened, raised mass of hardened tissue, often formed after an injury but sometimes a normal feature, e.g. the glandular wart-like structures on the labellum of some orchids, or in grasses, the hardened, usually hairy base of the dispersal unit, usually a floret or whole spikelet. pl. **calli**, **callosities**. adj. **callose**

calyx: the outermost floral whorl usually consisting of sepals or a calyx tube and calyx lobes. adj. **calycine**

calyx lobe: one of the free upper parts of the calyx which may be present when the lower part is united into a tube

calyx tube: the tubular, often cup-shaped or bell-shaped, fused part of the calyx when it is free from the corolla. *cf.* **floral tube**

campanulate: a 3-dimensional shape; bell-shaped

canaliculate: with a longitudingal groove or channel. *cf.* **channelled**

capitate: growing together in a head. Also meaning head-like as in some stigmas

capitulum: a dense cluster of sessile, or almost sessile, flowers or florets. pl. **capitula**

capsule: a dry fruit formed from two or more united carpels and dehiscing at maturity to release the seeds. adj. **capsular**

carpel: an organ, generally believed to be leaf-derived, which bears 1 or more ovules, a stigma and sometimes a style. Often much modified in a syncarpous ovary. adj. **carpellary**

cartilaginous: hard and toughened, but still able to bend

caryopsis: the grass fruit, which has the seed coat united with the ovary wall

cataphyll: a scale-like leaf

cauline: borne on the aerial part of a stem. *cf.* **radical**

chaff: debris often associated with species of the Myrtaceae that consists of sterile particles from infertile or non-fertilised ovules

channelled: a long groove running along the length of an object. *cf.* **canaliculate**

chartaceous: papery

ciliate: fringed with hairs

clavate: a 3-dimensional shape; club-shaped; thickened at one end

claw: the conspicuously narrowed basal part of a flat organ

clay: fine-grained soil that usually combines hydrated silicates of aluminum with traces of metal oxides and organic matter, plastic when wet, drying very hard

cleistogene: a plant producing cleistogamous flowers, common in Poaceae

climber: growing more or less erect by leaning or twining around another structure

clone: a group of plants derived from a single parent by vegetative means, thus genetically identical

coccus: one of the segments of a distinctly lobed fruit which becomes separate at maturity. Sometimes called a mericarp. pl. **cocci**

colliculate: covered in small, rounded protuberances, giving a hilly appearance

column: a structure formed by the united style, stigma and stamen(s), as in Apocynaceae, Stylidiaceae and Orchidaceae; or by twisted awns as in Poaceae

combinational dormancy: a type of seed dormancy whereby seeds possess a physiological block to embryo growth (i.e. physiological dormancy), coupled with the presence of a impermeable fruit or seed coat layer (i.e. physical dormancy)

complicate: folded upon itself. *cf.* **conduplicate**

companion spikelet: a paired spikelet in Poaceae that may be infertile and dissimilar in appearance to its pair

compound: consisting of two or more anatomically or morphologically equivalent units

compressed: flattened

concave: hollow, like the inside of a bowl

concolorous: uniformly coloured, as in upper and lower surfaces. *cf.* **discolorous**

conduplicate: folded together lengthwise, the upper surfaces parallel and facing each other

confluent: coming together and merging

conic: a 3-dimensional shape; cone-shaped, attached at the broader end

connate: fused to another organ of the same kind. *cf.* **adherent**, **adnate**, **coherent**

convex: with a rounded surface, like the outside of a bowl

convolute: an arrangement of petals in a bud where each has one side overlapping the neighbouring petal

cordate: a 2-dimensional shape; heart-shaped in outline, i.e. having the base broad and distinctly notched

coriaceous: leathery

cork: protective tissue replacing the epidermis on older stems of some plants, usually impervious to water

corolla: the floral whorl inside the calyx, usually consisting of petals or a corolla tube and corolla lobes. adj. **corolline**

corymb: an inflorescence, usually a raceme, in which the flowers, through unequal pedicels, are in one horizontal plane. adj. **corymbose**

cotyledon: the primary ('seed') leaf, each seedling having one or two, rarely more

crenate: with obtuse or rounded teeth which either point forwards or are perpendicular to the margin

crenulate: minutely crenate

crisped: very strongly undulate; margins which are wavy in a vertical plane, the waves themselves with secondary waves. *cf.* **undulate**

crustaceous: hard, thin and brittle

cucullate: hooded

culm: the aerial stem of grasses, sedges, rushes and other monocots, bearing the cauline leaves and the inflorescence

cuneate: a 2-dimensional shape; obtriangular, i.e. wedge-shaped

cupular: cup-shaped

cylindric: a 3-dimensional shape; tubular or rod-shaped with a length:breadth ratio of 2:1 to 3:2. *cf.* **narrowly**, **broadly**

cyme: an inflorescence in which each flower, in turn, is formed at the tip of a growing axis, further flowers being formed on branches arising below. adj. **cymose**

D

decompound: more than once compound, usually applied to three or more orders of branching

decumbent: spreading horizontally with the apex growing upwards, referring to stems. *cf.* **ascending**, **procumbent**

decurrent: having the leaf base prolonged down the stem as a winged expansion or rib

decussate: having paired organs with successive pairs at right angles to give four rows

dehiscent: breaking open at maturity to release the contents. *cf.* **indehiscent**

dendritic: branching from a main stem or axis; resembling the branching of a tree

dentate: with sharp, spreading, rather coarse teeth standing out from the margin

denticulate: finely dentate

depressed: flattened as if pressed down from the top or end; applied to 2 or 3-dimensional shapes, e.g. ovate and obovate or ovoid and obovoid which have a length:breadth ratio of less than 2:3

dimidiate: divided into two parts, one reduced so only one obvious

dioecious: having male and female unisexual flowers on different plants. *cf.* **monoecious**

disarticulating: falling apart into its component parts or separating at a joint or abcission zone

disc: the usually disc-shaped receptacle of the head in Asteraceae. Also the fleshy nectariferous organ which is sometimes annular or lobed and developed usually between the stamens and ovary. Also used for the enlarged style-end in Proteaceae

discoid: resembling a disc; a solid structure with two convex faces. Used in the Asteraceae to describe a head which has only disc florets present. *cf.* **disciform**, **radiate**

discolorous: having two colours, e.g. the lower leaf surface distinctly different in colour from the upper. *cf.* **concolorous**
disk: an enlargement of the floral receptacle or ovary, often secretes nectar, often cup-shaped or forming a ring, often displays the stamens
dissected: dividing into segments
distal: end of any structure farthest from the point of attachment. *cf.* **proximal**
divergent: spreading apart
divaricate: widely spreading
domatium: small cavities, most common on the lower surface of leaves, often in axils of the veins, and often linked to ants or mites. pl. **domatia**
dormancy: a characteristic of a seed that defines the environmental conditions that must be met for germination to proceed. A dormant seed is one that does not germinate under external environmental conditions that are normally suitable to support germination
dorsal: upper surface (= **adaxial**) or object
dorsiventral: with two distinct surfaces, upper and lower (usually used when different in appearance)
drupe: a 1-celled fruit with one or two seeds enclosed by a stony layer (**endocarp**) which is embedded in succulent tissue (**mesocarp**) surrounded by a thin outer skin (**epicarp**). adj. **drupaceous**
dry after-ripening: see **after-ripening**

E

ebracteate: without bracts
echinate: bearing stiff, stout, prickly hairs
eciliate: lacking cilia
eglandular: without glands. *cf.* **glandular**
ellipsoid: a 3-dimensional shape; elliptic in outline and with a length:breadth ratio between 3:2 and 2:1. *cf.* **narrowly**, **broadly**
elliptic: a 2-dimensional shape; oval in outline and with a length:breadth ratio between 3:2 and 2:1. *cf.* **narrowly**, **broadly**
elongate: lengthened; stretched out
emarginate: having a broad, shallow notch at the apex. *cf.* **retuse**
embryo: rudimentary plant contained within a seed, consists of cotyledon(s), radicle and plumule
embryo to seed length ratio (E:S ratio): the ratio between the length of the embryo and the length of the seed (measured within the internal wall of the seed coat)
embryo 'type': an embryo classification system used to describe the shape, position, size and developmental state of the embryo within a fruit/seed. Globally, 13 different embryo types are recognised
enation: outgrowth of one organ from another or epidermal outgrowth
endemic: having a natural distribution confined to a particular geographical region
endocarp: the innermost layer of the fruit wall, derived from the innermost layer of the carpel wall. *cf.* **exocarp**, **mesocarp**, **pericarp**
endosperm: the storage tissue that surrounds the embryo. The endosperm is digested and used as a nutritional source during the development of the embryo, during the process of germination, and in early seedling establishment
entire: without any lobes, incisions or teeth
ephemeral: short-lived
epicalyx: a whorl of bracts, just below or joined to the calyx, resembling a second calyx
erect: upright; perpendicular
erose: irregularly toothed, with a nibbled appearance
exocarp: the outermost layer of the fruit wall, derived from the outermost layer of the carpel wall. Sometimes called **epicarp**. *cf.* **endocarp**, **mesocarp**, **pericarp**
excentric: not in the centre, off-centre
exotesta: the part of the seed coat that develops from the outer surface of the outer integument
exserted: protruding beyond some enclosing organ, e.g. referring to stamens which project beyond the corolla or perianth, or to valves which extend beyond the rim of a capsular fruit. Sometimes the stamens are not actually longer than the corolla or perianth but are exserted due to the spreading of the segments. *cf.* **included**
exstipulate: without stipules. *cf.* **stipulate**

F

falcate: sickle-shaped
felted: a matted indumentum of intertwined hairs, felt-like in appearance
ferruginous: rust-coloured
fertile: provided with functional sexual parts which are capable of fertilisation and seed production. *cf.* **sterile**
fibrous: composed of fibres; including fibres
filament: the stalk of a stamen below the point of attachment to the anther

filiform: thread-like

fill/filled (fruit/seed): a term used to describe the internal cavity of a fruit or seed where all the tissues are present that are required for germination. A filled fruit/seed is not necessarily alive; however, in most cases fruit/seed fill is used as a proxy for 'viable' (e.g. after an x-ray fill assessment is carried out)

fimbriate, fimbriolate: fringed

fissured: cracked with deep slits (usually refers to bark)

fleshy: succulent, swollen with water

flexuose: zig-zagging, often referring to a stem

floccose: covered with tufts of soft woolly hairs that usually rub off readily

floral bract: the first bract below the flower

floral tube: the part of the flower enveloping an ovary and formed by the fusion of the calyx tube, corolla tube and sometimes stamen filaments and receptacle. Variously referred to elsewhere as **calyx tube**, **perigynium** or **hypanthium**

floret: one of the small individual flowers of the Asteraceae or the reduced flower of the grasses, including the lemma and palea

flower: the sexual reproductive structure of the angiosperms, typically consisting of gynoecium, androecium and perianth or calyx and/or corolla and the axis bearing these parts

foliate: having leaves, or leaf-like

-foliolate: used with a number prefix to denote the number of leaflets

follicle: a dry fruit, derived from a single carpel and dehiscing along one suture

fruit: seed-bearing structure formed from the ripened ovary after flowering

funicle: stalk of the ovule or seed, attaching to the placenta

furrowed: with longitudinal grooves or channels

fusiform: a 3-dimensional shape; spindle-shaped, i.e. is circular in cross-section and tapering at both ends

G

germination: the process by which a plant grows from a seed, defined as the point when the radicle (i.e. primary root) first emerges from the fruit/seed coat

germination potential (= germination percentage): the proportion of a seed batch that has germinated in a set period of time, normally expressed as a germination percentage or proportion. Occasionally, germination potential is also referred to as 'germinability'

germination stimulant: a chemical or biological agent that is known to promote germination. Common stimulants include the smoke-derived chemical karrikinolide (e.g. KAR_1), gibberellic acid (e.g. GA_3), and nitrates (e.g. KNO_3)

germination test: an evaluation of germination potential under standard conditions. Typical conditions include incubating seeds under appropriate temperature, light, and moisture conditions (e.g. in a temperature and light controlled incubator) on Petri dishes containing a substrate (e.g. filter paper, sand, or solidified water-agar). Germination is then assessed over a set time period (usually for up to 4 weeks)

glabrescent: becoming glabrous

glabrous: without hairs

gland: a secretory structure, e.g. a nectary, extra-floral nectary or a gland tipped, hair-like or wart-like organ. adj. **glandular**. *cf.* **eglandular**

glaucous: blue-green in colour, with a whitish bloom

globular: a 3-dimensional shape; spherical or orbicular; circular in outline

glume: one of the two bracts at the base of the grass spikelet, called the lower and upper glumes, due to their position on the rachilla. Also used in Araliaceae and Cyperaceae for the very small bracts on the spikelet in which each flower is subtended by one floral glume and in which there are often several empty glumes at the base. adj. **glumaceous**

gnarled: twisted, mis-shapen

grasslands: vegetation where grasses are the dominant life-form, trees usually absent or scattered, shrubs may be present or absent

grain: see caryopsis

granite: intruded, igneous, crystalline rock

granular, granulate: a surface covered with small grains

gravel: a mixture of rock fragments or pebbles, often rounded by water

gynoecium: the female part of a flower; a collective term for the carpel or carpels

H

habit: the growth form of a plant, comprising its size, shape, texture and stem orientation

halo: a pale margin found on the seeds of some *Acacia* species

head: a short, dense inflorescence. *cf.* **capitulum**

hemispherical: shaped like a half-sphere

herb: a plant which is non-woody or woody at the base only, the above-ground stems usually being ephemeral. adj. **herbaceous**
herbaceous: herb-like; often applied to bracts, bracteoles or floral parts that are green and soft in texture
hermaphrodite: bisexual, bearing flowers with both androecium and gynoecium in the same flower. adj. **hermaphroditic**
hilum: present on a grass caryopsis as well as seeds of the Fabaceae and Malvaceae, visible on its exterior and represents the attachment of the funiculus of the ovule to the inner ovary wall
hirsute: with long and rather coarse or stiff hairs, the hairs being less rigid and erect than when referred to as hispid
hispid: with stout, rigid or bristly, erect hairs
hoary: covered with a greyish layer of very short, closely interwoven hairs
hyaline: almost transparent
hybrid: the offspring of the sexual union of plants belonging to different taxa
hypanthium: see **floral tube**
hypocotyl: embryonic stem in seed or seedlings, located below the cotyledons but above the radicle

I

imbibition test: a test carried out to determine if a fruit or seed has the capacity to take up water; primarily used to determine the presence or absence of physical dormancy. Imbibition, or water uptake, is measured as the percent increase in fruit/seed weight following placement on a moist substrate or immersion in water
impressed: a nerve or vein that is usually shallowly below the surrounding surface
incised: cut jaggedly with very deep teeth
included: not protruding beyond some enclosing organ, e.g. stamens which do not project beyond the corolla or to valves which do not extend beyond the rim of a capsular fruit. *cf.* **exserted**
incumbent: lying on, and closely parallel to an object
incurved: curved inward; curved towards the base or apex
indehiscent: not opening or splitting to release the contents at maturity. *cf.* **dehiscent**
indumentum: the type of hairiness commonly found on external parts of plants. *cf.* **vestiture**
indurate: hardened, often the hardening developed only at maturity
indusium: tissue covering the sorus of a fern. Also used for the modified style end or pollen-cup of some Goodeniaceae (including Brunoniaceae). adj. **indusiate**
inequilateral: with two unequal sides
inferior: a flower in which the ovary is fused with the floral tube and the sepals, petals and stamens are inserted above most of the ovary. *cf.* **superior**
inflated: enlarged and hollow except in the case of a fruit which may contain a seed. *cf.* **swollen**
inflorescence: the arrangement of flowers in relation to the axis and to each other
infructescence: the fruit-bearing part of a plant derived from an inflorescence
internode: the part of an axis between two successive nodes, joints or point of attachment of the leaves
intra-carinal veins: fine veins between keels or a keel and a margin
inverse: with reference to grass glumes, the lower glume is longer than the upper
involucre: a large bract or whorl of bracts surrounding a flower or an entire inflorescence
involute: with the margins inrolled on the upper (adaxial) surface, referring to a leaf or other flat organ
ironstone: the name given to various hard iron-ores containing admixtures of silica, clay, etc

J

juvenile: young or immature, used here for leaves formed on a young plant which are different in morphology from those formed on an older plant

K

keel: a boat-shaped structure, with a prominent longitudinal ridge, or, in the Papilionaceae, the part of the corolla formed by the fusion of the lower edge of the two abaxial petals. adj. **keeled**. *cf.* **standard**, **wing**

L

lamina: the usually flattened blade of a leaf or frond. Also used here for the limb, the flattened part of the ligulate floret of Asteraceae. pl. **laminae**
lanate: covered with long hairs which are loosely curled together like wool
lanceolate: lance-shaped, much longer than wide, the widest point below the middle
laterite: a red, porous, ferruginous rock. adj. **lateritic**
leaf: a chlorophyll-bearing, lateral outgrowth from a stem (well differentiated)
leaflet: one of the ultimate segments of a compound leaf
leathery: a tough but flexible surface
legume: pods of the legume family (Fabaceae), derived from a single carpel, often opening along two suture lines
lemma: the lower of two bracts of a grass floret, usually enclosing the palea, lodicules, stamens and ovary
lenticular: shaped like a biconvex lens

ligulate: small and tongue-shaped or with a little tongue-shaped appendage or ligule, *q.v.* Also applied to the florets of the Asteraceae which have a single small tongue-shaped corolla lobe

ligule: outgrowth from the inner junction of the grass leaf sheath and blade, often membranous, sometimes represented by a fringe of hairs. Also a small, membranous, triangular organ on the adaxial side of the fertile leaf base in *Isoetes* or the narrow, upper part of a reduced petal in some Malvaceae or the single corolla lobe of a floret of Asteraceae. *cf.* **ligulate**

limb: the expanded portion of a flat organ above the claw. Also used here for the expanded portion of the calyx tube or the corolla tube, e.g. in the Araliaceae and Solanaceae

limestone: a soft or hard sedimentary rock, composed mainly of calcium carbonate

linear: a 2-dimensional shape: narrowly rectangular with parallel sides and a length:breadth ratio of at least 12:1

loam: a soil composed of roughly equal parts of sand, silt and clay

lobe: a usually rounded or pointed projecting part, usually one of two or more, each separated by a fissure or sinus. The terms calyx lobe, corolla lobe and floral lobe are used here for the projections of the calyx, corolla or floral tube. *cf.* **segment**

lodicule: minute scales found between the lemma and the fertile flower parts in grasses

longitudinal: refers to seeds held longitudingal on the pod. *cf.* **oblique**, **transverse**

loose: lax or not crowded; with distant or scattered units

M

mallee: a growth habit in which several to many woody stems arise separately from a lignotuber; e.g. usually applied to certain low-growing species of *Eucalyptus*

margin: (in leaves) the edge of the leaf blade

matted: closely intertwined, forming a carpet

mealy: with many small, floury grains

membrane: thin, film-like, usually flexible, often translucent. adj. **membranous**

mericarp: a 1-seeded portion of an initially syncarpous fruit which splits apart at maturity, e.g. in the Araliaceae, Malvaceae. Sometimes called a **coccus**, *q.v.*

-merous: used with a number prefix to denote the basic number of the 3 outer floral whorls, e.g. a 5-merous flower may have 5 sepals, 10 petals and 15 stamens

metamorphic rocks: rock that has been transformed to another kind, usually under the influence of heat or pressure

midrib: the primary vascular supply and support structure for a leaf, leaflet or bract

midvein: the main vascular supply of a simple leaf blade or lamina

minni-ritchi bark: reddish bark exfoliating in narrow, curled strips. Usually only applied to certain species of *Acacia* or *Eucalyptus*

moisture content (seed): the amount of water within a seed, normally expressed as a percentage of the dry-weight or fresh-weight of the seed

moniliform: usually applied to pods, which are constricted around the seeds so as to form a regularly spaced pattern, cylindrical then constricted, like a string of beads

monoecious: having both male and female unisexual flowers on the same individual plant. *cf.* **dioecious**

morphological dormancy: a type of seed dormancy classified by the presence of an underdeveloped embryo that requires a period of growth inside the fruit/seed prior to germination occurring

morphophysiological dormancy: a type of seed dormancy classified by the presence of an underdeveloped embryo (i.e. morphological dormancy) coupled with a physiological block to embryo growth (i.e. physiological dormancy)

mucro: a stiff or sharp, short projection of the midvein abruptly terminating an organ. cf. cusp

mucronate: abruptly tipped with a mucro, sometimes with a small amount of leaf tissue. *cf.* **cuspidate**, **muticous**

mucronulate: with a very small mucro; a diminutive of mucronate

multicellular: usually applied to hairs that consist of multiple cells

multistriate: with many striations

muricate: with numerous short hard outgrowths. *cf.* **papillose**

muriculate: with numerous minute hard outgrowths; a diminutive of muricate

muticous: blunt, lacking awn or point. *cf.* **mucronate**

N

narrowly: having a length:breadth ratio between 3:1 and 6:1; if the ratio is more than 6:1 then the shape is described as **very narrowly**, except in the case of **very narrowly oblong** which is termed **linear**

nectariferous: having one or more nectaries

nectary: a secretory organ producing nectar; commonly in a flower, sometimes on leaves, fronds or stems

node: the area of a stem where a leaf attaches

non-dormant: to lack a dormancy stage or mechanism

notched: with a nick or cut in an edge that is otherwise entire

nut: dry indehiscent 1-celled fruit with a hard pericarp

O

ob-: prefix meaning inversely or opposite to
obconic: a 3-dimensional shape; cone-shaped, attached at the narrower end
obcuneate: obversely wedge-shaped
oblanceolate: narrowly obovate and tapering to an acute apex
oblique: leaves with two sides of the blade unequal at the base; seeds at an angle to the main plain of the fruit
oblong: a 2-dimensional shape; rectangular with a length:breadth ratio between 3:2 and 2:1. *cf.* **narrowly**, **broadly**
obloid: a 3-dimensional shape; short parallel sides and rounded ends
obovate: a 2-dimensional shape; similar to ovate but attached at the narrower end and with a length:breadth ratio between 3:2 and 2:1. *cf.* **narrowly**, **broadly**
obtriangular: a 2-dimensional shape; triangular, attached at the apex and with a length:breadth ratio between 3:2 and 2:1. *cf.* **triangular**
obtuse: blunt or rounded at the apex, the converging edges separated by an angle greater than 90 degrees. *cf.* **acute**
-oid: suffix denoting a 3-dimensional shape
opaque: not translucent
operculum: a lid or cover becoming detached at maturity by abscission, e.g. in *Eucalyptus*, a cap covering the bud and formed by fusion or cohesion of sepals and/or petals. adj. **operculate**
opposite: describing leaves or other organs which are borne at the same level but on opposite sides of the stem; of floral parts, on the same radius. *cf.* **alternate**
opposite-decussate: opposite leaves in alternating pairs at 90 degrees to each other
orbicular: of circular outline
orifice: an opening or aperture such as that of a cup-shaped receptacle or the summit of a leaf sheath in grasses and sedges; also used for the 'pollen cup' at the apex of the indusium in Goodeniaceae
ovary: the basal portion of a carpel or group of fused carpels, enclosing the ovule(s)
ovate: a 2-dimensional shape; resembling a section through the long axis of an egg, attached near the broader end and with a length:breadth ratio between 3:2 and 2:1. *cf.* **narrowly**, **broadly**
ovoid: a 3-dimensional shape; egg-shaped; ovate in outline
ovule: the organ in the ovary which becomes a seed after fertilisation. adj. **ovular**

P

palea: the upper of the two bracts of a grass floret, usually enclosing the lodicules, stamens and ovary. pl. **paleae**. adj. **paleal**. *cf.* **lemma**
palmate: describing a leaf which is divided into several leaflets which arise from the same point. adj. **palmately**
panicle: a compound raceme; an indeterminate inflorescence in which the flowers are borne on branches of the main axis or on further branches of these. adj. **paniculate**
papilla: a small protuberance on the surface of an organ being an extension of one epidermal cell. pl. **papillae**. adj. **papillose**. *cf.* **muricate**
pappus: a tuft (or ring) of hairs, bristles or scales borne above the ovary and outside the corolla in Asteraceae and possibly representing the calyx; often persisting as a tuft of hairs on a fruit. adj. **pappose**
-partite: divided almost to the base into segments, the number of segments given as a prefix
pedicel: the stalk of an individual flower. In a single-flowered inflorescence, may refer to a peduncle, or both peduncle and pedicel when undifferentiated. Also used here for the stalk of a spikelet in Poaceae. adj. **pedicellate**
peduncle: the stalk of an inflorescence. adj. **pedunculate**
perennating: to survive from one growing season to the next, often with a period of reduced or arrested growth between seasons
perennial: with a life span extending over more than two growing seasons. *cf.* **annual**, **biennial**
perfect flowers: with both male and female floral parts
perianth: the outer floral whorl or whorls of a monocotyledonous flower, sometimes divisible into outer and inner perianth segments. Sometimes used by other authors to describe the floral whorl of a dicotyledonous flower when it is uncertain if the whorl is calycine or corolline in origin or used as a collective term for **calyx** and **corolla**
pericarp: the wall of a fruit developed from the ovary wall. Composed of the **exocarp**, **mesocarp** and **endocarp**
persistent: remaining attached; not falling off. *cf.* **caducous**
petal: free segment of the corolla. adj. petaline. *cf.* **lobe**
petiole: the stalk of a leaf. adj. **petiolate**
petiolule: the stalk of a leaflet. adj. **petiolulate**
phenology: the study of the relationship between seasonal climatic changes (e.g. rainfall and air temperature) and the reproductive life stage of plants such as budding, flowering, fruit development, and fruit/seed dehiscence
phyllary: a bract surrounding the capitulum in Asteraceae, forming the involucre. pl. **phyllaries**
phyllode: a leaf whose blade is much reduced or absent and whose petiole has assumed the functions of the whole leaf. adj. **phyllodineous**. *cf.* **cladode**

physiological dormancy: a type of seed dormancy whereby germination is prevented due to an embryo and/or seed coat-imposed dormancy. The low growth potential of the embryo cannot overcome the mechanical restraint of the seed or fruit coat. For germination to proceed a period of embryo maturation is typically required

physical dormancy: a type of seed dormancy classified by the presence of a water-impermeable fruit or seed coat, with the barrier to water uptake typically made up of palisade, or palisade-like, cells

pilose: with slightly stiffened distinct hairs ascending from the surface

pindan: vegetation consisting of a lower storey of grasses, a middle storey of dense *Acacia* thickets and a sparse upper layer of trees up to 15 metres high

pinnate: with the same arrangement as a feather; divided into pinnae; once-compound

pinnatisect: lobed almost to base or midrib

pistil: the combination of carpel, style and stigma

pitted: with small depressions

pleurogram: U-shaped or elliptic fracture line around the areole on the face of legume seeds

plumose: like a feather; with fine hairs arising laterally from a central axis

plumule: the shoot bud of the embryo

pod: a dry 1-many-seeded dehiscent fruit. Commonly applied to the fruits of Fabaceae

pollen: powder-like fertilisation agent, developed in anthers, containing a single set of chromosomes

pollen presenter: a specialised section of the style in Proteaceae that holds the anthers

pre-treatment: any form of treatment that is applied to fruits/seeds to render them non-dormant and to promote germination (e.g. nicking)

prickle: a pungent outgrowth from the epidermis, detaching readily (e.g. a rose 'thorn')

procumbent: trailing or spreading along the ground but not rooting at the nodes, referring to stems. *cf.* **ascending, decumbent, erect**

prostrate: lying flat on the ground

pruinose: with a frosty surface; having a thick, waxy, powdery coating

pseudostaminode: appendage between the stamens which is not a modified stamen

puberulent, puberulous: covered with very short fine straight erect hairs

pubescent: covered with short, soft hairs

pulvinus: an enlargement below the base of a leaf or leaflet in which changes of turgidity enable the leaf or leaflets to droop or fold. Also a glandular swelling in the axils of the inflorescence axis and primary branches of some grasses. adj. **pulvinate**

punctate: marked with dots, e.g. the clear-coloured oil glands of a leaf of Rutaceae and Myrtaceae

pungent: ending in a rigid, sharp point (also a sharp smell)

purity: the proportion of fruit/seed material relative to foreign debris (e.g. floral parts, chaff, plant/leaf fragments, or soil particles. Purity is expressed as a percentage

pyramidal: shaped like a pyramid; when applied to an inflorescence, usually an inverse pyramid

Q

quadrangular: with four angles

quality (of seeds): a general term that can encompass the purity, viability, germination capacity, and/or seedling vigour of a seed batch

quartzite: an extremely compact, granular rock, consisting essentially of quartz

R

raceme: an indeterminate inflorescence with a simple, elongated axis and pedicellate flowers. adj. **racemose**

rachilla: the main axis of a grass spikelet

rachilla extension: a prolongation of the rachilla beyond the uppermost floret in Poaceae

rachis: the main axis of the spike or other inflorescence of grasses; the axis of a pinna in a bipinnate leaf

radiate: arranged around a common centre; used here especially to describe an inflorescence of Asteraceae with marginal, female or neuter, ligulate ray-florets and central, perfect or functionally male, tubular, disc florets. *cf.* **disciform, discoid**

radicle: the primary embryonic root of a germinating seed

ray: the marginal portion of the inflorescence of Asteraceae and Apiaceae when distinct from the disc. Also, the axis or branches of a compound umbel in some Araliaceae, Cyperaceae and Loranthaceae

receptacle: the region at the end of a pedicel or on an axis which bears one or more flowers. In the Asteraceae the receptacle may be quite large. adj. **receptacular**

recurved: curved outward or downward

reflexed: bent or turned downward

resinous: producing resin; often sticky

reticulate: forming a network or reticulum

retrorse: bent or directed downwards or backwards. *cf.* **antrorse**

retuse: with a very blunt and slightly notched apex. *cf.* **emarginate**
revolute: with the margins inrolled on the lower (abaxial) surface
rhizome: a creeping stem, usually below ground, consisting of a series of nodes and internodes with adventitious roots. adj. **rhizomatous**
rhomboid: 4-angular, with obtuse angles
rib: a distinct vein or linear marking, often raised as a linear ridge
rigid: stiff
rim: the margin, often thickened and raised in *Eucalyptus* fruit. adj. **rimmed**
rootstock: underground stems and/or roots
rosette: a tuft of leaves or other organs resembling the arrangement of petals in a rose, ranging in form from a hemispherical tuft to a flat whorl. adj. **rosetted, rosulate**
rotate: wheel-shaped; applied to a corolla with a very short tube and a broad upper part which is flared at right angles to the tube. *cf.* **salverform**
rotund: 2-dimensional shape, nearly round with slightly compresed sides
rudimentary ovary: basal, small, non-peripheral embryo; relatively undifferentiated; endosperm copious

S

saccate: pouched
sagittate: arrow-shaped, with a pair of large, acute or rounded, usually overlapping lobes
sand: a loose granular soil derived from the erosion of siliceous and other rocks, coarser than silt
sandstone: sedimentary rock consisting of cemented sand and quartz grains, often red, yellow, or brown
savannah: a grassland ecosystem with widely spaced trees and an open canopy supporting a herbaceous layer dominated by grasses
scabrous: rough to the touch
scabridulous: minutely scabrid
scale: a small peltate disc. adj. **scaly**
scarious: dry and membranous
schizocarp: a usually dry fruit which splits longitudinally into indehiscent or tardily dehiscent parts (**mericarps** or **cocci**)
scurfy: covered with small, dandruff-like scales
secund: arranged on one side only
sedge: a plant of the family Cyperaceae
seed: reproductive structure produced from a fertilised ovule with an embryo, endosperm, and usually a seed coat
seedling: a plant produced from a seed, with the term typically used to described a plant that has emerged from the soil surface
segment: a free or almost free part or subdivision of an organ. Calyx and corolla segments are called sepals and petals respectively. In the case of undifferentiated dicotyledonous flowers, the segments are called floral segments. *cf.* **lobe**
sepal: free segment of the calyx. adj. **sepaline**
septum: a partition or cross wall. Also a term for the sterile structure separating the seeds in a mature fruit of *Banksia.* pl. **septa**. adj. **septate**
septifragal: fruit dehiscent along the carpel septa, valves falling, axis persisting
sericeous: silky; covered with close-pressed, fine, straight silky hairs
serrate: toothed so as to resemble a saw; with regular, asymmetric teeth pointing forward
serrulate: finely serrate
sessile: without a stalk
seta: a bristle or stiff hair. pl. **setae**. adj. **setose, setaceous**
shale: fine-grained, stratified, sedimentary rock
sheath: a tubular organ enveloping another organ (e.g. bracts and leaf bases)
sheathing: clasping or enveloping the stem
shrub: a woody plant usually less than 5 metres high and many-branched without a distinct main stem except at ground level. *cf.* **undershrub**
silicula: a broad, dry, usually dehiscent fruit derived from two or more carpels which usually dehisce along two sutures and which has a persistent partition after dehiscence. *cf.* **siliqua**
simple: not divided, e.g. applied to a leaf not divided into leaflets. *cf.* **compound**
skeletal (soils): very shallow, usually over rock
soil seed bank (topsoil): the accumulation of viable seeds in the soil profile over one to several years. A topsoil seed bank provides a source for plants to re-establish after disturbance (e.g. fire) and is a highly valuable resource for use in restoration sites (e.g. topsoil re-spreading after mining operations cease)
solitary: usually used to describe flowers which are borne singly, and not grouped into an inflorescence

spathe: a large bract ensheathing an inflorescence or its peduncle. adj. **spathaceous**
spathulate: spoon-shaped; broad at the tip and narrowed towards the base
spike: an unbranched inflorescence of sessile flowers or spikelets. adj. **spicate, spiciform**
spikelet: the grass flower head, generally composed of 2 glumes and one or more florets. Also used for the small spike-like inflorescence or inflorescence units commonly found in Apiaceae, Cyperaceae and Restionaceae
spindly: thin (usually refers to a single-stemmed plant with well-spaced branches)
spirally alternate: and arrangement whereby leaves or branchlets are alternate on a stem, but form a spiral
spur: a pouch-like short to slender, usually hollow extension of some part, usually of a flower
stamen: one of the male organs of a flower, consisting typically of a stalk (filament) and a pollen-bearing portion (anther). adj. **staminal, staminate**
staminal cup: a modified **disk** found in Amaranthaceae (e.g. *Ptilotus*)
staminode: a stamen without pollen; often lacking an anther, often reduced in size, sometimes elaborated in structure. adj. **staminodial**
standard: the usually adaxial petal in the flower of Papilionaceae. *cf.* **keel**, **wing**
stellate: star-shaped, usually referring to hairs with radiating branches
stem: the main axis or a branch of the main axial system of a plant, developed from the plumule of the embryo and typically bearing leaves
stem-clasping: leaves or bracts whose base is broadly attached to the stem, often enclosing the stem or nearly so. *cf.* **sheathing**
sterile: lacking any functional sexual parts which are capable of fertilisation and seed production. *cf.* **fertile**
stigma: the usually papillate or glandular part of the style which receives the pollen. adj. **stigmatic**
stipe: a stalk or support such as the petiole of a frond or the stalk of an ovary or fruit
stipel: stipule-like appendage at the base of a leaflet (in unifoliolate leaves, inserted on the petiole, not on the stem). pl. **stipellae**. adj. **stipellate**
stipitate: having a stalk or stipe, usually of an ovary or fruit. *cf.* **sessile**
stipule: one of a pair of leaf-like, scale-like or bristle-like structures inserted at the base or on the petiole of a leaf or phyllode. adj. **stipulate**, *cf.* **exstipulate**
stolon: the creeping stem of a rosetted or tufted plant, giving rise to another plant at its tip, or in *Drosera* a vertical underground stem connecting the tuber with the above ground parts and bearing adventitious roots. adj. **stoloniferous**
straggling: growing irregularly and untidily. *cf.* **gnarled**, **spindly**
stramineous: a straw-like colour, very pale, dull yellow
stratification (warm): the process of dormancy loss under warm, moist conditions. A type of pre-treatment used to alleviate physiological and morphophysiological dormancy whereby moist seeds are incubated at temperatures >15°C (usually at 25–30°C) for up to 10 weeks, then transferred to cooler conditions for germination. Stratification is an alternative treatment to dry after-ripening and wet-dry cycling
striae: parallel longitudinal lines or ridges. adj. **striate**
style: the usually narrowed, elongated part of a carpel or group of fused carpels, between the ovary and stigma
sub-: a prefix meaning nearly or almost, as in subcapitate or subequal
subshrub; undershrub: a small, usually sparsely branched woody shrub less than 1 metre high. *cf.* **shrub**
subspecies: taxon differing in minor morphological characters such as size or shape of parts, and either partially or completely isolated by means of geographic, ecologic or other barriers
substrate: underlying rock
subulate: narrow and tapering gradually to a fine point
succulent: fleshy, juicy, soft in texture and usually thickened
superior: a flower in which the ovary is free and mostly above the level of insertion of the sepals, petals and stamens. *cf.* **inferior**
surmounted: placed on top of
suture: a line, mark or groove marking a natural division or union of parts of an organ
swollen: enlarged and solid. *cf.* **inflated**

T

tap-root: a primary root which goes straight down
taxon: a classificatory group of any rank, e.g. a family, genus, species or any infraspecific category. pl. **taxa**
tepal: a segment of the perianth, which is not differentiated into calyx and corolla; a sepal or petal
terete: circular in cross-section
terminal: at the apex or distal end
testa: a seed coat
throat: refers to the mouth of tubular flowers, at and below the point of fusion of the petals. *cf.* **tube**

thryse: a mixed inflorescence, the main axis racemose, the secondary axes cymose. Can also refer to a compact panicle with a cylindrical form
tomentose: covered with short cottony hairs, more or less felted together, shorter and less dense than lanate
torus: a ring-shaped cylinder. adj. **toroid**
toral rim: a modified portion of the style in many Proteaceae (e.g. *Grevillea*)
trailing: stems prostrate, but not rooting at the nodes
translucent: letting some light pass, but not transparent
transverse: broader than long; applied to 2 or 3-dimensional shapes, e.g. elliptic and oblong or ellipsoid and obovoid which have a length:breadth ratio of less than 5:6
tree: a woody plant usually over 5 metres high and with an unbranched lower axis
triangular: a 2-dimensional shape; 3-angled and 3-sided with a length:breadth ratio between 3:2 and 2:1. *cf.* **narrowly**, **broadly**, **deltate**
trifoliate: having three leaves
trigonous: obtusely 3-angled; triangular in cross-section with plane faces
truncate: with an abruptly transverse end as if cut off
tube: the fused portion of an organ (e.g. calyx, corolla, bract)
tuber: a stem, usually underground, enlarged as a storage organ and with minute scale-like leaves and buds or 'eyes'. Some monocotyledons e.g. *Thysanotus* develop thickened roots called tuberous roots. adj. **tuberous**
tubercle: a wart-like protuberance. adj. **tuberculate**
tuft: a densely packed cluster arising from an axis. adj. **tufted**
turbinate: top-shaped; inversely conic
twining: coiling around an object
U
umbel: an inflorescence in which the pedicels originate from one point on top of the peduncle and are usually of equal length. adj. **umbellate**
umbonate: more or less round, with a small elevation in the centre
uncinate: hooked at the apex
undulate: with an edge or edges wavy in a vertical plane; may vary from weakly to strongly undulate or crisped. *cf.* **crisped**
V
valve: one of the parts produced by the splitting of a capsule when ripe or a part of the specialised opening of a capsular fruit
variant: any definable individual or group of individuals which may or may not be regarded as representing a formal taxon after scrutiny
variety: a classificatory rank below that of subspecies
vein: a strand of vascular tissue
vernicose: with a shiny, varnish-like surface
verrucose: warty
versatile: anthers that appear to be 'hinged' on the filament
vestigial: the remaining trace or remnant of an organ which was probably fully developed in some evolutionary predecessor
viable/viability: seeds that are alive and have the ability to germinate if provided with the correct external environmental conditions
villous: with long, soft hairs sometimes lying on the surface, finer than in pilose
viscid: sticky on the surface; coated with a thick, syrup-like secretion
W
wet-dry cycling: a type of pre-treatment used to alleviate physiological dormancy and morphophysiological dormancy. Wet-dry cycling involves incubating seeds under warm, dry conditions (e.g. 30°C / 50% RH), interspersed at regular intervals (e.g. every 2–4 weeks) with short periods (e.g. 1–2 days) of exposure to moist conditions. Repeated cycles of this treatment have been shown to alleviate seed dormancy in some species
whorl: a ring-like arrangement of similar parts arising from a common point or node
wing: any flat, often membranous expansion or flange, e.g. on a seed, stem or one of the two lateral petals of a papilionaceous flower or one of the petal-like sepals of Polygalaceae. *cf.* **keel**, **standard**
woolly: dense, matted, long, curled hairs. = **lanate**
wrinkled: a surface with irregular wrinkle-like lines
Z
zygomorphic: having only one plane of symmetry, usually the vertical plane, referring to a flower, calyx or corolla. *cf.* **actinomorphic**.

Index

www.ingramcontent.com/pod-product-compliance
Lightning Source LLC
LaVergne TN
LVHW070947210926
848482LV00002B/122